P9-DHR-949

LIONS IN THE BALANCE

LIONS
IN THE BALANCE

Man-Eaters, Manes, and Men with Guns

CRAIG PACKER

The University of Chicago Press Chicago and London

Craig Packer is professor of ecology, evolution, and behavior and director of the Lion Research Center at the University of Minnesota. He is the author of *Into Africa*, also published by the University of Chicago Press.

The University of Chicago Press, Chicago 60637
The University of Chicago Press, Ltd., London

Printed in the United States of America

24 23 22 21 20 19 18 17 16 15 1 2 3 4 5

ISBN-13: 978-0-226-09295-9 (cloth)
ISBN-13: 978-0-226-09300-0 (e-book)
DOI: 10.7208/chicago/9780226093000.001.0001

Library of Congress Cataloging-in-Publication Data
Packer, Craig, author.
Lions in the balance : man-eaters, manes, and men with guns / Craig Packer.
pages ; cm
ISBN 978-0-226-09295-9 (cloth : alk. paper) — ISBN 978-0-226-09300-0 (e-book) 1. Lion—Conservation—Tanzania—Serengeti National Park. 2. Lion—Ecology—Tanzania—Serengeti National Park. 3. Lion—Conservation. 4. Lion—Ecology. 5. Packer, Craig—Diaries. I. Title.
QL737.C23P32 2015
599.75709678′27—dc23
2014049420

♾ This paper meets the requirements of ANSI/NISO Z39.48–1992 (Permanence of Paper).

Both bad and good, and much of both, must be borne in a lifetime spent on this earth in these anxious days.

BEOWULF, CA. 1010

And it seemed as though in a little while the solution would be found, and then a new and splendid life would begin; and it was clear . . . that they had still a long, long road before them, and that the most complicated and difficult part of it was only just beginning.

ANTON CHEKHOV, 1899

Contents

Undertow

NAIROBI, DECEMBER 1999

The night had been cool, and Susan was lying curled on her side, our backs touching under the coffee-colored comforter. The yellow morning light filtered through the curtains; the sun was already up, but it was still early. We were in Barbie Allen's house in a leafy green neighborhood called Spring Valley. Barbie's father was a first-generation settler from the United Kingdom; he had been given land by the colonial government for service in World War One, and Barbie had famously retrieved a pistol from underneath her pillow and chased off an intruder after being shot in the arm during the Mau Mau Rebellion of the early 1950s. For the past twenty years, Barbie had graciously hosted my family whenever we passed through Nairobi.

Spring Valley is one of the poshest parts of Nairobi; most homes are surrounded by high walls and guarded by Kenyans in security uniforms. Barbie has lived here for thirty years; she only has one night watchman, who goes off duty when her African cook starts work first thing in the morning. Her picture-windowed living room is decorated with landscape paintings, and her shelves are filled with hundreds of books on wildlife and Africana.

Susan and I were sleeping in one of the downstairs bedrooms; a few birds chirped in the trees outside, but the morning was otherwise quiet.

Barbie called from the hallway, "Craig, there's trouble."

I rose quickly, pulling on my clothes, and started toward the door. Susan sat up, covering herself with the comforter, and asked, "What's wrong?"

Before I could answer, Barbie and her cook were pushed into the bedroom by two men with guns.

I can clearly recall Barbie's expression of anger and frustration, her cook's frightened resignation, Susan's cowering behind the covers, and the two pistols pointed straight at us. But of the following few moments, all that remains was the one impotent thought: "Don't kid yourself; you have no idea what is going to happen next. You have no idea. Don't kid yourself."

The two men wanted our valuables. And there was nothing more important in that moment than making sure they got every scrap of value in the room. Placate them. Win them over. Keep everyone alive.

Standing a foot taller than the cook and the two thugs, I was the obvious focal point. Any attempt at heroism would be in my hands—but my hands were empty, and they had guns, so I avoided looking them in the face and tried to make myself seem as small as possible.

I handed them the Pelican case with the video camera from the Discovery Channel ("Send us any footage that you take of your lions or the Serengeti." "But I'll be on my honeymoon and working with my students . . ." "That's, OK," they persisted, "anything interesting you see . . ."). I hand over my own camera in its case along with our passports and all our cash.

Placate them, win them over: "Here is our money as well as our camera."

Susan was confused at first, pushing a gun away from her face, thinking they were the law and she was being taken as a criminal. But now she is starting to panic.

Barbie, angrier than I've ever seen, scolds the gunmen, "This is her first trip to Africa!"

One of the men has worn a nasty scowl since first entering the room. He raises his pistol to Susan's forehead and says, "Welcome to Nairobi."

Just like a scene from *Pulp Fiction*.

Barbie groans in disgust. The second crook focuses on the loot; he sees Susan's new training shoes, starts looking around for a way to gather the goods.

I offer one of our suitcases.

It occurs to me that really only Mr. Psycho is intent on keeping us in line. The way he is holding his gun makes me want to get a better sideways look; maybe it is a dummy.

You have no idea; don't kid yourself.

They order Barbie and me to leave the room. Mr. Psycho escorts us out the door, leaving terrified, naked Susan alone with Mr. Loot.

Oh god, now they are going to rape my wife on our honeymoon, on her first day in Africa. This is Kenya during the height of the AIDS pandemic. I have to gain control; it is just Mr. Psycho and me for the moment, one-on-one.

But Mr. Loot suddenly appears with Susan; they had lagged behind long enough for her to get dressed and to assist him with our luggage; he wasn't interested in her, he never touched her, he just needed help opening the Pelican case.

Mr. Psycho asks, "Where is a room with a lock?"

They hustle us inside the downstairs bathroom; Barbie sits to my left on the edge of the tub and leans sideways against the sink; Susan sits to my right. I can't remember where the cook was at that point, maybe behind us in the bottom of the tub. Susan seems convinced this is where they will shoot us, the blood of four bodies running collectively down the drain; she rests her head on my lap and waits for the end.

The gunmen stand just outside the bathroom door; they are having some sort of discussion in Kikuyu—I have no idea what they are saying.

I ask Barbie to cool off and Susan to stay calm. I want them to keep their eyes down, stare directly at the floor. I'm thinking that maybe if the raiders realize we couldn't recognize them in a lineup, they'll let us go. If they had wanted to kill us just for the fun of it, we would already be dead.

The robbers close the door from the outside before realizing that the door only locks from the inside. They wave us out into the hall, demanding a room that can be locked from the outside. Barbie leads us toward her bedroom upstairs. As she passes a light switch at the base of the stairs, she flicks a switch—intending to activate the silent alarm but, instead, turning on the roof siren—*Rrrrrrrrr*.

"It's OK," Barbie reassures them, "I do that all the time; it only blew for a second. No one ever comes."

Mr. Psycho is apoplectic.

The gunmen push us up the stairs and into Barbie's room. Mr. Psycho's expression is intense, focusing all his hatred on this one moment of existence.

He raises his gun, arm stretched at full length, pointing his pistol at Barbie's head, looking for all the world like that famous black-and-white photo of a Vietnamese execution.

He says, "And now we will kill the Mama."

Susan and I sag halfway down to our knees, moving toward him, crying, and screaming. There is no way we are going to let him kill her.

We are only a few feet away, trembling, pleading, and his scowl grows even more grotesque, then he turns on his heel, walks purposefully out to the hall, clangs shut the iron door and locks us all inside the room.

He rejoins Mr. Loot in another room, but we aren't sure if it is safe to breathe until we hear them leave the house, then Barbie picks up the phone and calls the law.

The Nairobi police would love to come help, but it is a Saturday morning, and they don't have a car. Could she please come fetch them?

TSAVO, APRIL 2001

This is my last day in Kenya. We've been here for ten days, and I have to drive the two lion dummies back to Tanzania tomorrow. My graduate student, Peyton West, has spent nearly three years trying to learn why lions have manes, and a couple of Caputos have come out to document the grand finale.

Phil Caputo, the writer (his memoir, *A Rumor of War*, is a classic on the war in Vietnam) and journalist (he won a Pulitzer in 1972), has developed an obsession for man-eating lions; he orchestrated this expedition so we could contemplate the historical forces that spawned the Man-Eaters of Tsavo. Bob Caputo (no relation to Phil) is a *National Geographic* photographer; Bob is even taller than me, and we've known each other for nearly thirty years. I like to tell people that I gave Bob his first camera, but he is always quick to point out that I *sold* it to him.

Phil is substantially shorter than either of us, gruffly self-conscious about his stature and the sparseness of his remaining hair (the state of which we can view all too clearly from above). He also seems determined to transform our time together into something worthy of Hemingway. Whiskey flows each night at camp, reminiscences focus on war (Phil rode around Nam in a jeep with Hunter S. Thompson one day when the gonzo journalist claimed they were being swarmed by pterodactyls), but the heat of the day drives Phil into his canvas tent for a nap while Bob, Peyton, and I usually head off to the nearby lodge for a swim in the pool and a respite from the responsibility of meeting Phil's he-man aspirations.

Tsavo is a hot scrubby place; the lions here look like buzz-cut Serengeti specimens: sleepy and unlikely to view any of us as food. But Peyton and I are intrigued by the shortness of their manes—which pretty clearly reflect the intense heat and humidity of a coastal equatorial climate. We brought an infrared camera with us, and we can see from the thermal imagery that these lions are heat stressed. We want to know what lions "think" when they encounter a male capable of sporting a bulky dark mane.

Peyton's most important tools for tackling this problem are two life-sized toy lions with fake fabric manes: by setting out two dummies with contrasting manes, she can see how real lions respond. Back in the Serengeti, she found that the females adore a male with a darker mane—they don't seem to care much about length, but give them a choice between light and dark, and they approach the dark-coiffed dummy like a mistress to her Latin lover. In fact, we've given the manes Mediterranean nicknames to reflect their essential characteristics: short blond is Lothario, short dark is Romeo, long blond is—what else?—Fabio, and long dark is Julio.

These experiments can border on the hilarious: we set out the dummies with their contrasting manes then broadcast the noises of feeding hyenas to attract nearby lions looking to scavenge a free meal. As soon as the lions move in, we turn off the tape and let the lions focus on the dummies. Females can be quite shameless, coming up to the dark-maned dummy, flirting outrageously, sometimes taking out their frustration by tugging the dummy's tail. Disappointed by the dummy's indifference, they eventually sniff underneath its tail and realize their mistake.

Males are intimidated by dummies with longer manes—they approach in slow motion—literally pussyfooting through the tall grass so as not to make a noise—before choosing the shorter-maned invader as the less fearsome opponent.

Peyton has experimented enough to know that females prefer darker manes and care little about length. She is also certain that males fear longer manes, but she needs one more test to know if they are similarly intimidated by darkness. She performed dozens of these experiments in the Serengeti, and she literally used up every lion in our study area—once a lion has taken that sniff beneath the dummy's tail, the scales fall from its eyes, and every accessible male has seen this trick before.

So when Phil invited us here to talk about man-eaters, I told him we would join him if we could bring along the thermal camera and arrange to expand Peyton's sample size. And it has worked out wonderfully. We've done two other tests the past week—in both cases the males avoided the longer-maned opponent—but we need one more test of dark versus blond.

We've seen two sets of males in the past ten days—one quartet conveniently split up for a few days so we could test one male by himself and the other three the next day. The trio was especially satisfying because they not only approached the small-maned invader, but they also attacked it in front of Bob's camera. Score one for *National Geographic*.

But we need to find that other pair of males we glimpsed a few days ago, and there is really no good reason to think we can succeed. Tsavo's coastal humidity and ninety-degree heat forces us to rise at dawn to look for lions before they all disappear into the shade by 9 or 10 A.M.

This morning, Bob and I went out by ourselves, each hoping to find anything at all—male or female. But Peyton took Phil through the heavily thicketed area toward the lodge that was the last place anyone had seen the other pair of males, and she's desperate for a last chance to test the males' fear of darkness.

Phil has pretty much given up on me. He likes to talk about ancient Greece, and he finds my evolutionary slant on animal behavior and human morality a bit disheartening. He wants life to have Meaning. I mostly care about survival value and reproductive advantage. He wants the Man-

Eaters of Tsavo to have had mystical significance. I mostly want to know what happened in the 1890s to cause the Tsavo lions to broaden their diet to include humans. He likes to evoke a shiver up his own spine when he hears a lion roaring at night. I like to step out farther into the darkness and soak up the sound of a full-throated roar; it brings me a sense of safety because it means the lions aren't being pursued by poachers or poisoned by angry cattle herders. We even clashed over his taking Larium to protect himself against malaria. "That stuff should be banned," I practically shouted, "It's way too powerful to use as a prophylaxis—it has horrible side effects like paranoia and heart palpitations!"

I feel like I've driven these same sandy tracks every day of my life, but I'm truly torn between calling it quits and changing my schedule to stay on in Kenya long enough for Peyton to get her last data point.

The soil here is salmon pink, the low bushes have tough green leaves, the grass is faded, and the shadows beside the trees are starting to darken as the sun beats down. Too late for lions today, so I head back to camp.

Peyton's already back, and she tells me that she saw the elusive pair of males just before they disappeared into the bush. Bob returns empty handed an hour or so later, and we decide to see if we can retrieve Peyton's males out of the thicket. We need to mend the short-maned dummy after its manhandling the other day, so I poke the stuffing back inside while Peyton darns a few rips and tears. Bob photographs us at work, and we all laugh again that people have actually paid for us to come out here and play toys with lions. Phil sits by his tent, writing up his notes while we transform Fabio into Romeo.

It is especially hot this afternoon, and Bob and Peyton and I take frequent swigs from the big water jug in the middle of camp. Drinking tepid water is a poor substitute for another dip in the pool, but there is no time for a swim today.

Our dummies are mended, dark versus blond, and we're ready to roll. Peyton is behind the wheel, and she is beside herself. She has spent years on this project; she wants to get this thing done. She still can't believe that her results are going to add up to enough to earn a doctorate. When we first met on the Yale campus, I could see immediately that she was unlike any other graduate student I had ever worked with. Peyton is a Manhattanite; she worked for a publishing company in San Francisco, but she has a profound sense of curiosity about the natural world that is matched by insight, enthusiasm, and creativity. It was her idea to bring infrared technology to the project—she has a million ideas and exudes a contagious sense of excitement.

But today she is resigned to whatever might happen. We drive into a labyrinth of stunted thorn brush at the bottom of a shallow gorge a hundred meters wide.

"They were heading through there, up the slope but looking for shade," she says pointing above the steering wheel as we slow to a stop.

"How on earth did you see them in here?"

"They had just crossed the road about a quarter mile away, and we managed to keep sight of them as they trekked through here. I'm not exactly sure where they are right now, but if we crank the amp, we might draw them back down."

We get out of the car and set to work, moving quietly back to the horse trailer, opening the door, and hauling out the dummies.

"I'll aim the speaker so if the lions reappear, they'll see the dummies from that clearing over there," she says, pointing to a grassy area about eighty meters away.

Bob pulls up with Phil, "Where do you want us?"

She directs them so that their car extends the line of a semicircle behind the trailer.

Dummies positioned, speaker wired, and amp powered on, she frowns, sighs, and says, "Let's see if we can wake them up."

She plays the roar of a Serengeti female. A strange female roaring in a males' territory is sure to grab the attention of any red-blooded male within earshot.

No sign of anything.

"So it goes," she says biting her lower lip then reaching back to turn off the amp.

I start to ask if she wants to try again, but then she gasps as the two males emerge off to our left, walking to our right and away from us.

"OK," her confidence blooming, "I'll let them get beyond the clearing then try again."

The lions are making a large circle somewhat away from us when the roar booms forth again, and they make the equivalent of a spiral arc back toward us—exactly along the path Peyton had picked when she placed the dummies in line with the clearing. It is like watching a master fly fisherman reeling in the lions from a perfect cast.

And here they come, one of these boys has a darker mane than the other, and as often happens, the dark-maned male leads the way. They can see the dummies, and their approach is slow, deliberate; caution is written all over their faces.

Forty meters away, then twenty, and finally, at ten meters, their choice is clear.

"Oh, my god! Oh, my god!" she is almost bouncing up and down. The dark-maned male sniffs beneath the tail of the blond-maned dummy; his blond companion joins the investigation. The last piece of the puzzle falls into place.

Black manes are a turn-on for the ladies but intimidating for rival males.

And elation combines with relief in a way that I wish I could keep with me for the rest of my days. I'm so utterly thrilled that I hardly notice what Peyton says or does until the moment when the lions realize they've been duped by some stupid toy and wander off into the sunset.

We retrieve the dummies and share our excitement and relief with Bob and Phil. "Do you know how many years it has taken to reach this moment?"

And they beam at us like parents at the school play.

"Hey, it's our last evening together; we need a picture with the dummies." So we clown around and add a few shots in infrared for the eventual *National Geographic* story. But when we look at ourselves on the screen of the infrared video, none of us notice at the time anything odd about the color of Phil's nose. The warmest parts of our faces, mostly on our foreheads, look white or yellow in infrared, whereas cold patches are rendered in blue. While most people's noses are cool, Phil's blue patch extends out toward his cheeks and up toward his forehead.

Dummies reinserted into the trailer, dusk has settled, and we wind our way home in the fading light, arriving back at camp shortly after dark.

The last supper. Whiskey is poured, wine uncorked. Toasts given and received. High fives all around. Food arrives, and we reach a satisfied state of companionship.

Bob sits to my right; Phil is across the table.

We are all focused on our food but then I catch sight of Phil staring straight ahead at me, all lights gone from his eyes. He makes a groaning, croaking sound, and then he is dead.

"Phil!" I shout frozen in my seat, but trying to engage his attention. "Phil!"

Peyton reaches over to his stiffened corpse and yells in his ear. Bob leaps up, bounds around the table, and starts to give him the Heimlich maneuver. We're all shouting, "Phil! Phil!"

Peyton checks his wrist, "He doesn't have a pulse."

His body sits rigid in his chair, his glazed eyes wide open. No pulse, no breath, no sign of life.

Peyton is crying inconsolably.

Then like the weird croaking of some toad from hell, Phil comes back to life, groaning loudly as he fills his lungs with air, his wild eyes start to relax, and his face takes on a human expression.

After a minute or so, he is able to talk, "That was strange," he says in a surprisingly understated manner.

Meanwhile, I've got Bob's satellite phone, standard equipment for every *National Geographic* photographer on assignment, and I'm talking to the Flying Doctors in Nairobi. We discuss the possibility of landing a light aircraft on an unlit airstrip in Tsavo or immediately driving Phil up to Nairobi hospital.

The desk operator at Flying Doctors asks a series of questions: Does the patient have a history of heart disease? Food allergies? Could there have been a bee sting or snake bite?

No, no, no. None of the above.

Then she asks, "What medications is he taking?"

"Phil, what meds are you taking?"

"Only the Larium."

"Larium!" I'm practically yelling at him.

"Yes," he said, "I stopped for a few days after our conversation about side effects, but then I got so paranoid around the three of you that I decided you were all trying to manipulate me somehow."

So the nurse says, "And you say his last name is Caputo? He's probably just had a seizure from the medication. It is especially common in men of Mediterranean descent. Has he been drinking alcohol?"

"Yes."

"And has he been getting enough to drink otherwise?"

"He mostly drinks alcohol in the evening; we were out all day today . . . He got pretty badly dehydrated."

"Well, if he has another seizure tonight, you'd better drive him up right away. Otherwise he can wait until the morning. We see cases like this every few months."

OK.

"But you should keep an eye on him all night, just in case it happens again."

So I hang up and communicate the news to the troops. But in mentioning the need for a nightlong vigil, I'm quick to say, "You know, I'm really sorry, but I can't look after him tonight. I have to drive back to Tanzania by myself in the morning."

Bob chimes in, "Yeah, and I'm way too tall to fit in his tent."

So Peyton agrees to be his roommate for the night.

After the ups and downs of the day, I stumble back to my tent to pack up and get some sleep. Nothing breaks my night; there is no moon, and the inside of my tent fits like a glove. Sleep is the easiest thing in the world.

Getting up in the morning, I find Peyton sitting by the campfire with a cup of tea.

"How was the patient?"

"Ugh! He woke up at 2:00 A.M. and told me he wanted to kiss me."

"So did he die again?"

"I sure hope so."

NGORONGORO CONSERVATION AREA, DECEMBER 2003

Dennis Ikanda rolls to a stop beside a dung-flattened surface surrounded by a ring of thorn branches (a *boma*) that somehow serves as a restraint against wandering livestock and a flimsy barrier against nocturnal meat-eaters.

Looking like outsized Twinkies, simple dung-daubed huts form a dotted circle around the boma, which is itself maybe twenty meters wide. Women and children soon appear next to the car; the smaller kids are either naked or clad in a few tattered rags; younger teenage girls carry some of the babies—though to look at, you'd swear these child minders were only eight or nine years old. Everyone here is stunted by malnutrition, and the girls' ages can only be inferred by the modesty of their dress: bright blue cloth draped over their shoulders and a near absence of jewelry means they are old enough to be decorated but a year or so too young for marriage.

A few old women appear, decked out with the stiff O-shaped bead necklaces that rest on their shoulders and jut out from their collarbones perpendicular to the empty breasts that form skinny V's dangling below their corrugated chests.

I've learned not to react to the swarm of flies that settle on our faces the instant we get out of the car. Dennis says a few formal greetings to the old women in Maasai before switching to Swahili and asking to meet with the *mwenyakiti* (chairman) of the village.

These flies take some getting used to. They just cling to your face; if you wave at the air an inch or so in front of your eyes, they hang on until the disturbance is over. You have to wipe your face, swat your cheeks and forehead, breathe through your nose (careful not to inhale too sharply!), and squint your eyes as they crawl into your tear ducts.

The kids are beyond all that. They are amazed at the sight of a white man, intrigued by a black man who can't speak more than a few words of Maasai. But they remain utterly oblivious to the flies that treat their watery eyes like wading pools and their crusty lips like parking lots.

The *mwenyakiti* appears, and Dennis shows the appropriate deference to the old man who is dressed in a faded red blanket over one shoulder and a second cloth wrapped around his waist. His sparse hair is barely long enough to show white on his black skull; his expressive face could have been carved from a weathered stump. But there is an unmistakable glint of pride in his eyes as he is the clear center of attention of these two strangers who drove all the way to his village to see him again.

Dennis is enrolled in a Master's program at the University of Dar es Salaam, and I am his external adviser. I fund his tuition and fieldwork, and he comes here about once a month to ask about livestock, about any problems with lions, hyenas, and leopards. We are in the Gol Mountains, a remote part of the Ngorongoro Conservation Area, midway between the eastern border of the Serengeti National Park and the western flanks of the Ngorongoro Crater. The Maasai have lived in this valley for a hundred years. Unlike most other tribes in Tanzania, they have steadfastly refused to modify their lifestyle to suit the twenty-first century. The land is harsh. The soil is mostly made up of volcanic ash; grass grows with abandon during the short rainy season, but if you were to plow the ground, the dust would all blow away. So the best way to make a living is to let the cows, goats, and sheep turn the grass into blood, milk, and meat.

But this whole area is a wildlife paradise: the Serengeti and the Ngorongoro Crater are two of the most productive ecosystems in the world. All those wild ungulates support a remarkable abundance of large carnivores. And as long as the wild prey are around, the carnivores pose no real threat to the people who live here—for, uniquely, the Ngorongoro Conservation Area (NCA) is a national park with one more species than usually found in a national park: Maasai are viewed as an integral part of this ecosystem, so they are allowed to live anywhere in the NCA other than the floor of the Crater itself—allowed to live anywhere in the NCA, that is, as long as they continue to follow their traditional lifestyle.

Which means life in a dung-lined hut; water from a spring at the base of a mountain or from a seasonal water hole; no electricity or material wealth—other than as measured in the traditional way: a rich Maasai has more livestock than his neighbors.

It also means virtually no education. Maasai elders dismiss educated young villagers as *wasomi*—people who just read. They are not keen on sending their kids to school. Boys should herd livestock. Girls should help raise babies.

The nearest hospital is a two-day walk; the nearest school is three days away. Oh, and the school. It is a boarding school for kids between the ages of eight and twenty-four: one coed dorm with half as many beds as students and no adult supervision after class each afternoon.

It isn't enough to say there is no running water, no electricity, and no basic sanitation—watch your step: those long skinny turds don't all come from dogs—or that there is no real appreciation of education or health. Hang around at night, and there is no light. No kerosene lamps, no candles. Just darkness.

And you are huddled with your siblings and cousins in the dark and maybe you've all been asleep, and then something starts the dogs barking outside, and maybe one of the dogs suddenly goes quiet, while the rest of the dogs really start to make noise, and your livestock are snorting and worried, and in their panic they storm through those thorn walls and you are suddenly outside now, listening to the hooves thumping against the hollow-sounding soil, feel the pounding through the soles of your bare feet. And somewhere in all this chaos there is a hyena or a leopard or a lion. On the one hand, the sudden silence of the dog might mean a leopard—a leopard caught one of the dogs last month and carried it up the mountain—or a hyena—hyenas are demons sent by witches, and they always manage to escape. But, on the other hand, it might be a lion.

If you are a *morani* (warrior), you have your spear at the ready—you could be the hero, but you will have to wait until the morning light before you can go out and prove yourself. If it is a lion, you want to be the first to spear it—and if the lion turns on you, make sure it mauls you on your chest or stomach, on your face, shins, or throat. Any place where you can show your scars with pride, show the incontrovertible evidence of courage. A scar on your back would be a permanent reminder of cowardice, an ineradicable trace of shame.

Some kids are heroes before they become warriors. Yesterday we went to a village near Olduvai Gorge to see the now seven-year-old herd boy who, a year ago, had come off a hill about three kilometers from home when a large male lion trotted toward him. The child didn't hesitate; he started yelling and waving his stick, then picked up a couple of rocks and threw them at the lion, who, for his part, didn't miss a step, picked up the boy like a rag doll, shook him around in annoyance, then spat him out and ran over to the goats, snatched one on the run, and took off into the bush.

The child was upset that the lion caught one of his goats, but his father forgave him and spread the story of the boy's courage.

We talked to the boy for about half an hour, and I took pictures of his scars. He was confused by his father's quiet pride, and he seemed relieved when we left them alone together in the glow of the evening sun.

Few stories end this happily. Cows can wander off while boys are messing around or generally distracted from their daily chores. Losing a goat is one thing, but the loss of a cow is unacceptable. Some negligent boys refuse to go home again for fear of being beaten after losing a cow, so every year another boy seems to stay out until he either starves to death or feeds the monsters of that crushing darkness.

*

Dennis is still working on his list of questions with the village elder, and my Swahili is only adequate enough to get a general outline of the discussion. So far nothing seems too far from the norm, so as usual I'm distracted by the visuals out here. The air is so clear that the trees on the far slopes of the Ngorongoro Crater highlands feel within easy reach. The lime-green grass stretches from our feet to the near hills.

The kids stare at me like I'm from a distant planet. I try to look past the flies walking over their faces and engage with them like any other kids in the world, brighten up the occasion, clown around a bit. It doesn't take much to make them laugh. I squat down to engage them at eye level, tip my hat, point at Dennis and the village elder, act stern for a moment and mime their conversation. Some of the kids look around to see if it is OK to laugh, but others burst forth with spontaneous glee. A toddler staggers over from the pack and holds onto my shoulder. One of those girls who look no more than eight or so comes up to retrieve the little boy. She is all business, hoists him up, and ties a sort of sling to keep him steady against her back. He looks back at me while she hauls him over to the women inside one of the huts.

We have reached some sort of equilibrium until I take out my camera—instantly shattering the peace. I make the universal gesture of asking if someone wants their picture taken followed by another signal indicating that they will be able to see the picture themselves.

A couple of the bolder girls stay put, and, as promised, I turn on the viewfinder so they can admire their portraits. A kind of magnetic pandemonium ensues, with kids bunching toward the middle of the heap to look at their friend's pictures and then pose for the next shots.

These images will stay with me forever, but the kids only have a few seconds to see themselves before the screen moves onto the next sequence of faces.

But that seems to be more than enough. Most of the kids quickly tire of the photos, and the few persistent show-offs are eventually persuaded that the camera has all been used up. But we've reached a new stage of our relationship, and there is a sense of something closer to trust rather than of speculation or suspicion.

Dennis tells me that he has finished, asks if I have any other questions for the old man.

"Yes, yesterday we met the seven-year-old who tried to chase off the lion all by himself. What does the *mwenyakiti* think about this story?"

Dennis translates, the old man looks at nothing in particular and rat-a-tats back his reply.

"He thinks that boy was very brave."

"I do, too. Incredibly brave! But that wasn't really what I meant. What does he say about relying on such a small child to protect livestock? Wouldn't it make more sense to send out the morani? I mean, warriors would be far more effective than little kids. They are the guys with the spears, and I thought they used their spears to protect their livestock."

Translation, pause, a more measured response.

"He says that there aren't enough morani to go out with every herd."

But then Dennis alters his tone a bit, "That's what he says, but there are plenty enough morani to do the work. What he doesn't want to tell you is that he doesn't want these kids to go to school. If they leave here for school somewhere else, they won't come back."

"OK, ask him about sending out a small child by himself. Wouldn't it be better to have them go out as a group so they can protect themselves—safety in numbers and all that?"

"I'll ask, but he won't tell you the answer you want to hear."

"Why not?"

"Because if these kids go out in a group, they'll spend most of their time showing off and playing with each other rather than herding. It's when the kids get distracted that the lions can catch any cow that wanders off into the bush."

We have three more villages to cover today, so I thank the old man for his time and wave goodbye to the kids. But they've already forgotten me, and as much as I hate to admit it, I'm struck by how everyone seems so happy—or, if not happy, at least content—with their lot. Many of these kids will die before they reach eighteen, and most girls will get pregnant as soon as they attain fertility. But nobody asks for anything other than a few liters of water, nobody finds anything interesting or desirable about wearing clothes or driving off in a Land Rover. Nobody minds that their portraits only persisted for more than a few seconds in a stranger's camera.

The afternoon light does its magic, intensifying every color in the world as we drive past the cliffs and tree-studded slopes of the Gol Mountains with white-barked trunks standing out from the green and yellow foliage like freestanding piano keys.

People are heading home before dark. Kids traipse a few paces behind rafts of goats and sheep. Women haul stacks of firewood on their backs.

"Dennis, where are all the cattle? And the morani?"

"The rains only started last week, so all the morani are off watering their cattle at Ndutu." Ndutu is the site of the nearest lake, about twenty-five kilometers away.

"So the morani do herd livestock . . ."

"Yes, but mostly when they take them far from home—and especially when they take them west."

"Ah, because of the Sukuma."

"Yes," he says. The region is dominated by the Maasai to the east, but the dominant tribe to the southwest of the Serengeti is the Sukuma.

I say, "And the Sukuma have guns . . ."

"AK-47s."

According to tradition, every cow in the world was given to the Maasai by their god, Lengai. Mount Lengai is an active volcano that looms up behind us at the north end of the Ngorongoro Crater highlands, a towering reminder of innate entitlement. So the Maasai have been engaged in a never-ending quest to retrieve all of their cattle, some of which might have been under the temporary care of another tribe. Of course, no one else subscribes to this particular interpretation of Maasai scripture, and the Sukuma have developed very effective methods for retaining their herds—and they no longer restrict themselves to just playing defense.

"The Maasai elders got together last year and told the morani to stop stealing cattle from the Sukuma—too many morani were getting killed."

"So they decided to give up cattle rustling rather than buy guns to fight back?"

"Yeah."

"Seems almost enlightened."

"Not really. The Maasai are very, very conservative. They don't want to change one single thing."

It's hard to conduct a coherent conversation in a noisy Land Rover, so I can only guess his meaning. Perhaps the warriors prefer their spears to guns?

The sun has fallen below the crest of the Gol Mountains, so the green plain has gone gray. About half a dozen people are trekking the last few hundred meters before home when we see a woman waving at us to stop.

Dennis pulls up next to the woman and her two companions.

She speaks no Swahili, so an older man explains, "She hurt her hand."

She holds up her right hand to show a grotesque swelling between her fingers and wrist. The skin is purplish and oozing pus; it is so taut that her fingers look starved of blood.

"She was bitten by a snake while collecting firewood. Do you have any medicine?"

Dr. Packer takes charge of the situation: "We don't have any antivenin. Does she know what kind of snake it was?"

"The snake bit her ten days ago."

"Oh, OK, so there's no worry about the poison from the snake anymore. But it has gotten infected. She needs antibiotics."

"Can you help her?"

I reach into the back of the car, pull out my bag, "Yes, but I only have enough medicine for the next two days. Here, take these pills, but tell her she needs a full course of antibiotics for ten days or so. She needs to go to hospital."

He translates from Swahili to Maasai.

Rather than any sort of relief, she sets her face and shakes her head angrily.

"What's wrong?" I ask in surprise.

"She won't go to hospital."

"Why not? Her hand will get worse. She might lose it if she doesn't get the right treatment. It could even poison her blood and kill her."

Again, no go.

"We can take her tomorrow. She needs a proper doctor."

No, no, no.

"Why not?" I'm almost frantic.

"Her husband is away at Ndutu with their cattle. She can't go without his permission."

In English, "Oh, forget about her husband." Then in Swahili, "Her husband won't be happy if she loses her hand or dies. We can take her to hospital in the morning and bring her back before her husband gets home."

No, no, no. Even though her pain must be excruciating, she is more frightened now than when we first met her.

Dennis starts the car, "There is nothing else we can do. You gave her your pills."

The sun has set; the evening is gray; the car starts to roll.

"We did nothing."

1 Fools Rush In

MINNEAPOLIS, 6 MARCH 2008

I almost never dream of animals. But when I was awakened by the alarm at half past five this morning, I was still annoyed by that damned lion in the gray-green light, sniffing around outside my tent, trying to come inside. All night, he kept finding a weak spot in the canvas or a loose zipper, and he would come in with me again and again. I had to keep getting up and leading him back out. He wasn't hungry, and I wasn't afraid. He just wanted in, like a huge house cat wanting in to the bedroom.

It was a snowy Minnesota morning. My first thought after fully wakening was, "Keeping lions out of my tent all night long—that's why I'm always so tired in the morning!"

It took a few moments to summon up the courage to leave my warm bed and walk barefoot across the cold floor. Then in the middle of brushing my teeth, standing on the white bathroom tile, the penny finally dropped, and I felt like Freud.

The dream. It meant something.

I've spent decades studying lions, most of my adult life. But for the past few years I've kept the lions on the proverbial back burner.

This morning I was flying off to a conference in Santa Fe where I was supposed to give a talk about lions again—and nosy Simba here had been eager to remind me that I had been neglecting him.

So OK, big guy, you can have my undivided attention for a few days now, but there has been a perfectly good reason for setting you aside. Something I have to do.

Tanzania's population is currently thirty-eight million people—about the same as California—in an area as large as California, Nevada, and Oregon combined. It is one of the poorest countries in the world, despite being the home of the Serengeti, Mount Kilimanjaro, and the spice island of Zanzibar. Maybe as much as 80–90 percent of the population lives in poverty, yet nearly a third of the country has been set aside for natural resource management: national parks, game reserves, and forest reserves. But ask most rural Tanzanians what they think about wildlife, and they'll probably tell you that if they can't eat it, they'd just as soon eradicate it.

Elephants and lions? The two most hated species in the country.

Tanzania has the last great populations of wildlife on the continent. Over a million wildebeest in the Serengeti alone; maybe half the lions left in Africa; one of the largest populations of elephants. Extraordinary biodiversity of birds, mammals, reptiles, amphibians, butterflies, and plants.

Riches beyond compare? Depends on whom you ask. Some people are infatuated with large animals; others live in fear.

And what to make of the Tanzanians themselves? A lot of people just look on rural villagers as pests. An ocean of poverty that consumes bushmeat and converts wild land to cotton fields and subsistence crops. A vast incubator of emerging infectious diseases and a potential source of economic migration, bringing crime and instability to the rest of the world.

It's not that people haven't tried to alleviate the miseries of Africa; Tanzania receives a remarkable amount of aid money from Europe, Japan, and the United States. But you'd be hard-pressed to see much sign of progress.

Much sign, that is, if anyone had been keeping track. I've worked in Tanzania for the past thirty-six years, and while some things have certainly gotten better in many of the larger cities—Dar es Salaam, Arusha, and Mwanza—the rural villages still largely consist of mud huts and makeshift shelters, emaciated cattle, and puny babies. Roads to make an SUV weep, decrepit railway carriages, and sanitary practices blissfully unaware of germ theory and the miracles of modern hygiene.

I remember someone in the 1980s talking about economic growth in the third world, saying that beyond the distinction between developed nations and developing nations, economists were starting to wonder if Africa required a third category: the never-to-be-developed nations, despite unrivaled natural resources and all of the goodwill in the world.

When I first went to Tanzania in 1972, the country had fewer than ten million people. Now there are nearly forty million, and if population growth continues at 3–3.5 percent for the next forty years, there will be 160 million by the year 2048.

And if that happens, you can kiss all the lions goodbye—even the dream lions that sniff around the tent in my bedroom in Minneapolis.

So where to begin? How about with Dennis Ikanda in the Ngorongoro Conservation Area sometime after that trip to Tsavo in 2001 and a few months before 9/11.

Dennis was born in Arusha, a middling city about sixty miles from Mount Kilimanjaro at the foot of Mount Meru on the edge of Maasailand. His parents went to school in Arlington, Texas, when he was twelve years old. They went to study the gospel, and Dennis acquired a wonderful grasp of vernacular English—a skill that made me realize how superficially I had appreciated the intelligence of most Tanzanians. He had worked as a tour-company manager in Arusha after graduating from college, and he had frequently spotted Peyton and another lion assistant, Grant Hopcraft, when they came into town for supplies. He asked them about opportunities and e-mailed me for a job interview.

I hired him to take over from another Tanzanian student, Bernard Kissui, who had just finished his Master's project in the Ngorongoro Crater, and it soon became clear that Dennis should also consider a graduate degree. But I didn't want him to conduct research on a topic that would only be interesting to professional ecologists—with his language skills and ease with people, he should take on a project with a more direct relevance to wildlife management.

Dennis based himself in the drivers' quarters at a posh lodge on the rim of the Ngorongoro Crater, but he commuted regularly to the Serengeti to discuss lions with Grant and Peyton, and on this particular day he decided to take a shortcut back to the Crater through the Gol Mountains.

He was pretty much in the flat middle of nowhere when he picked up the signal of a radio-collared lion that normally lived near the center of Serengeti National Park, over forty kilometers to the northwest. Approaching the strengthening series of beeps, he spotted a party of six Maasai morani, covered in blood, two of them barely able to walk. They waved frantically at him to stop and told him that they had just speared a lion.

He drove them back to the scene, and there was a dead female with tail and claws removed, her radio collar still around her neck.

The Maasai explained that they had tracked her from over thirty kilometers farther to the east—that she had killed their cattle last night.

Dennis didn't contradict them, knowing that she would have been impossibly far from any livestock depredation, if it had even occurred last night or any other night in the past month. He merely asked how they knew this was the cattle killer, and they told him how they had tracked her to this spot, where she was suddenly accompanied by a dozen other lions. Dennis asked for details, and they eagerly lifted their bloody shins to show where she had lashed out, clawing and biting before losing her life at the end of a half-dozen spears.

*

Thus Dennis ended up studying the Maasai in the flyblown villages of the Ngorongoro Conservation Area. He had originally hoped to watch lions in the same areas, but they were too elusive and too scarce. So he had to rely on the Maasai's own accounts of livestock losses and close encounters, and he was immediately struck by the consistencies in people's stories. If one village told of a break-in to their boma and a loss of two goats or a lion attack on a morani in the middle of a heavy rainstorm, the neighboring villages knew the same basic details. The bush telegraph worked fast; these were major events in people's lives.

But the Maasai were suspicious of Dennis. He drove a vehicle; he wore khaki; his ancestry was Bantu. Maybe he worked for the NCA authority. Tribalism is not as serious in Tanzania as in, say, Kenya (Luo versus Kikuyu) or Rwanda (Hutu versus Tutsi) or even South Africa (Xhosa versus Zulu), but Dennis was definitely not Maasai, so he couldn't be trusted.

At the beginning, Dennis held a fairly low opinion of the Maasai. He didn't especially empathize with their stubborn reliance on pastoralism ("They don't have to be this poor; they could always sell a few cows"), their vulnerability to predators ("Man, all these kids you see here should be in school. The morani don't have anything better to do; why don't they guard the stock?"), or their squeaky-clean reputation as guardians of wildlife ("Last week they speared a buffalo over there; a few weeks ago, they killed a couple of elands").

For several years, my trips to Tanzania were dominated by excursions into Maasailand with Dennis. One year we went to a high-relief area north of the NCA. We looked over the escarpment of the rift valley, and Dennis said, "Here's where they drive the wildebeest over the cliff. Dozens at a time."

"But they don't eat wildebeest."

"No, they just do it for fun."

"Who? The kids?"

"Kids, yeah, and the morani."

"But I don't believe they just do it for fun. The Maasai have to avoid the wildebeest because of the malignant catarrh—the disease kills their cattle."

"That's true, but they don't have to stick around here when the wildebeest come with their calves."

Malignant catarrhal fever is caused by a herpes virus. It is harmless to wildebeest, but it causes disastrous infections in cattle. Adult wildebeest hardly shed any viral particles, but newborn calves are like malignant-catarrhal-fever factories; any contact between livestock and infant wildebeest (or even their afterbirth) is disastrous. The wildebeest calve in the

Crater highlands at the height of the rainy season, grazing on grasses enriched with calcium and phosphorous from the volcanic soils and spilling countless viruses on the fresh green lawns.

The migratory pattern of the wildebeest prevents the Maasai from grazing their cattle on their home turf during the short period of green grass. But "pattern" presumes predictability, and everyone's climate has been in a state of flux over the past decade or two—and when you are poor, very poor, and your livelihood depends on scratching out something from nothing, there is a horrible temptation to gamble away your last chance.

In an average year, the Maasai would have a good idea of the near future: if the clouds build up in November, the wildebeest will arrive by December and pop their babies here at the end of January. So the herders would normally have a month or two to ready themselves for their own migration to the Rift Valley or up near the Kenyan border, keeping their precious cattle as far from the malignant catarrhal fever as possible.

While some parts of the world are becoming hotter or dryer with each passing year, Tanzania's weather is becoming more erratic. Some years the rains fail and the wildebeest never move out this far. Droughts are hard on everybody, and there is no way your starving livestock could survive a long cattle drive, so you stick around and wait for better times.

But sometimes the rains fail in November and December, and the heavens pour forth in January so that the wildebeest arrive before the lawns have turned the slightest bit green.

And if you gambled on a year without wildebeest, you've lost everything to malignant catarrhal fever.

There is no insurance against this sort of thing, no pastoralist pension plan, no savings in the bank. Nothing.

Meanwhile you see these flipping wildebeest everywhere, chomping away on your best pastures, and it is because of the wildebeest that your family can't even live in the Serengeti anymore. The Serengeti National Park was created to provide a refuge for the wildebeest, and the government moved your great-grandparents away from the sweet grasses along the rivers that hold water all year round, forcing you to live in this semidesert with the dusty soil and the grass that never grows more than a centimeter high.

So maybe the Maasai do drive the wildebeest off the cliff just for fun or maybe the reasons are a lot more complicated. But this was only the first year of Dennis's project, and he didn't feel any need to ask.

It is easy to get frustrated with the Maasai, easy to fall into the familiar disdain of the elite for the great unwashed—or even of the semipoor for the ultrapoor. But the Maasai's worldview is so intensely focused that sometimes you just have to laugh.

Grant Hopcraft, part-time Crater lion assistant during the late nineties, employed a Maasai guard on the rim during his weeklong camping trips from the Serengeti. Grant is a low-key, gentle soul. He is the man I most wanted my own son to grow up to be like. Grant thinks before he speaks. He considers the opinions of others, and he cares about people.

After a long day of looking for lions down on the Crater floor, Grant would wind his way up the narrow twisting road to his camp and spend the evening in the company of his Maasai guard, Olokoko. Grant grew up in Kenya, so he is fluent in Swahili, and his relaxed manner made it easy to sustain long complicated conversations about the world at large.

They would talk about things the guard could never imagine, having never left the NCA. He knew about internal combustion engines, of course, and he could see aircraft flying overhead. But he'd never seen an ocean or a train, and one night Grant looked up at the full moon and said, "You know, people have been to the moon."

"You are lying to me. No one can go to the moon."

"No, no, it's true!" And so Grant explained about rockets and space suits.

And Olokoko asked, "How long does it take to get there."

Grant wasn't too sure, but he said, "A couple of days, I guess."

"Now I know you are lying!"

"Why?"

"Everybody knows the moon goes away in the daytime."

"No, no, it's true. Let me show you."

So he went into his tent and came back with a grapefruit and a tomato and gave an impromptu lecture on celestial orbits.

The Maasai listened intently, watching Grant manipulate his fruit in the moonlight.

"OK, so the moon is always there."

"Yes," Grant replied, relieved that his lesson had taken root.

"Let me ask something."

"Go ahead!"

"Are there cows on the moon?

"No, there are no cows on the moon."

"Then why bother to go there?"

I took my kids to the NCA a couple of times during those years. My daughter, Catherine, had never particularly enjoyed Africa or the Serengeti or the animals in the park, but she was transformed by her experience with the Maasai. It wasn't anything that you could easily pin down; she still doesn't especially want to come back to Tanzania. But maybe it was her realization that the red blankets and beads didn't come off when people went home at night. Or that the women and children weren't free to make their own decisions about anything important in their lives. Or that the human condition included almost no limit to its squalor. Whatever happened, she was altered by the experience, and her attention to the Maasai's untarnished humanity turned my own thoughts to new directions that took a few more years to fully sink in.

My son, Jonathan, in contrast, was always in heaven whenever he came to Tanzania. Everything was an adventure, full of snails, pails, and a chance to drive off-road in a stick shift. His viewpoint tended to echo Dennis's initial skepticism, seeing the way that an old Maasai named Pakipuni would always manage to find us no matter where we went and somehow manage to hit me up for a few shillings in some dubious exchange. Jonathan, however, was the only member of the family to try out Pakipuni's barbecued goat that obviously, I tell you, obviously, had been sick with anthrax. And if you get sick, young man, it's going to be your own damned fault; I don't care if you do have a full supply of Cipro . . .

But the goat meat didn't sicken my adventurous son, though his lesser illnesses during various trips to Africa ultimately inspired him to aim for medical school rather than a career watching animals.

There came a point, somewhere in the third and final year of Dennis's fieldwork, when he finally earned the trust of the Maasai elders, and they told him things they had been holding back. It was only legal to kill a lion in retaliation for livestock depredation—like in the doubtful case of our collared lion from the Serengeti. But after all this time, the elders decided that Dennis was not a policeman. He obviously didn't work for the NCA; he wasn't a spy.

So they told him things. They told him about specific cases of lion killing going back forty years. They told him where the lions had been killed, the years they were killed, the seasons. But even now, they were careful not to tell him why.

They no longer bothered to claim that each lion killing followed a cattle killing. But they still wouldn't admit to *Ala-mayo*, the ritual killing of lions that makes a young man a hero.

Dennis nevertheless managed to assemble a more complete picture than they would have guessed. Most of the livestock depredation occurred in the dry season, when the wildebeest and zebra were up in Kenya, and the local lions had nothing else to eat but livestock. But most of the lion killings occurred in the wet season, when the unsuspecting tourist lions from the Serengeti had followed the migratory wildebeest into the NCA—lions that were too stupid to run from a Maasai approaching on foot with a spear.

But it wasn't all one-sided. Lions really did kill Maasai cattle, and the dry-season culprits were often killed in retaliation, too.

Dennis went farther afield, interviewing Maasai from other parts of Tanzania and even a few from Kenya, and discovered that no small number of morani would make the pilgrimage to the NCA each wet season to conduct their *Ala-mayo* in groups of ten to fifteen warriors. The Gol Mountains were a well-known Mecca for the spear-throwing set.

But these itinerant heroes came from areas where lions also ate cattle, so the economic wounds of lion depredation were still fresh. Motives for revenge were happily merged with the joy of the hunt, never mind if the target had actually perpetrated the crime.

Grant Hopcraft has the appearance of a Clark Kent looking for the right moment to slip into the nearest phone booth. He is a mensch: quiet, observant, and smart enough to stay away from trouble. He was a mainstay in the Serengeti for many years. As a Master's student, he went to every location where lions had been observed with a kill in the previous thirty-plus years, and he found that the lions didn't catch much food in the open areas, where the prey could easily detect a large yellow cat trying to sneak up on them. Although this may seem somewhat obvious, the opposite conclusion wasn't at all obvious: lions feed best in areas where prey animals are almost never found. Lions need cover to stalk and surprise their prey, but this means they are rarely close to the big herds. So choosing habitat on the basis of prey *accessibility* rather than prey *availability* is now known as the Hopcraft Effect. At least among the Serengeti research community.

In the course of discovering the Hopcraft Effect, Grant would return to the precise spot where a lion had once been seen feeding, then he would step out of his Land Rover and measure the height of the grass and the availability of any other cover. This may sound hazardous: seek all the places where lions have been known to catch their dinner then potentially put yourself on the menu. But this was the Serengeti, and these lions have no notion of people as food. Man-eating lions are mostly restricted to the southern part of Tanzania where the natural habitat has been converted to subsistence agriculture. So Grant's biggest concerns were the risks of poisonous snakes lurking in the grass and the occasional conflicts between his housemates.

*

The Serengeti lion project is based at a cinder-block research house in the middle of the national park. The lion house has three bedrooms, a dining room, living room, bathroom, kitchen and larder. Grant and Peyton each had their own rooms, and a third student, Karyl Whitman, would visit on occasion from her field camp just outside the park.

Karyl had come to Tanzania to study the impact of trophy hunting on lion populations in the Maswa Game Reserve, located on the southwest border of the Serengeti. Trophy hunting is big business in Tanzania, and lions are the most important single species for the industry, but shooting an adult male has far-reaching consequences. A mature male lion is someone's daddy, and if something bad happens to dad, he may be replaced by a rather monstrous stepdad who will want to mate with mom again right away—and the best way to ensure a quick roll in the hay is to kill all her kids as quickly as possible so she'll stop nurturing her ex's brood and immediately resume sexual receptivity.

I had met Karyl a few years earlier; she was eager to study the impacts of lion trophy hunting, but I was worried that she might not get the necessary permissions—the Tanzanian government distrusted anyone who wasn't an insider in the trophy hunting industry. So we discussed potential alternative research projects in case her hunting study fell through, maybe even doing something with the lion's mane—although I had another student in the pipeline who would probably end up on that project. . . .

The prospects for Karyl's hunting project remained uncertain for nearly two years. Meanwhile, Peyton West joined our PhD program, and it started to look like the mane project would have to be split in two. So Peyton and Karyl circled each other warily, and it was clear that Dodge City wasn't going to be big enough for the two of them. In the end, Karyl got all the permits she needed to work in the hunting blocks, but relations with Peyton were often still tense.

Grant had been navigating minefields for months on the fatal day when Karyl drove up to the lion house from the Maswa, covered in dust from head to toe. Peyton, for her part, had been stressing out over her mane experiments: she had set out two dummies a few days earlier, turned on the tape deck, and a big male lion emerged from the river, just as planned, but then a bull elephant came out of nowhere and chased him away before he could choose dark versus blond. Even though the lion never reached the dummies, he got close enough to see that it was just a large goofy toy, and he would never respond again. She was running out of males.

Karyl lived in a modest tent in the hunting concession, cooking on a camping stove and spending hours each day in a dusty old Land Rover, which constantly broke down. For her, a visit to the three-bedroom lion house was a spacious break from her cramped grungy lifestyle, a time to socialize and a chance to enjoy a kitchen and an oven.

Peyton had found an ostrich egg a day or so before Karyl's return and graciously offered it to Karyl before disappearing into the bathroom to wash her own hair.

Karyl felt touched by the rather surprising and appropriate peace offering—a single ostrich egg is the equivalent of at least two-dozen hen's eggs, and it tastes best when baked in a quiche or a cake. Standing at the kitchen sink with her clean wet hair draped over her shoulders, she felt as close to domestic bliss as she could imagine.

The sun was shining; music was playing. Free from the confines of her tent, Karyl could stand up to her full height on a hard floor, and there was even running water. Grant had started an early dinner, and she had all the pans laid out on the kitchen counter. They opened a couple of beers, and when Grant stepped into the living room to change the music, Karyl began looking for something to crack open the egg.

He was only about ten feet from the kitchen door when he heard an explosion, a scream—and according to some reports—the sound of Peyton cackling in the bathroom.

Grant burst back into the kitchen and encountered an overwhelming stench, the sight of Karyl drenched with runny, rotten egg, and her perfect silhouette at the back of the kitchen where her body had stopped the stinking contents from plastering itself on the wall.

Professional hunters don't have much tolerance for an attractive young woman running around on her own in their game reserves; it dampens the sense of danger. Hunting clients were strictly off-limits to Karyl, but she got the sense that a lot of them were almost ludicrously gullible. The usual macho chatter around a campfire can sound like gospel to a city slicker who has just paid a fortune for an exotic adventure in the company of a world-famous professional hunter (PH). So if someone really wants to feel frightened by the darkness, there really are a lot of fangs, tusks, horns, and claws out there.

Everyone has a good story; that is just summer camp stuff.

But Karyl also heard of PHs who allowed their clients to exceed their lion quota by shooting a more impressive individual and tossing away the weedy specimen they had shot the week before. She heard about clients who arrived with nineteen-year-old hookers and unlimited quantities of alcohol. She heard of clients with such bad aim, whether from alcohol or adrenaline, that they couldn't hit a barn door.

Some clients are hunters—who cherish the stalk and the experience of the hunt and may never even pull the trigger—but most clients seem to be shooters—inexperienced, impatient nouveau riche who spray bullets at anything that moves. Why take twenty-one days to hunt big game? Get it over with ASAP so you can go home and get back to work.

The PHs despise the shooters, but it is not their job to be judgmental.

Lions and leopards are invariably shot at sites where PHs have provided bait. A client that orders a twenty-one-day safari gets a license for a lion and maybe a leopard, too. A few days before the client's arrival, the camp staff and PHs go out and shoot a buffalo or two, then check for lions every day. If they attract a male, they build a blind nearby and shoot another buffalo or maybe even a hippo to keep the lion around until the client arrives. So the greenhorn client gets his adrenaline rush around the campfire the night before, and the next morning the PH takes him to the blind and tells him, "How lucky! A lion on the first day of your safari!"

It's a bit like shooting a fish in a barrel. But after all the build-up and excitement and the sheer rush of events, the client may be literally shaking in his boots. The setup at least ensures a reasonably clean shot, and if the client messes up, the PH is standing right beside him, armed and loaded, cool as a cucumber, ready to put the poor lion out of its misery.

A decision has been made somewhere, sometime that a lion must die to generate revenue for the government of Tanzania to justify setting aside 300,000 square kilometers for wildlife, so the death should at least be humane. Better to euthanize a lion at a bait site than to risk a shot to its gut and a miserable lingering death.

But there was something else Karyl learned that we would never have guessed. A real hunter wants an animal that is a real trophy, and trophy characteristics are determined by the size of its horns, tusks, antlers, or whatever. The Serengeti lions are famous for their manes; most of the park is above five thousand feet—the weather is cool for Africa—so these well-coiffed lions rate a premium price from wealthy clients.

But the hunters kept insisting that there were all these maneless males in the Serengeti.

Manelessness is a physiological response to a hot humid climate. Peyton had to venture down to the stifling coastal heat of Tsavo to find a maneless male—that is why we took the infrared camera—and even the Tsavo males eventually grow a modest mane; it just takes a few extra years. The mane is an honest advertisement of a male's overall health; every male has the genetic predisposition to grow a mane—and mane hair continues to grow as long as the male maintains reasonably good health. But only the best can produce enough testosterone to grow a black mane, a thick-haired mane that traps the heat. If a dark-maned male is ever wounded in a fight with another male, he survives longer than a wounded blond; the cubs of dark-maned males are more likely to reach maturity. Mane coloration varies through a male's life, depending on his condition; coloration is the key variable in a lion's love life.

But in cooler climates, even the weediest male can grow a weedy blond mane.

Provided they live long enough.

So when the hunters told Karyl about the maneless Maswa males, I asked if someone could send some photos—and here were all these "trophies," and they really were maneless.

Yep, just like any other preadolescent male who doesn't need to shave. These males were only two to three years of age. A male lion doesn't reach his full dimensions until he is about five or six, so the big brave hunters were shooting the equivalent of sixth graders.

Karyl endured two long miserable years in Tanzania, trying to find enough females and cubs in the hunting areas to see whether the local lion populations were harmed by the continuous removal of adult males and by the frequent takeovers and infanticide by replacement males, who would live only until the next hunting season, then be shot and replaced in turn.

At one stage, Karyl came down with amebic dysentery and malaria at the same time. She didn't have the right meds and became so disorientated that she convinced herself she'd be fine if she just got plenty of bed rest. She stayed in the middle of nowhere until she grew too weak to stand, drink a glass of water, or think straight. An emergency airlift finally evacuated her from the Maswa and flew her up to Nairobi Hospital.

Her project had to be changed in the end because of the sheer impossibility of trying to obtain accurate counts of gun-shy lions in a hunting area. She spent the majority of her time in the Serengeti National Park, validating methods for estimating lion numbers in our known population. In southern Africa, researchers broadcast lion roars and hyena calls and count the number of lions that show up at the loudspeakers. Karyl discovered that not only did these "call-up" techniques underestimate the true number of Serengeti lions but also the error varied by different degrees in different circumstances: lions might eagerly respond in some areas but remain hidden in others. This means that no one may ever know how many lions exist in most parts of Africa, since there are few places where lions can be studied as closely as the Serengeti.

And if no one can count the number of lions in a hunting reserve, they can never set a scientific quota. If someone ever tells you that they never shoot more than 3 percent of their population in a year, you should ask, "Three percent of what?"

After coming home at the end of her final field season, Karyl worked with a computer whiz in Minnesota to construct a lion population simulation model (SimSimba) that created a virtual lion population with separate prides and mothers and cubs and male coalitions that moved around the silicon landscape just like in a fancy videogame. The fake population pretty much lived and breathed like the real thing, except whole decades of births and deaths flew by in microseconds.

When conditions mimicked the ecology of the Serengeti, the model produced a stable population with about fifty adult females, twenty to thirty adult males, and maybe a hundred cubs and subadults—about the same number that might occupy a decent-sized hunting block.

Then she imposed a variety of hunting strategies for harvesting males from the virtual lion population. Most harvest strategies are based on an annual quota, so she ran hundreds of computer simulations where different numbers of males were removed each year. Every pride contained a resident coalition, and, depending on the quota, one or more males in the coalition might be removed at random each year. Nomadic males searched for weak resident coalitions, challenged them, and, if successful, replaced them, killed their cubs and started a new reproductive sequence with the females.

Then she did something very clever: in some runs, she defined "adult male" so as to restrict harvests ("offtakes") to animals that were at least six years of age; in others, adults were at least five, at least four, or as young as three. Judging from what she saw in the hunting concessions, shooting three-year-olds seemed to be common, and in areas with heavy hunting, no males could ever survive long enough to make it to six or more.

The outcomes were startling: if offtakes were carefully implemented so that hunters never shot males younger than five or six years old, their harvests made no serious impact on the viability of the simulated populations. However, if harvests included males as young as four, the population was severely compromised, and if harvests included three-year-olds, the populations would invariably go extinct.

So after all her torments and frustrations in Africa, she had landed on her feet after all. She had a story that people needed to hear. Trophy hunting is a polarizing issue, inciting powerful emotions among Bambiologists and gun-toting crazies. But some of the emotion could be defused by accentuating the positive: trophy hunting is not inherently damaging to lion populations—provided that hunters take care to let the males mature and wait to harvest them after their cubs are safely reared.

Over a quarter of Tanzania's surface area is devoted to trophy hunting—most of the land is dull, hot, and filled with tsetse flies. In short, it is no place for a photo safari. So hunting could well provide the best possible incentive for conserving vast tracks of land. Lions occupy the top of the pyramid. If hunters take care of entire ecosystems—the land, the plants, and the herbivores—they would be rewarded with healthy numbers of lions.

But trouble was brewing. Botswana had banned trophy hunting of lions in 2000, partly on the grounds that infanticide would devastate harvested populations. A move was afoot to up-list the lion to appendix 1 of the Convention for International Trade in Endangered Species (CITES). Appendix 1 status might lead to a de facto ban on lion hunting all across Africa—reducing hunting companies' interest in conserving them and eliminating any financial incentives for rural Africans to tolerate them.

Then, with the next CITES convention on the horizon, yours truly—Mr. Tact—had the delicate task of coaxing Karyl to run more simulations on SimSimba in the waiting room of the pediatric hospital. A couple of years after finishing her fieldwork, she had married a fellow graduate student at Minnesota, moved to Alaska, and gave birth to a boy, Miles. Though he seemed perfectly healthy, their pediatrician noticed something odd during a routine check-up, and intensive exams revealed Miles to have a serious heart condition. He underwent repeated open-heart surgeries before he was six months old.

I applied gentle persistence, paced by long periods of patience, but, truth be told, Karyl needed something to occupy her mind during the long recovery period following each surgery. Months went by, Miles came home, and Karyl finished everything. I polished off the manuscript, and we sent it to *Nature*, where it was published in 2004.

The response was immediate. The press was amazed to think that trophy hunting could actually benefit lion conservation, until I reminded them of Ducks Unlimited, one of the most successful conservation programs in America, which had done more to preserve wetlands than anyone else—all so that duck hunters could preserve their hunting heritage for their children and grandchildren.

But, of course, there are others who can never accept that the means justify the ends—those that think killing a single lion could never be justified for the sake of a population, that conservation means protecting Leo no matter what. A population is vague, anonymous—better to preserve the bird in hand than protect two, unseen, in the bush.

I'm not disinterested in this debate. I work at a university. Animal-rights extremists occasionally break into our medical school and liberate lab animals—animals with no immune system that serve as models for disease or cancer research, animals that will be fatally infected the moment they reach the freedom of the great out of doors. My office in Saint Paul is across the street from a plant genomics building that was firebombed because genetic engineering of food crops was considered immoral.

I find it hard to pity the poor plants when so many people are starving, hard to feel comfortable with liberating animals that descend from a lineage kept in captivity for generations and have no clue what a cat or a hawk or an on-coming car can do to you.

At the same time, we knew perfectly well that trophy hunters were no angels. But the simple truth of the matter was that they controlled four times as much lion habitat in Africa than was protected by the national parks. So 80 percent of the lions left in the world were in their hands—regardless of the blood on those hands.

There was no choice but to accentuate the positive. Trophy hunting could become an essential tool for lion conservation, provided that hunters only harvested fully adult males and left the youngsters alone until they had safely fledged their first cohort of offspring.

ARUSHA, 5 JULY 2008

"There has been a plane crash," Andrew tells me as I enter the living room of our house. "Four people from the Wildlife Division." He hesitates for a moment. "One of them was Miriam." Another brief hesitation, "Miriam Zachariah."

Not quite a chill at the back of my head; more like a numb muddle of sensations. Miriam had once been my enemy, as Andrew well knew, and maybe a year ago, I would have been angry, foolish, or heartless enough to blurt out something harsh.

"What happened?"

"They took off from Monduli yesterday and crashed into the mountain."

I had shaken hands with Miriam a week earlier; she had seemed relieved that our battle was over. She was forty-ish and rising up in the Tanzanian government—she had recently been appointed deputy director of wildlife. Miriam was still full of life; her absence was palpable. She had done a lot of the dirty work for Emmanuel Severre (pronounced: seh-VERY), director of the Wildlife Division in the Ministry of Natural Resources, but her sin was loyalty, and after the strongman left office, she seemed just as loyal to the forces of reform.

I first met Severre in 1983 when Tanzania was at low ebb after being shunned by the international community for invading Uganda and overthrowing Idi Amin. The nation was bankrupt, but Tanzania's socialist policies had created an enormous Soviet-style bureaucracy where people pretended to work and the government pretended to pay them. Severre was an eager young wildlife biologist stationed in the Selous Game Reserve—an area the size of Switzerland with a couple of dozen rangers and himself. He wanted to find out what warthogs ate in the Selous, and he collected plant specimens every year, but the samples would rot before they could be analyzed because his research budget was so small.

I bought him an extra drum of fuel, and years went by before I saw him again. By 1999, he was stationed in Dar es Salaam (known simply as Dar by the locals) at the headquarters of the Wildlife Division or "the WD," as we always call it. He was the CITES officer, keeping track of trophy exports and presumably helping to set hunting quotas.

Severre had become the protégé of a director of wildlife named Muhidin Ndolanga, who systematically institutionalized corruption in the WD—Ndolanga encouraged his inner circle to become business partners with the hunting companies, and Ndolanga himself had developed a taste for the high life. "Hunting blocks" are tracts of land that vary between five hundred and two-thousand square kilometers in size. Each block is leased to a private company for a number of years, and the rent is ridiculously low: a few cents per acre. The WD then allows its well-chosen pals to sublease their blocks for ten to twenty times as much—a patronage system that keeps nearly a third of Tanzania's surface area out of the hands of subsistence farmers. Consequently, many of the donor agencies like the US Agency for International Development (USAID), the World Bank and WWF remained uncomfortably silent about the corruption that putatively conserved so much forest and wildlife.

During our reunion at the CITES office, Severre was still as friendly and energetic as I remembered, but I had come to the WD that day to meet with Ndolanga's recently appointed successor, Bakari Mbano. Bakari had been the principal for Tanzania's wildlife college, and he later worked as a consultant in Botswana. He was a genuinely nice man who was also hopelessly honest and open. The WD had made mincemeat out of him, and he seemed barely alive when I came to ask permission for Karyl Whitman to start her study of trophy hunting in the Maswa Game Reserve—a potential exposé that Ndolanga never would have allowed.

Bakari was barely five feet tall, and he looked like a child behind his enormous desk, chained to a hopeless task while his heart was elsewhere. When I walked in, it was clear he was glad to see a friendly face, but the lights in his eyes soon faded as we discussed the complexities of lion hunting, and I was sorry to have to ask him for a favor.

Bakari had only remained in office for another year or so. Severre contrived some sort of dispute over travel allowances or something equally trivial, Bakari was forced into an early retirement, and the prodigal ascended to become the next director of wildlife. Appointed by the president, Severre assumed one of the most powerful positions in the government in 1999. He outranked the director general of Tanzanian National Parks (TANAPA), the conservator of the Ngorongoro Conservation Area Authority, the principal of the College of African Wildlife Management, Mweka, and the director general of the Tanzanian Wildlife Research Institute (TAWIRI).

In fact, Severre became more powerful than any subsequent minister for natural resources, even wielding more power over the fate of Tanzania's wild lands than the country's president.

Ndolanga had only been ousted after he had been called to testify before a parliamentary committee. He had lost his temper and attempted to strangle one of the members of Parliament (MPs) in full view of the security agents. He nevertheless remained in office until he refused to show up to his trial for assault and battery. Bakari Mbano had maintained his integrity and his good name during his tenure as director of wildlife, but his subordinates had laughed at his naïveté and actively frustrated his attempts at reform. Severre had been handpicked by Ndolanga and had only been bypassed for Bakari as a result of the little fracas in Parliament. But after Bakari's retirement, Severre's appointment was inevitable, and he took power with a vengeance.

Severre's story was all too common in Africa—he abused his power for nearly a decade and caused a tremendous amount of harm—but he was a complex character.

Some time ago, Severre and his wife took their kids to the beach just north of Dar. The kids played in the tide pools and scampered across sandbars while their parents snoozed on shore. The afternoon wore on, and they awoke to find that the tide had come in and their kids were gone.

They searched frantically, but it was too late, and they knew it was their own fault.

Severre's wife suffered a nervous breakdown and had to be institutionalized. Severre never remarried—there were no more children—and he became legendary for his work ethic: he never seemed to leave his office before about ten o'clock at night. He *became* the WD—making every decision, controlling every aspect of every operation, awarding the leases, setting the quotas, writing new versions of the legislation that would give him ever more power.

In June 2004, I met Severre for the first time since his appointment as director of wildlife and asked him to consider a new approach to hunting lions, using an age minimum instead of quotas. I had phoned from the Serengeti, and, much to my surprise, he quickly agreed to a meeting.

I flew down to Dar two days later, and found that Severre had invited two men to my presentation: Sheni Lalji, a big-time hunting operator with business interests all over the country, and Michel Mantheakis, another operator and the head of the Tanzania Professional Hunters' Association. Sheni and Michel are both Tanzanian citizens, but Sheni is a Baluchi, whose family originally came from Pakistan, and Michel is a rather short, wiry Greek with a British accent.

We met in the boardroom at the headquarters of the WD, just the four of us. The introductions were brief, and I was barely paying attention anyway since I was so focused on trying to win them over—and my arguments were compelling, surely. In the long term, there is no conflict between business and conservation. Lions are like a crop. Look after them properly, and you can harvest more of them, making lots more money. Just be patient and let the males grow up, find a pride, rear a family, send the kids safely off to college, then—bingo—it's harvest time!

But if you're impatient, you'll condemn your populations to frequent infanticide as the dads are replaced by the next in line.

And not only is an age minimum good for your population, your males will be proper trophies with big manes—and, as Karyl had taught me to say, proper trophy males can command much higher prices by safari hunters.

"Now the only difficulty with this system is being able to estimate the lions' ages." My audience clearly agreed and looked forward to the next slide. "In the Serengeti, we've followed lions for so long that we know the precise birth date for every individual in our study area. But long-term records aren't available anywhere else. So we have been looking for a good indicator of age that could be used for a lion that has never been seen before."

Sheni and Michel watched intently.

"Here," I said, pointing, "you can see how the lion's nose—this triangular bit—starts out pink in young animals, but then it gets more and more freckled and then turns completely black by the time they're ten years old. So if you only shoot males with noses that are over half black, you can shoot as many as you want—you wouldn't even need to have quotas anymore."

And thus was established the pattern that would haunt me for the next four years.

I come into a room full of strangers whose livelihoods are on the line in ways I can only dimly appreciate. I start out telling them things they need to hear—even desperately *want* to hear—then I forget to shut up. My penchant for saying too much will ruin everything over and over again, and still I don't learn. It has cost me two years of cooperation from the hunting industry, a million dollars or so, and possibly even my sanity.

And there I was with my old pal the director of wildlife and two of the major figures in the hunting industry, and instead of sticking to science and estimating the ages of lions, I had to talk about policy.

They interrupted, not very politely. Mantheakis was especially aggressive. What did I know about business? Why was I trying to interfere with established traditions?

"The great thing about this system is that you could change your approach entirely. Instead of quotas—which limit your revenues—you could have as many clients as you wanted. It would be like a lottery—sell thousands of tickets, and some lucky guy gets to shoot the five- or six-year-old male.

"You wouldn't be selling trophies so much as selling opportunities—just like selling air. A real trophy should be rare and precious—a real prize that was only won by the best hunter."

Mount Mantheakis erupted. When a company markets a lion hunt, the sale depends on a high probability of success. If word got out that Tanzania sold ten times as many hunts as they had lions, everyone would go to Botswana or Zambia instead.

The last thing he wanted was a reputation for selling air.

Sheni's reaction was a bit more nuanced, or maybe, like me, he felt like a member of a persecuted minority—him the Asian in Africa, me the bloodstained scientist in a tank full of sharks. But despite his apparent empathy, he was no more pleased than Mantheakis.

Severre was in control, and he had what he needed. He saw how his operators would respond, and he also saw that I had just saved him hundreds of thousands of dollars.

Even though I was an idiot for trying to sell air to trophy hunters, I had made a much larger point: quotas were essentially arbitrary since no one knew how many lions they had in their hunting blocks. Trying to count lions was impossible, especially in areas where people liked to shoot them. The Convention for International Trade in Endangered Species would be considering a proposal to reclassify the lion on appendix 1 in October 2004—just a few months away—a move that might end all lion trophy hunting in Africa. The proposal had originated from Kenya, a country that had banned all trophy hunting in the 1970s, and they had support from Botswana, which had banned lion hunting in 2001.

To counter the Kenyan proposal, the WD might have been forced to conduct a nationwide survey of all the lions in Tanzania and so would every other country that wanted to preserve lion hunting. Karyl and I had aimed our paper for a high-profile journal like *Nature* because we had wanted to spread the word that it would be impossible to overhunt lions if clients only shot males that were at least five or six years of age. Imposing an age minimum meant that no surveys would be necessary after all.

Severre saw that right away—and he knew that CITES would see it too. A few weeks later, Severre invited me to speak to the Tanzanian Hunting Operators Association—TAHOA—at their annual meeting before the start of the 2004 hunting season.

This time, I stuck to the science of harvesting older males. The talk went over much better. Lion hunting was sustainable! No one needed to know how many lions they had in their blocks. Science had saved the day.

The Tanzanian hunting companies would dry up and blow away without lions on their menu, maybe 40 percent of revenues came from clients who were required to purchase twenty-one-day safaris for the privilege of hunting a lion. That meant adding a couple of buffalo, a hippo, kudu, or eland and spending a fortune on one grand expedition.

Most surprising was when eighty-year-old Gerard Pasanisi, the father of the Tanzanian hunting industry, came up to me with an enormous smile on his sun-damaged face and thanked me in his heavy French accent. Pasanisi was in a league of his own—Mantheakis referred to him as God—and nothing could happen in the Tanzanian hunting industry without his approval. Pasanisi had been close to every Tanzanian president; his clients included Giscard D'Estaing (former president of France), the first President Bush, Stormin' Norman Schwarzkopf, the owner of Real Madrid Football Club (the same guy who bought Beckham!), and dozens of minor royalty throughout Europe and North America.

So I left the TAHOA meeting composing in my head all the things I was going to tell Karyl—about how, despite all her frustrations, illnesses, and tragedies, she had started something of lasting importance. Some day there would be more lions in the world because of her simple insight about waiting to shoot males until after they had completed their paternal duties.

Severre and I were allies working toward a common goal: preserve the status quo on the lions, and incentives would remain in place to keep all that land safely set aside for conservation. We met again just before I went home to Minnesota, and I had an idea how to help him even more.

"The Kenyan proposal claims that lions are disappearing all over Africa, but hunting clients shoot hundreds of lions each year in Tanzania's hunting blocks—and our lions in the Serengeti are at an all-time high, people are losing livestock to lions everyday throughout Maasailand, and you have all those man-eaters down around Selous, right?"

"Of course," he said as we sat around a coffee table in his office at the WD with a few members of his inner circle.

"I've seen that report by Rolf Baldus," I charged along, trying not to react to his grimace at Rolf's name, "If the rest of the world realized how many Tanzanians were being eaten by lions each year, they'd think twice about ending lion trophy hunting."

"We have too many lions in Tanzania, not too few," Severre extended his hand toward his crew, who all nodded.

"Exactly!" says I, rolling along. "Why don't we document the extent of the man-eating problem before October, and you can use the results at CITES. I have a couple of students who've been studying cattle killing up in the north; they could use the same techniques to learn why people are being eaten by lions down south."

And that is how Dennis Ikanda became the world's expert on man-eating lions. He had completed his fieldwork on the Maasai in Ngorongoro and discovered several factors that made cattle herds particularly vulnerable to lion predation. Lions were most likely to attack herds that were tended exclusively by small children, and they also preferred large herds with relatively few herders. Not exactly rocket science, but valuable for trying to mitigate the problem—since there is nothing like quantitative data to get people to realize the consequences of their behavior, even if they've behaved the same way for a thousand years.

In the meantime, Bernard Kissui had started similar studies around Tarangire, and he was finding out that lions were far more vulnerable to retaliation for cattle killing than were hyenas or leopards—lions tended to stand their ground long enough to be speared whereas hyenas and leopards vanished before the Masai could catch them.

The wildlife authorities had asked me to help with their man-eating problem near the Selous back in the 1980s, but I had no idea how to approach the problem. I was only familiar with well-fed lions within the Serengeti and Ngorongoro Crater—cats that would run away at the first sign of a perpendicular biped. In the Serengeti, lions would occasionally walk past the veranda while we were enjoying the evening; they carried on untroubled as long as we sat still and watched in silence. My vehicle got stuck in the mud right next to a dozen lions when my daughter Catherine was three years old; I simply hoisted her on my shoulders and strode determinedly past the whole pride.

But the lions around the Selous were a different beast altogether. They treated people as food. Rolf Baldus had come to Tanzania in the mid-nineties as part of a German government aid program to help the WD raise revenues and improve management practices. Rolf was a hunter himself, and he loved the romance of the Selous. When I first met him in his office at the Ivory Room (WD headquarters), he handed me a lion skull and asked how old the animal had been.

"Probably only three years old or so."

"Are you sure?" he asked, disappointed.

"Yes, look at how sharp his canines are—and his teeth are still white."

"What about that big tooth at the back? He had an abscess—it must have hurt like hell. He probably couldn't catch normal lion food with all that pain."

"What's his story?"

"The villagers called him Osama. He killed forty-three people."

"How long did it go on?"

"Over two and a half years."

"Oh, then he didn't do it all by himself—he's hardly three years old. He would've been only six months old at the beginning—still a cub with his mother."

"He wasn't driven to eat people by his tooth?"

"If he was around for the first few cases, he wouldn't even have had his permanent teeth yet. So he didn't do it because of his tooth. He certainly didn't kill all those people by himself."

In addition to his interest in feline dentistry, Rolf had asked Severre's staff to extract all the records at the WD of man eating around the Selous over the previous ten years. So when Dennis and Benny went down south, they were able to interview the survivors or victims' families, and discover the details of the attacks.

And the devil was in the details. These were not cases of big cats defending themselves against angry pastoralists; these lions know what they are doing: they want to eat *you*, and your little dog, too.

These cases included attacks in the middle of town and lions breaking into people's houses—pushing through the walls, digging down through thatched roofs, snatching babies out of the arms of nursing mothers, grabbing people as they went to the outhouse at night.

Over a hundred people a year.

Lindi District, located on the coast of the Indian Ocean and not far from the Mozambique border, averaged nearly one attack a month for the past *fifteen years*. Lindi District is about the same size as Rhode Island.

So opinions are divided over the merits of lion conservation. Most rural people in southern Tanzania would just as soon be rid of every lion in the world—although not everyone felt this way. Villagers in these areas often distinguish between lions and man-eaters by referring to the latter as Simba-*watu*—literally, "lion people" but the phrase is usually translated as "spirit lions."

Dennis explained it to me like this:

"Down south, a lot of people think man-eaters aren't animals, but spirits sent to torment them by their enemies. Some people are so convinced these aren't real lions that they won't even try to catch them.

"They believe that if someone in their family was attacked or killed by a spirit lion, it was punishment for committing a crime against someone in the community—like sleeping with their wife or stealing. So the family feels disgraced by the crimes of the relative. The disgrace can be so serious that they might not report the attack to anyone and move away to another village.

"They usually tell the witch doctor first—asking him if it was a real lion or a spirit lion. So long as he tells them it was a spirit lion, they don't fight back. But if the outbreak goes on, the witch doctor will change his mind, and the whole village goes after the lion. The men go out with fishing nets and sticks. The men with the nets form a semicircle around the spot where the lion was last seen. The men with the sticks beat on cans and drums and drive the lion towards the nets, the nets are pulled closer together, and if the lion is in there, they attack it with the sticks."

Guns are virtually nonexistent in these villages, although someone might show up with a homemade muzzle-loader every now and then. These are not high-quality nets and hitting at a lion with a stick is not the most efficient way to dispatch an angry 300–450-pound carnivore.

All of the wildlife in the country belongs to the government of Tanzania, so villagers are not supposed to kill man-eaters on their own. The WD is meant to work in conjunction with the district game officers, who may need two to four days to locate a functional firearm, the appropriate ammunition, and a working vehicle to travel to the affected village.

Even if the district game officer shows up quickly, these lions are good at their craft. They hide by day and roam by night. One man-eater in southern Tanzania became known as Simba *Karatasi*, "paper lion," because his movements were as erratic as a piece of paper blowing in the wind.

While there are also people down south who dismiss the notion of Simba-watu, everyone is fed up with being lion food. Their MPs complain in Parliament. Newspapers report the lurid details with all the enthusiasm of a supermarket tabloid. And successful counterstrategies have become the stuff of legend. One man discovered his half eaten wife when she didn't come back from the outhouse. He laced her remains with rat poison. Another man discovered the bottom half of his mother-in-law. Rat poison again finished the job.

Lots of lions were eating lots of people. Why should Tanzanians have to live with these dangerous animals? There were plenty of trophy hunters who came to the Selous and presumably helped to control lion numbers. Why would anyone want to ban lion hunting?

When I got back to Minnesota, I worried that Severre might need help in convincing the rest of the world to vote down the Kenyan proposal. I started calling people at Safari Club International (SCI). I sent out copies of our *Nature* paper, the TAHOA PowerPoint, and my point-by-point rebuttal to the Kenyan proposal. Safari Club International seemed somewhat dubious, but John Jackson, who was director of Conservation Force, recognized my sincerity in wanting to preserve lion hunting, and I flew to Washington, DC, in late September to meet him in person. It quickly became clear that I would have to choose one or the other—John had once led SCI, but he established Conservation Force as a rival to the powerful organization located just a few blocks from the Capitol Building.

I liked the fact that John wanted to act fast. He knew Severre personally, and within a week I was an official member of the Tanzanian delegation to the CITES convention in Bangkok, and a few days later I was in Thailand courtesy of Conservation Force. I joined thousands of people, representing every country in the world. Presentations were simultaneously translated into a dozen languages—just like at the United Nations, but these talks were about turtles and birds and trees and rhinos and elephants.

Of all these species, the lion was expected to be the star of the show. A huge coalition of animal-welfare organizations had lobbied hard to uplist the lion to appendix 1 so as to ban lion trophy hunting once and for all. Earnest young Brits, Europeans, and Americans handed out lapel pins of round-faced smiling lions; the Born Free Trust hosted a fancy reception with a colorful array of Thai cuisine and exotic fruits. Speeches proclaimed the plight of the lion. Lion numbers were falling dramatically across Africa, diseases were ravaging remnant populations, and trophy hunting was the final straw.

And these people were all living safely in London or Amsterdam or Los Angeles, and they had convinced themselves that the only way to save the species was to protect the lion from the evil hunter. But the arguments in the Kenyan proposal were flat wrong. First, the ghost writers of their report (who were hired by the International Foundation for Animal Welfare or IFAW) had exaggerated the risks of infanticide at least tenfold and conveniently ignored the fact that the incoming males only eliminated their step-cubs so that they could quickly replace them with their own offspring. Second, they had put together a very superficial set of estimates about the remaining population of lions across Africa that somehow claimed that Tanzania had virtually no lions outside the Serengeti—whereas our man-eating survey plus all the trophy-hunting records clearly showed that lions still occupied over at least a third of a very large country. Third, they implied that lions faced imminent extinction from a virus, feline immunodeficiency virus (FIV), that infected most of the large populations still extant in Africa.

The FIV argument was the most annoying of all. We had worked with the American lab that had discovered the virus in 1990. Our collaborators at the National Cancer Institute had looked at FIV in the lions from a variety of angles, and there were no worrying consequences of infection. In fact, every single adult lion in the Serengeti and Ngorongoro Crater had been infected with FIV for the past twenty years (and genetic studies suggested they'd been infected for thousands of years), yet the populations were doing just fine.

But the author of the Kenyan report had made extravagant claims about the health impacts of FIV using data that were flimsy to nonexistent, and no one in the scientific community took him seriously—though his claims had caught the attention of a website that set up an online prayer circle to save the lions from AIDS.

So as soon as we convened in the vast conference hall in Bangkok in October 2004, the official delegation of the United Republic of Tanzania prepared our rebuttal. I typed out a bunch of statements, and Severre supervised the revisions, specifying the appropriate language for CITES. The director general of the Tanzanian Wildlife Research Institute, Dr. Charles Mlingwa, contributed his perspectives, and we sought the approval of the Honorable Zakia Meghji, the minister of natural resources and tourism, who had accompanied us as a sign of the seriousness with which the Tanzanian government viewed the future status of the lion. Mama Meghji was dressed in a traditional Zanzibari gown, and she cut an impressive figure as we entered the dining area and sat down to lunch at our own table, far from the Kenyans, who were clearly unsettled by our magnificence.

The lion proposal was due for debate in three days, so once we had polished our rebuttal, we circulated it to the pro-utilization faction (hunters, ranchers, loggers) and to the antis (as John Jackson called them), and after an appropriate interval, I went to see the Kenyan delegation.

The Kenyan minister and his top deputy were otherwise engaged, but I had the full attention of their younger counterparts—who went pale once they realized the weaknesses in their ghost-written proposal.

I then went into a lecture hall to give a presentation to fifty or so antis. The American and Royal Societies for the Prevention of Cruelty against Animals were there, along with Born Free and IFAW. It was essentially the same talk I'd given to TAHOA in Dar—with the same message: shoot older males, and infanticide is not a problem. But this time I hammered home the messages that the trophy hunters had always taken for granted. Lions were mean, vicious, terrible, horrible, awful animals; local people hated them. Man-eating lions killed over a hundred people a year; cattle killers damaged the livelihoods of thousands of impoverished pastoralists.

Far more land is set aside for trophy hunting than for national parks. Take away the incentive for hunters to grow a healthy crop of lions, and the king of beasts would be eliminated from most of its remaining range.

Love it or hate it, lions needed trophy hunting as much as trophy hunting needed lions.

And by the end, the air had left the room; the antis knew they had lost.

But there were some sharp minds in the audience, and they made themselves clear.

"You may be right that a well-regulated hunting industry would be good for lion conservation. But do you really think they'll do what you say?"

"Well, that's always the question," I began, "and there's every reason to be skeptical. But all I can say is that the Tanzanians seem genuinely worried about their lions, and the Tanzanian Hunting Operator's Association—TAHOA—unanimously approved a new policy in July to restrict lion hunting to males that are at least six years old."

"How do you know they'll follow through?"

"Well, the six-year rule would save them a lot of money on surveys, and they're desperate to keep their lions on quota, so if they don't follow through, they'll have a lot of explaining to do." I stood up to full height, "After all, the flip side of our simulation model shows that if they shoot too many underage males, the impact would be devastating, and they know they're going to be closely watched from now on."

One person was already watching me. The head of a British animal welfare organization, who I will call Sara Bambara. Sara earned her PhD studying bat-eared foxes in the Serengeti, but she always cared more about the foxes' well-being than the details of their behavior or ecology—she fed raisins to starving foxes each dry season; she rescued a sick fox pup and nursed it in her tiny house until it died of rabies. She nurtured a monkey with a broken arm. Whenever she sat on her verandah, mongooses scurried at her feet, a dozen species of birds hovered around her like a halo. After finishing her degree, she left academia to follow her heart.

In the early days of the Iraq War, Sara called me about the lions in Uday Hussein's private menagerie near Baghdad. Saddam Hussein hadn't been captured yet, civilian casualties were rapidly climbing, and Iraq's infrastructure had been shattered; the air conditioning no longer worked at the Hussein family zoo. Several animal organizations wanted to rescue the lions, and the question was whether to ask the Royal Air Force to airlift them to South Africa or leave them in the heat of Iraq.

I recommended that they just shoot the lions now and get it over with.

I never found out what happened, and I wasn't going to ask her at the convention. Sara is German, and to call her "intense" is kind of like saying that the sun emits the occasional photon. But she is also a scientist and far more rational than her peers, and she helped me navigate through the various antis, although her message was relatively simple, "Don't trust any of them. They never keep their word. They are just looking out for themselves, and each one of them wants to be the one to take credit for saving the lion."

But before I left the conference room, I made one more comment to the antis: "Hey, don't feel too bad if your proposal fails this time. The hunting industry is scared to death they'll lose the lion. I had tried working with them in Tanzania for years, and they never would have listened to me if you hadn't put together that proposal. While your arguments may be flawed, I agree that the trophy hunters should be kept on a tight leash. So keep watching, and if they don't behave themselves, you'll win next time."

Then came the European Union delegation, which wanted to broker a deal between Kenya and Tanzania. We were supposed to meet the Kenyans the night of the big CITES reception, but they failed to show. The following day, the Dutchman in charge of the rapprochement gathered together a quorum of Kenyans, and we all sat together at a long table in a distant wing of the Bangkok convention center.

Severre proceeded to lecture the enemy about the fact that Kenya and Tanzania were both members of the recently reformed East African Community and that protocol required the Kenyans to obtain community approval for any document impinging on Tanzania's sovereignty, specifically informing him in person of any attempt to influence international law in a way that would affect Tanzania's wildlife policies. Mama Meghji nodded in approval, and the Dutch intermediary strained to find some way to prevent the Kenyans from having looked naive.

The European Union wanted the Tanzanians to acknowledge that the Kenyans had backed down, so therefore couldn't the Tanzanians show similar good grace?

No.

And why should they? The lion proposal was a mess, promulgated by the animal rights organizations, and the Kenyans had been willing partners in an attempt to spread the antis' ideology to the rest of Africa. Severre had spent a full day in a meeting with the Southern African Development Community representatives, and the Mozambicans, Zambians, South Africans, Zimbabweans, and Namibians were just as irate over the lion proposal as Tanzania—even Botswana no longer seemed convinced that a lion ban was such a good idea.

So we left our meeting expecting the Kenyans to back off completely.

But while they had withdrawn their original proposal, the next afternoon, the Kenyans slipped in an official request for a lion "review," and the Tanzanian delegation cried foul. Charles Mlingwa made it halfway across the hall before someone persuaded him to calm down.

I shared my colleagues' frustration, but I was also intrigued that a CITES review might mean that someone might actually collect real data on lion-hunting offtakes. Each African country would now be required to put forward an official statement on the status of lion populations in their parks and reserves, and CITES would review the lion's status again in the next few years.

Safari Club International had snubbed me in Bangkok because I'd hung out with John Jackson at breakfast and introduced him to all my Tanzanian colleagues. But the sheer size of SCI gave them influence with the CITES secretariat, and SCI largely guided the lion review process the following year, arranging for regional workshops in Cameroon and South Africa.

I was next invited to a meeting in Kasane, Botswana, in March 2005, where it would be decided whether to reopen lion hunting after a three-year ban. Botswana was the erstwhile home of the author of the Kenyan CITES proposal, but he didn't show up and was instead represented by Derek Joubert, the *National Geographic* filmmaker as well as a close friend and business partner of the presumptive next president of Botswana, Ian Khama.

Joubert has made a career out of filming Botswana's lions, presenting himself at times as a scientist in his photo essays in *National Geographic* magazine. Joubert's films have been hugely successful and highly controversial, edited in ways to tell a largely fictional narrative for the sake of dramatic impact rather than to represent the natural world.

On the first day of the Botswana workshop, Joubert set up his cameras in the middle of the conference room and started to film the proceedings. When asked what he was doing, he replied that he was on assignment from the National Geographic Society—a claim that the society later refuted.

Botswana had publicly banned lion hunting because of increasing levels of problem-animal conflict. Botswana was the richest country per capita in Africa; it had vast mineral wealth and a tiny population. Rural Botswanans traditionally practiced pastoralism, and their love of livestock was similar to that of the Maasai—cattle remained the gold standard of family wealth. With all the money in the resource-rich country (diamonds, gold), and the government's paternalistic policies, even the poorest family in Botswana could afford such luxuries as television, and job opportunities had drawn most rural youth to the capital city of Gaberone.

So cattle husbandry had deteriorated to where people opened the gates to their bomas in the morning, went inside their huts to watch TV, and the cattle wandered off untended for the day. In the evening, the villagers closed the gates on any cattle that had come back to spend the night in the boma, and the stragglers were left out to fend for themselves in the dark.

A graduate student from Oxford, Graham Hemson, worked in one of these areas and saw how the lions would catch a wildebeest or zebra or cow in the middle of the night. As daylight broke, the lions would continue feeding on their wild prey but quickly abandoned any half-eaten cow, as if the lions knew they shouldn't be eating beef. Lions with a sense of guilt?

In the 1970s, Botswana had promoted the health of the nation's livestock by fencing vast areas to stop wildlife migrations that might expose cattle to foot-and-mouth disease, leading to massive die-offs of wildebeest and other native ungulates—and an international outcry from conservationists. But ecotourism had grown to such an extent in the 1990s that the country attempted to promote conservation by compensating cattle herders for their losses to lion predation, thereby hoping to discourage the extent of retaliatory killings. But compensation schemes always seem to fail. Whether in Montana or Scandinavia, ranchers love to be paid for losing livestock, and if they can only be paid for depredation, their cow must surely have been eaten by a wolf or lynx rather than have died from starvation or disease.

In Botswana, if a villager could only be compensated for lion depredation, their cow must only have been eaten by a lion and never by a hyena or leopard.

Even with their vast mineral wealth and tiny pastoralist population, the Botswana government quickly ran out of funds to pay off all the compensation claims.

With all this controversy and with cattle being eaten while villagers watched their TVs, the government had to justify why it was OK for rich white people to come shoot lions in the game reserves while local people were supposed to let lions get away with killing their cattle.

So the government banned all killing of lions for any reason.

At least that was the official story.

The hunting industry told its own version. Derek Joubert was close with Ian Khama, the son of Botswana's first president and the heir apparent. Joubert and Khama were partners in a photo-tourist company, and they wanted to shut down hunting for business reasons. Hunting blocks in Botswana were immediately adjacent to tourist blocks in a crazy-quilt pattern all around the country—tourists wanted to take photos of tame animals that they could approach to within a few meters. Botswana was the only country in Africa where hunters stalked their quarry on foot, tracking them in the sands, and once a wild animal had been habituated to people in vehicles, it became an easy trophy. Joubert had repeatedly complained of hunters who shot his tamed lions when they wandered into neighboring hunting blocks.

Joubert had also played a role in the Kenyan CITES proposal in exaggerating the impacts of infanticide and the fiction of FIV dooming the lion to extinction. So no one was particularly comfortable when he set up his cameras and started filming every word we had to say.

I gave my usual talk about lion hunting, several other scientists reported on their estimates of Botswana's lion population size, and the government representatives were clearly leaning in our direction—exhausted as they were by the complaints of uncompensated villagers and the pressures from the hunting companies.

Joubert gave the final presentation at the workshop, claiming that it was much more difficult to estimate a lion's age than Karyl and I had suggested. We had emphasized the darkness of a lion's nose, and sometimes even a small cub might have a very dark nose. He went on to say that, well, he wasn't a virologist, but he knew that when he had a virus, he wasn't well, and since almost every lion in Botswana had FIV, they couldn't be feeling too well either.

I sat on my hands until he was finished, but then I stood up and said that of course small cubs often had gray noses, but turned pink by the time they were half grown. No hunter came to Africa to shoot a lion the size of a tabby cat. As for FIV, we had looked long and hard for fifteen years for any evidence of harm, and we couldn't find anything—and, if we had, we would have been the first to report it. Some species of monkey carried SIV perfectly well; given enough time, a host species can evolve mechanisms to withstand chronic infection from a retrovirus, and lions had had FIV for eons.

The feeling was somewhat the same as teaching evolution to a large class of freshmen in the Midwest: knowing that no matter what you say, no matter how extensive or clear the evidence, there will always be about half the students who will refuse to accept anything that threatens their faith in Genesis.

Joubert must have had another agenda as well, but I still don't know if he was acting out of business interests or if he had a similar worldview as the antis at CITES. Either way, his arguments were overruled, and Botswana reopened lion hunting a few months later.

But all was not well in Tanzania. I had attended an SCI convention in Reno in January of 2005, repeating my TAHOA PowerPoint about lion aging for the professional hunters, meeting with operators at their display booths, and generally trying not to freak out at the undercurrents of the slaughter on display. There were taxidermists with stuffed baboons screaming at stuffed leopards; stuffed bears standing on their hind legs; stuffed lions galore. And there wasn't a single stuffed lion that could have been older than about three or four years of age.

The most popular booths showed videos of beautiful animals in brilliant sunlight that crumpled into the dirt after lead slugs had pierced their hearts. Or videos over the shoulder of the PH whose modus operandi was to provoke a buffalo or hippo to charge him before he put a bullet through its brain at about twenty feet.

There were booths from Europe, Africa, Canada and the United States, South America, and Asia. Some combined fishing with big game; some were family owned and modest. Others took testosterone to the next level. The worst were from Spain and Argentina, which proudly displayed photographs of smiling hunters who had shot five hundred to a thousand birds in a single day.

The avian massacres stopped me in my tracks.

"That's really impressive," I said. "I mean, that's an incredible number of animals. I used to go duck hunting when I was a kid, we'd get five or six on a good day, and we'd always eat them all. But who could eat a thousand ducks?"

"Oh," the vendor stumbled, "they only eat a few of these themselves. They give the rest to the local communities."

John Jackson attended every hunting convention, so he introduced me to various key figures in the hunting industry in Botswana and French West Africa and encouraged everyone to see my talk, but it was only sparsely attended. The PHs were out looking for business, and the operators only had these few days to fill the bulk of their schedules.

Of those who did attend, there was a consensus that the lion was in trouble throughout Tanzania. It was harder to find good lions, and the one company that had already tried to follow our six-year minimum sometimes found no suitable lions at all. There was also the rather awkward point that hunters found it hard to get a good enough view of the lion to estimate his age. The Tanzanian hunting companies all shot their lions at baits, and the blood of the carcass could obscure the lion's nose. Besides, it was one thing to look at a sleepy lion's nose in the Serengeti, but hunting-block lions were jumpy creatures that would run away if they saw you—and they generally saw you as soon as they pointed their noses at you. The PHs needed other age indicators that they could use even when the lion was pointing in the opposite direction.

But there was a far more profound and pervasive problem: there might only be four or five honest operators out of fifty-plus companies in Tanzania. Only a tiny minority was willing to take the longer view of conservation; the rest were fly-by-night operations that only cared about short-term profits. They had no incentive to ensure healthy populations in the distant future. If they overshot the lions in Tanzania, they could move operations down to Zambia or Zimbabwe the following year.

And the whole system was driven by the subleasing of the well-connected politicians who rented their blocks season by season; a matrix of institutional corruption that ran straight through the WD. Rolf Baldus was persona non grata with Severre because he had tried to link European aid to concrete reforms in the Wildlife Division, particularly by trying to impose an open and transparent method for assigning the hunting blocks. Rolf worked for the German equivalent of USAID, an organization called Deutsche Gesellschaft für Technische Zusammenarbeit, and he had a particular interest in the diets of Osama and Simba Karatasi because his agency's aid projects were near the Selous. But he had also become all too familiar with the sordid details of corruption in the hunting sector. His own office was less than twenty meters from Severre's, but relations were so bad between the two men that he hadn't set eyes on Severre for several years. While we were all in Reno, Rolf was packing to leave Dar after the German government had given up all hope of reforming the WD.

George Hartley works for one of the only honest companies in Tanzania. Tanzania Game Trackers (TGT) is owned by a wealthy family in Houston, and they have spent millions of dollars on antipoaching and the ecological restoration of their hunting blocks. The main complaint of the rest of the industry has been that TGT doesn't have to operate within the economic constraints of a true hunting company. No *real* hunting company could afford to engage in such philanthropy.

Nor, apparently, would any "real" hunting company pass up an opportunity to shoot a lion no matter how young and scruffy.

George had sent me photos of all the lions shot by TGT over the past few years; TGT had even fired a couple of PHs who allowed their clients to shoot obviously underage lions. Tanzania Game Trackers proudly advertised the quality of males they were now harvesting, after having allowed their young males to grow up and develop full physiques. Business was booming, and their bookings stretched far into the future.

But their blocks were flanked by unscrupulous neighbors who, like Joubert's neighbors in Botswana, would shoot TGT's finest males by setting baits close to the boundaries. So TGT wasn't always able to reap its own harvest.

"I'm telling you, this can't work unless everyone follows the same rules," George said over the phone when I was in Minnesota one winter's day.

"But TAHOA all agreed to the age minimum."

"Doesn't matter. Half those guys that voted on the age minimum don't even hunt those blocks themselves. They're subleasers, and they don't care what gets shot in their blocks. Most of the rest will tell you one thing and do another."

"There must be some way to enforce it," I insisted.

"Not with Severre in charge."

"Why not?"

"He works for them."

My office is always cluttered. I have three large desks in a small room with two or three computers running, and papers scattered over every surface. I'm up on the fifth floor, and on a clear day I can look beyond the Capitol Building in Saint Paul all the way to Wisconsin.

But it was gray day in January 2005, and I was trying to absorb as much light as I could, looking for any bit of color on my shelves. I remember wracking my brains, trying to atone for my impetuosity in giving a corrupt industry carte blanche to hunt lions to extinction.

I also remember the precise moment when the solution presented itself. I was staring at the CD cases on my office shelf, just behind the Play-Doh mobile of colorful fishes that my daughter Catherine had made when she was about eight years old. I was mesmerized by the long row of vertical CD labels, each a different color, and the thick black strings supporting tiny red, blue, and yellow fishes.

Something clicked, and I said to George, "What about some sort of independent entity that could serve as a certification system? You, know, like with tropical forestry—forest certification. Everything is getting certified these days. Why not certify hunting trophies?"

And that is the moment I left the safety of man-eating lions and leapt headlong into ruin.

I came up with the name Savannas Forever a few days later at our kitchen table in Minneapolis. Catherine was still home from college for the holidays, and she was checking her e-mail when I told her I wanted to set up a system for "conservation hunting" and asked if she liked the sense of the words as well as the repetitive *s*'s, *v*'s, and *r*'s. She gave her seal of approval, and I wrote it out in large letters in my newest PowerPoint.

I went to the Dallas Safari Club convention the next day to give a talk on lion hunting, and it was at a dinner of steak and potatoes that I first informed John Jackson about my idea for hunting certification. John is a lawyer by profession; he wears a lawyer's glasses, but his white hair and stiff smile make him look a little like John McCain. He gets nervous about last minute surprises, and I was afraid that he might pull the plug on my talk the next day. But we were dining with my old friend from Berkeley, Laurence Frank. Laurence was the first to apply scientific techniques to the Maasai-lion problem; his research had inspired Dennis Ikanda's project in Ngorongoro, and Laurence was an intense advocate for the resumption of trophy hunting in Kenya. Kenyan wildlife policies appeared to be run entirely by the International Foundation for Animal Welfare, and Kenya had been home to a lot of famous safari hunters back in the day. Kenya was the Holy Grail of SCI and all the other hunting organizations—any chance of reopening hunting in Kenya would be a huge victory against the antis.

Laurence was quick to voice his support for Savannas Forever, and John followed in a surprisingly neutral tone—he was doubtful whether the idea would work, but he didn't dismiss it out of hand.

The next day, I talked at breakfast to at least two hundred hunting operators, PHs, and taxidermists. I started out with pretty much the same story as at SCI in Reno, but then I took a new direction. The hunting industry was essential for the future of so many species. Hunters, after all, were the stewards of the land. They had accepted the responsibility for looking after the wild areas, for conserving the rare and valuable animals. They were the unsung heroes of conservation, but a few unethical companies could tarnish their reputation. These rotten apples provided ammunition for the antis. The antis wanted to ban hunting, and the antis spent most of their time and energy trying to influence public opinion. The hunting industry had no mechanism to counter these claims, so countries like Kenya had banned hunting and were controlled by an animal welfare organization; Botswana had banned lion hunting for a couple of years, and we nearly lost the lion at CITES. The whole industry was at risk.

With an organization like Savannas Forever, the hunters' good works could be objectively evaluated and celebrated. Governments could be reassured; public opinion could be altered as surely as Ducks Unlimited had done for waterfowl hunting in North America.

And there I was in a room full of armed men in Dallas, talking about how some of them were truly bad eggs who needed to be exposed and driven out of business, but no one took aim or even scowled at me. John Jackson was dizzied by the whole performance, but he didn't try to turn me around. I later wandered the display booths with their photos of happy hunters and fishermen, and the head mounts of deer, elk and big-horned sheep and the occasional stuffed Cape buffalo, leopard and lion, and I was amazed by the number of vendors who stopped and thanked me for wanting to try to do something for their image and to help clean up their industry.

My wife Susan had been looking for something more fulfilling than serving as vice president of marketing at yet another Minnesota business, and she also wanted to spend time in Africa. We had taken an uneventful trip to South Africa a few years after the horrific morning in Nairobi, and she was eager to see Tanzania when she wasn't suffering from post-traumatic stress disorder.

It seemed like the perfect merger of science and business. We would work with hunting companies to help maximize their harvests scientifically and sustainably and promote their conservation practices with the hunting public, possibly even enabling them to charge a premium as a top-tier company.

Susan researched certification programs in forestry, aquaria, and whatever else, and we discussed the relevant dimensions of a system for trophy hunting. Not only should they shoot male lions of a safe age, but they should also adhere to higher standards for Cape buffalo and leopards. Leopards were a special appendix 1 species at CITES—they had been given such high status because of the once extensive market in leopard-skin pillbox hats (among other fashion items). Otherwise, leopards were everywhere, eating guard dogs in Nairobi and Arusha and generally skulking around in any patch of forest. Leopards were high-prestige trophy animals, but hunters were only supposed to shoot males above a certain minimum body length, and leopards were just as infanticidal as lions. Cape buffalo were the meat loaf of the hunting industry. Clients shot two to three of them on a twenty-one-day big-cat safari, and we had heard complaints that fully mature bull buffalo were becoming scarce in many parts of Africa.

But hunters had more responsibilities than just sustainable harvest strategies. The Tanzanian government required them to assist the WD with antipoaching and to implement community development projects in nearby villages. So we were thinking of ways to help the hunters improve their antipoaching efforts and to select community projects that would actually improve the lives of neighboring villagers—increase the hunters' return on investment, as a businessman might say.

I flew to South Africa on Christmas Day 2005 with Catherine, who had enrolled in a workshop on conflict resolution at the University of Cape Town. We spent a couple of days together climbing Table Mountain, watching penguins at The Boulders, and wandering the Cape of Good Hope on New Year's day 2006 before I left her to attend an SCI-sponsored lion meeting in Johannesburg.

And there was John Jackson again, along with Bob Byrne, a kinder, gentler representative from SCI, a couple of my old friends from the Wildlife Conservation Society, the International Union for Conservation of Nature, and CITES, and delegations from countries stretching from Sudan, Ethiopia, and Somalia down to Namibia and South Africa.

The Tanzanian elections in December 2005 had led to a reshuffle in the government. Charles Mlingwa was now an MP, and the new director of the Wildlife Research Institute was my former postdoc from the Serengeti, Simon Mduma. Mama Meghji had been appointed finance minister by the new president, and the new minister and the new permanent secretary in natural resources and tourism were two of Severre's most outspoken enemies, so he had stayed in Dar to preserve his hold on power. Severre sent Miriam Zacharia as his personal representative, and she was joined by Julius Kibebe, the CITES officer in the WD.

I had been asked to evaluate the trophy-hunting trends across Africa, and my presentation was far more critical of the hunting industry than the TAHOA/CITES/SCI versions.

The Convention for International Trade in Endangered Species publicly distributes all data on the international trade of endangered species, including the number of trophies exported from each country. There were seven countries with consistent exports of lion trophies: Botswana, Central African Republic, Mozambique, Namibia, Tanzania, Zambia, and Zimbabwe. The traditional view of trophy hunting is that quotas should remain lower than 3 percent of the target population. There were no definite population numbers for any of these countries, but a couple of continent-wide surveys had recently been completed, and judging from the annual lion offtakes compared to these guesstimates, the level of lion hunting wasn't particularly worrisome—except for Zimbabwe.

Whereas all the other lion-hunting countries seemed to be harvesting 1–3 percent of their lions, Zimbabwe often shot 10–14 percent of their lions in a year.

More worrisome was the fact that Zimbabwe was the only country in Africa to include females on quota. This just isn't sporting. Trophies are meant to be resplendent males. A population's breeding potential stems from its females.

I had been in near-conflict with another research group in Oxford, led by biologist Andy Loveridge, who had issued a report in 2004 stating that trophy hunting was endangering the lion population in Hwange National Park, located near the Botswana border in western Zimbabwe. We almost came to academic blows over our different perspectives—Andy's group felt that trophy hunting was nearly as bad as claimed by the Kenyan CITES proposal, and Karyl and I considered Andy's group to be alarmists.

As often happens, we were both right, but our respective positions were too blinkered by our own personal experiences. When Andy and I sat down together in Johannesburg, we realized that the Zimbabwean quotas were vastly higher than Tanzania's. Hwange Park held relatively few lions. The northern end of the park was immediately adjacent to a relatively lush hunting block, whereas the park itself was harsh and arid. So whenever the lions in the hunting block were harvested, the Hwange lions quickly moved next door to become next in line, and the Joubert/TGT bad-neighbor effect was having an impact on the entire park.

By now, I had collected a rather disturbing selection of photographs of trophy lions that were available from the Internet. The stuffed lion heads at SCI might have been from males that were only three or four years old, but those were often the best of the batch from each country. It was appalling to see rich white hunters delighted with themselves for having shot two- or three-year-olds with the merest wisp of a mane.

I was prepared to show the ruinously high offtakes in Zimbabwe and photo galleries of subadult trophies from Zimbabwe and Tanzania—and Tanzania had explicitly agreed to a six-year minimum.

But as I gave a practice run for my fellow scientists, Bob Byrne peered over their shoulders and called an emergency meeting of the Zimbabwean delegates.

I met with Bob and the Zimbabweans the next day at a restaurant that served traditional African cuisine. I ordered mopani worms and cow-hoof soup. The soup was gelatinous and gross; the fat black worms had been sun-dried so were rather tough.

I wanted to know why Zimbabwe's quotas were so high compared to every other country in Africa. It seemed that Zimbabwe had gotten used to shooting exceptionally high numbers of lions and decided to add females to maintain their offtakes. They also seemed to be meeting their quotas by shooting a large proportion of young animals.

They were mortified. Safari Club International had worked hand in hand with the Zimbabwean wildlife authorities for decades. Their flagship program for all of Africa, Communal Areas Management Programme for Indigenous Resources (CAMPFIRE), was often (though not universally) considered one of the most successful conservation efforts in the world.

They seemed reluctant to discuss the driving force behind their high quotas, but they promised to ban the hunting of female lions, effective immediately, and they had already lowered their quotas near Hwange in response to Andy Loveridge's analysis.

So next came the Tanzanians. I met with Miriam, Julius, and Simon after dinner and showed them the PowerPoint, and they panicked at the sight of all those subadults.

"You can't show that!" Miriam pleaded.

"I have to. TAHOA promised to impose an age minimum, and the WD was supposed to enforce it. Look, half of these animals were shot in 2005—you all approved the rule before the 2004 season."

"We did—it takes time—we will," she became desperate, "Look, you won't be speaking tomorrow on behalf of Craig Packer; you will be speaking on behalf of all Tanzania. Our delegation won't have a chance to speak for our country. You will be the only voice for thirty million Tanzanians."

"So you promise to enforce the age minimum?"

Three heads bobbed up and down furiously.

"What about Savannas Forever?"

"We'll require all the hunters to join Savannas Forever."

"But I need to show these pictures so that everyone understands why we need change."

Miriam brightened, "But you don't have to say where they came from, do you?"

I steered my talk the next day without setting off too many alarms. I explained that Zimbabwe's lion offtakes had been too high the past ten to fifteen years, but they had recently lowered their quotas and would no longer shoot females. I also confirmed that an age minimum was urgently needed, as seen by these recent trophies from several unnamed countries.

I was a member of a larger panel of scientists at Johannesburg who filled out a map of where lions still roamed, and it painted a striking picture. Across all of Africa, by far the most important area for lions was a more or less continuous zone from northern Mozambique throughout southern, western, and northern Tanzania up to the southern edge of Kenya. The species range had presumably once covered most of the continent south of the Sahara. What had happened to all those lions in most of Kenya, Zambia, Zimbabwe, Mozambique, South Africa, and Malawi, let alone Ethiopia, Somalia, and West Africa?

No one had been watching every square foot of Africa to document the retreat of every last lion, but the gaps in lion distribution obviously followed the land-use patterns of an ever-increasing human population. Vast tracts of agriculture (large scale in southern Africa, solid blocks of subsistence farmers everywhere else), high-intensity cattle ranching, and ever-expanding cities meant no lions. And around the edges of these areas, people killed lions for eating their livestock—only southern Tanzania and northern Mozambique had substantial numbers of man-eaters anymore—but the bottom line was the same: people had no tolerance for lions, and people were spreading.

How could an unregulated hunting industry possibly stem the tide?

2 Savannas Forever

DAR ES SALAAM, 21 JULY 2008

I have to renew my research funding every three to five years, and every new deadline looms like a tombstone with the arrival of each new grant. There is no congressional earmark for lion research. A jury of my peers at the National Science Foundation judges my lion proposals against competing projects, and the average success rate is less than 16 percent. So if I get complacent or have a run of bad luck, the Serengeti lion project could end forever.

So after twenty-three years of clinging to twigs, I seized on two new branches that the National Science Foundation (NSF) had established in early 2001. One program focused on the ecology of infectious diseases and the other on biocomplexity.

The disease ecology program was like a godsend. The Serengeti lions had been studied for nearly thirty years with only the occasional sign that they could succumb to anything other than the attack of another lion, the horns of a buffalo, or the spear of a Maasai. But then one early morning in February 1994, a group of tourists were riding in a hot air balloon above the Seronera River, wafting in the cool breeze of dawn, gliding a few feet above the fever trees, looking down on storks and geese in their nests. Fresh excitement: a male lion out in the open beside the river! But as they floated over, they realized something was terribly wrong. The lion was convulsing, flailing about, seemingly possessed by demons.

Their report took several hours to reach Melody Roelke, the chief veterinarian of the Serengeti National Park, who rushed to investigate along with a visiting ecologist, Ray Hilborn, and his wife, Annika. The lion hadn't moved far since morning, and Melody saw right away that he was seriously ill. He was disoriented, unaware of the approaching Land Rover, unable to hold his head steady.

Then he started grinding his teeth, working his neck up and down. He tried to stand, but then collapsed onto his side, and all four legs thrashed and flailed uncontrollably as he went into a grand mal seizure, foaming at the mouth, every muscle in his body flexing and twitching beyond the limits of endurance. The attack stopped after a minute or two, and he was completely exhausted, breathing heavily and trying to regain his balance sufficiently to lie on his chest and raise his head.

The reprieve lasted for maybe a quarter hour, then the whole cycle repeated itself again. Another brief reprieve was followed by yet another set of violent spasms—the pattern recurring again and again until he died and was eaten by hyenas.

I've watched Annika's videotape dozens of times, and it has become emblematic of the following weeks and months when another half-dozen lions were observed with the same symptoms while many, many more died unseen. Over a third of the Serengeti lions—more than a thousand animals—appeared to have suffered the same fate.

I was in Minnesota when I received a long description of the seizuring male and of three more dead lions. I called Melody on the phone in Serengeti and worked with the FIV and genetics lab at the National Cancer Institute to bring out enough veterinary supplies to launch a full-scale investigation.

April is the height of the rainy season in Tanzania, and the landscape had an eerie verdant beauty when I arrived in Seronera. The air was utterly transparent all the way from Nairobi to the Gol Mountains and the flanks of the Ngorongoro Crater highlands. The ground glowed green like the Land of Oz. When I landed in the Serengeti, Melody was out on the plains, dealing with a fourteen-month-old cub who had gone into convulsions in a shallow pool of water as his mother sat helpless nearby. Every lion in the park seemed to be sick or dying.

I arrived at the house of the Frankfurt Zoological Society representative, Markus Borner, as he watched the BBC on his satellite TV. The screen was filled with images of dead human bodies washing into Tanzania in the rivers flowing down from Rwanda. Hutu militias had just massacred a hundred thousand Tutsi.

April did, indeed, seem to be the cruelest of months, as T. S. Eliot had said. The green growth of spring was fueled by the bodies of the dead.

Melody was one of the best vets I've ever known. She was totally focused on her profession: nothing could get between her and a sick cat. On the one hand, she thought she recognized some of the symptoms of canine distemper—which she had once encountered in a captive tiger. But on the other hand, the blood smears looked like the lions were suffering from a massive infection of tick-borne disease.

I was doubtful about the blood parasites; I had organized a parasitological survey nearly ten years earlier, and the lions always seemed to be infected with something. What was so special about now?

Well, 1993 had been extraordinarily dry, and the Cape buffalo had started dying of starvation, so the lions had been feasting on buffalo for months. Could the lions have OD'ed on buffalo guts? Maybe the drought-stricken buffalo had started consuming the Serengeti equivalent of locoweed in a desperate search for fresh food, and any toxins might have concentrated in their livers. . . .

Melody politely ignored my demented ramblings and focused on collecting as many samples as possible while the animals were still showing signs of disease. We needed to find every pride in the Serengeti study area and sample two to three lions per day until we ran out of supplies. While most of our prides included a radio-collared female, about a third of the collared females had died in the outbreak, so I had to find the lions the old-fashioned way: through an inexplicable combination of intuition and blind luck. And I felt like a magician. I would drive into a flat featureless plain with no trees, no nothing for miles in any direction, and—voila!—there would be the pride we needed, lying around with their paws held out, waiting to have their blood drawn.

Melody would spring to work, and we spent hours darting, bleeding, probing, and testing. Samples were accumulated at a satisfying pace. But what I most liked about Melody's pride-side manner was her attention to each individual, continuously monitoring its temperature with a rectal thermometer, spraying its flank with water if it started to overheat, and rehydrating the lion with a saline drip.

The answer to the question of what was killing the lions came in June 1994, and the event was big news. Press coverage was extensive—*Time* and *Newsweek* ran stories in the same issues that they reviewed the release of Disney's *Lion King*, and smaller items ran in local newspapers all the way to Australia. The diagnosis came from wildlife veterinarian Linda Munson after she had examined bits and pieces from the drowned fourteen-month-old.

Linda was on the faculty of the vet school at the University of Tennessee, and CNN invited her to Atlanta for a live TV interview. The reporter asked about the cause of the Serengeti die-off, and Linda started to explain that the lions had been victims of a virus called canine distemper virus (CDV), which was normally associated with domestic dogs. This was the first report in wild lions, although there had been an earlier case in a captive tiger. . . .

But Linda saw that her interviewer was no longer paying any attention. The reporter was staring openmouthed over Linda's shoulder at a TV monitor where their own two faces had been replaced by a car chase on

Maasai child brings his puppy for vaccination against rabies and CDV.

the LA freeways involving about a hundred squad cars and a white Bronco driven by O. J. Simpson.

Sarah Cleaveland is a veterinary epidemiologist who first came to the Serengeti in 1991 to study the diseases that threatened the highly endangered African wild dogs. She was also the first Serengeti ecologist to venture outside the park boundaries and seriously consider the way that human lifestyles influenced the overall ecosystem.

Distemper is a directly transmitted disease, closely related to measles, and it infects a phenomenally wide range of species. During the seven- to twelve-day course of infection, the victim either dies or develops life-long immunity—just like measles. In even the most contagious diseases, an infected individual rarely manages to spread the pathogen to more than two or three other new victims. Consequently, a dangerous virus can only persist in a very large population of susceptibles.

Consider this: lions get CDV, and sometimes it kills them. But at most, there are only nine thousand to twelve thousand lions in all of Tanzania, and maybe a similar number of spotted hyenas and leopards. Now, how many domestic dogs live in Tanzania?

Sarah had gone into dozens of villages, counting the number of dogs in each household and determining how long they lived, and she estimated a total of five million domestic dogs in the country as a whole. Five million *unvaccinated* domesticated dogs.

Sarah's passion has never been CDV; she is the preeminent rabies expert for all of Africa. She initially came out to work with Frankfurt Zoological Society and Markus Borner to try to protect the wild dogs from rabies after the disease had annihilated a pack in 1990. The wild dog population had been declining for years; healthy packs would mysteriously sicken then disappear. Some people suspected CDV, others blamed rabies, but it had always been guesswork. When a diagnosis finally came in 1990, it was possible to establish a definite course of action.

But Sarah's wild dog project quickly morphed into something much larger. After looking at rabies in domestic dogs, she realized that published estimates of human rabies deaths in Tanzania were far too low. The World Health Organization and other international agencies had largely dismissed rabies as a significant human health threat in Africa, but Sarah found evidence for thousands of deaths each year rather than just dozens.

We joined forces after the 1994 lion die-off. At first, I was rather appalled by the public's greater concern for the lions' deaths than for all the untreated childhood diseases in the neighboring villages, but I wasn't going to turn down international offers of help, and I asked Sarah if there was any way to protect the lions from the dogs.

"Sure a mass dog vaccination program, but it would be very expensive."

We received a surprising number of donations from lion lovers in the United Kingdom and started Project Life Lion in 1995, vaccinating hundreds of domestic dogs in a half-dozen villages at the presumed epicenter of the 1994 outbreak.

But funding dwindled as the publicity faded, and Sarah was ready to move on to other problems until we saw the NSF announcement of the new disease ecology program in 2001, and we tossed together an audacious proposal to vaccinate every dog in a ten-kilometer band all the way around the Serengeti—approximately thirty thousand dogs per year for five years.

The National Science Foundation rejected our first application but encouraged us to resubmit, and we hit pay dirt after adding a theoretician, Andy Dobson, from Princeton. The National Science Foundation provided salaries for a large staff of Tanzanian vets and four graduate students. Sarah and her staff achieved the initial vaccination goals in 2003 and eventually increased coverage to fifty thousand dogs per year in the face of rapid human population growth around the Serengeti.

For rabies, the vaccination program was a stunning success. Serengeti wildlife are no longer exposed to canine rabies, and the elimination of the disease from dozens of neighboring villages has saved many human lives—mostly children since they are more likely to play with the family pets and more likely to be bitten on the head and neck.

However, our cordon sanitaire failed to fully protect the lions against CDV, which broke through the vaccination zone in 2006. The outbreak was minor, causing no deaths and infecting a smaller number of lions, but we had learned an important lesson. Controlling distemper would require a far more extensive vaccination program than might ever be possible. While we had successfully blocked rabies, CDV spreads so easily that a ten-kilometer buffer was too narrow. If we seriously wanted to eliminate CDV, we might have to vaccinate every dog in the country.

From a strictly scientific point of view, we were a success. All our students completed their PhDs, we learned a lot about disease ecology, and Chicago's Lincoln Park Zoo agreed to take over the Serengeti rabies vaccination program indefinitely. The World Health Organization accepted Sarah's contention that rabies is such a significant health threat to rural Africa that they solicited (and received) funding from the Gates Foundation for Sarah to conduct large-scale rabies control programs in other parts of Tanzania.

*

Linda Munson was diagnosed with breast cancer in the year 2000. She spent weeks at Sloan-Kettering, her cancer metastasized, and the treatment shattered her strength. She had no idea how long she had left, but the fire gleamed brightly through her hoarse slurred voice as she taught me over the phone how to perform a necropsy in March 2001.

A fatal disease outbreak had just struck the Ngorongoro Crater lions. A severe drought had just ended; the lions had been eating an unusual number of sick buffalo. Some of the lions had been seen passing blood in their urine, and many more were covered in ulcerative sores. Similar sores had been seen during another die-off way back in 1963, when people blamed biting flies, *Stomoxys calcitrans*, for fatally draining the lions of all their blood.

The Crater was verdant when I arrived in late March: purple flowers, yellow flowers, green, green grass dotted with tawny lions. Black clouds of *Stomoxys* flies rose a few feet into the air whenever a sick lion rolled over and then landed again on the lion's other side. The glistening red ulcers on the lions' flanks were certainly filled with flies, but the sores were all self-inflicted; the lions itched so badly that they had rubbed themselves raw wherever they could reach to scratch.

Golden lions in an emerald landscape once again littered with dead buffalo following a severe drought.

When Linda saw the pathology samples, she reacted a lot like Melody did back in 1994. The sick lions were overwhelmed by babesia, a tick-borne disease ordinarily associated with cattle that is commonly known as redwater fever. But the serology results once again revealed CDV infection, even in the smallest cubs.

I was initially disappointed that we hadn't discovered anything new, but the Crater outbreak finally made sense of the 1994 die-off.

"Melody was onto something when she noticed the blood parasites," Linda told me. "The lions might always be infected with tick-borne parasites, but CDV is immunosuppressive, like AIDS, and an animal seldom dies from the virus itself but from whatever else is around at the time."

By the end of 2006 we had evidence of seven different CDV outbreaks in the lions, but five could only be inferred from antibodies in blood samples. These were "silent" outbreaks where none of the lions had shown any sign of disease. The other two outbreaks caused horrifying illness and over 35 percent mortality.

So what was the connection with drought? As animals starve, their immune systems fade, and they can no longer fend off any low-grade chronic infections. The starving buffalo became infested with ticks—and also became breeding grounds for pathogens like babesia. The lions might have thought they were getting a free lunch with all those weak buffalo staggering about, but, well, we all know how that one goes. I inspected a half-eaten buffalo carcass one afternoon during the Crater die-off, and the ticks were jumping ship like rats.

It made a neat story, but it was only just a story until we found the grant money and the right person to test all three hundred lion samples that had been stored away in freezers from as far back as 1984. Linda's protégée, Karen Terio, ran a series of sophisticated molecular assays and confirmed that, while the lions were almost always infected with low levels of babesia, their levels of blood parasites skyrocketed immediately after the droughts of 1993 and 2000.

It was an astonishingly clean result. Some of the Serengeti prides lived in open plains where the buffalo never roamed—and the plains lions showed no mortality even though they all became infected with CDV in 1994. One woodlands pride suffered high levels of babesia after eating plenty of buffalo, but they escaped infection with CDV, and again no one died.

Thus CDV was only lethal in combination with a heavy load of babesia.

Large-scale rabies-control programs make sense: human lives are at stake and the intervention actually works. But CDV is only important insofar as it affects a few charismatic mega-vertebrates like lions and African wild dogs. So target the high-risk animals—with tick repellant, perhaps—when the risks climb sufficiently high and let nature run its course in everything else.

The second new source of NSF funding involved the emerging field of "biocomplexity," which emphasizes the connectedness between human activities and ecological processes, and, after the distemper/babesia die-offs, I was eager to bring a new perspective to the Serengeti.

Think about a villager's decision to keep a domestic dog. Domestic dogs carry infectious diseases that can infect lions and African wild dogs; a loss of these wild predators might increase the number of wildebeest, and more wildebeest might increase the risks of malignant catarrhal fever in the very livestock that the domestic dogs were meant to protect in the first place.

Now think about the continued growth in the human population around the park boundaries that increases the demand for bushmeat to the extent that the wildebeest population eventually collapses, causing a severe decline in international tourism and reducing the incomes of local people. Worsening poverty leads subsistence farmers to view added children as a vital form of insurance against economic hardship, and added population growth further worsens poverty in future generations.

Virtually nothing can change in the villages outside the Serengeti without an effect on wildlife, and vice versa—and many of these changes could lead to a nasty downward spiral.

So I assembled a second team of scientists to tackle the interrelatedness between local people and the Serengeti's plants and animals. I had only ever considered lions in isolation until the CDV outbreak, but carnivore populations were declining all around Africa, and lion conservation was starting to feel more important than lion ecology.

And I also felt Savannas Forever might serve to extend our study of biocomplexity to every major ecosystem in Tanzania.

DAR ES SALAAM, 26 JULY 2008

"Hey," said Susan, phoning from the Tanzanian Wildlife Research Institute (TAWIRI) office in Arusha. "I just talked to Simon Mduma, and he says we don't need clearance to do pilot work in Meru—we just needed permission for the health studies."

"That's great. We're finally set!"

And wouldn't it be pretty if it were so.

I'm in Dar es Salaam, staying at the Econo Lodge, an Indian-owned hotel off Libya Street near the Morogoro Road. This is the old part of town, and the Econo is easy walking distance to my ersatz office and my two favorite restaurants, the Hong Kong (with the best Chinese seafood anywhere) and Chef's Pride (with the best chicken *sekela*—a spicy tandoori). I'm in room 512, a corner room with a rooftop view of the new construction in town, five mosques, one church, the twin towers of the Bank of Tanzania, and the new lighthouse at the mouth of Dar harbor. Most of the older buildings are three or four stories with rows of black plastic thousand-liter water tanks on their roofs, along with satellite dishes, forests of TV antennas, and clotheslines hanging out various arrays of laundry, some items faded and pale, others colorful and crisp.

My ceiling fan keeps me cool, and I spend a lot of time here writing and listening to music on my iPod. But the street sounds pass easily through the open windows, and the muezzins from those five mosques wake me up during the night and early morning. One muezzin is gentle and melodic, but the worst sounds harsh and hectoring. Large cylinders of cooking gas are rolled noisily along the sidewalk, clanging together as they are aligned to take up the least amount of space at the fuel store across the street. Cars

Top, mobile salesman in Dar es Salaam with his supply of rat poison.
Bottom, entrance to the Econo Lodge in Dar.

are always honking in the background, with people shouting to each other above the din.

June and July are the mild season here. Walking to the office is quite pleasant at this time of year. I have an air-conditioned workspace in the coffee shop of the Kaminski-Kilimanjaro Hotel. I have Internet, and easy access to government offices, embassies, and all the big international aid agencies.

Tanzania was known for a long time as a failed socialist experiment. Its first president, Julius Nyerere, was idealistic to a fault, and he had grand plans to lead the country along its own path. He wanted everyone to have access to health care and education, so he started a policy of villagization in the rural areas (*Ujamaa*) to make every community large enough to sustain a clinic staffed with doctors and a decent school.

After winning Tanganyika in World War I, the British colonial government put German assets into a trust, which Nyerere used to establish the University of Dar es Salaam at independence in 1963. Nyerere was known as Mwalimu (teacher), and he sincerely worked to advance the welfare of his citizens rather than to enrich himself. Most remarkably, he left office voluntarily—establishing a policy of term limitations—then made sure that his successor followed suit. At any one time, Tanzania has about half of all the living ex-presidents in Africa who have left office voluntarily. When Nyerere left office, he was only modestly well-to-do.

In this, he may have been unique.

Nyerere's most important achievement was to forge a national identity through the adoption of Swahili as the country's national language and through a concerted effort to weaken tribal identities. Travel anywhere in the country, and you'll find that people see themselves as Tanzanians first and foremost. Whereas in Kenya, people will first tell you their tribe before admitting to being Kenyan.

These are the good things that Nyerere did, but the good is oft interred with our bones.

Alas, Nyerere found economics boring. When asked how he was going to pay for all those social services in the Ujamaa villages, he just shrugged. His socialist policies were also a powerful disincentive against high-powered economic activity, which kept the country far poorer than capitalist Kenya.

Tanzania's socialism disappeared sometime before the fall of the Berlin Wall, but the apparatus of the socialist state has taken on a life of its own. The government is everywhere, and there is nothing anyone can do unless the government gives its OK. A lot of foreign aid is currently being spent on getting the government to decentralize and otherwise relinquish its power, but that is a lot easier said than done.

When I first came back to Tanzania after unveiling Savannas Forever at the Dallas Safari Club in the summer of 2005, I went straight to Severre's office, and, much to my relief, he accepted the idea of hunting certification. He was still glowing after our success at CITES, and it seemed that I might still be useful to him.

I felt like I had taken another small step toward reforming the hunting industry. But when I told George Hartley at Tanzania Game Trackers, he was doubtful.

"You can't trust anyone in this business."

"But that's why there should be hunting certification."

"It won't work unless you can get everyone to sign on."

"I don't even know who owns most of these companies. Where do they live? How do I sign them up?"

"Come back to Reno; they'll be at the next SCI convention."

"What if they won't join?"

"There are a few key people you need to convince. Get them on board, and the rest will follow. We've turned Mantheakis around, and Danny McCollum will join for sure."

"What if no one else joins?" I insisted. "Why not make it mandatory? Severre's been positive so far."

"If it's mandatory, the government will control it, and it'll be useless."

"We'd keep it safely separate from the government. We'd be independent and report to TAWIRI, not Severre."

And all the time, I'm thinking, if Severre is so corrupt, why has he been so encouraging? He knows I wouldn't be any part of a cover-up.

So a few months before going to Jo-burg, I called him on the phone. After making sure that he was still in favor of Savannas Forever, I asked, "Sir, I'm curious what you see as the best reason for hunting certification?"

He told me that while the hunting industry was mostly just fine, there were a number of operators that had gotten their hunting blocks through the minister of natural resources, Mama Meghji, and they were a bad lot, so Savannas Forever would help him regain control of the southern part of the country.

But even with Severre's reassurances, I realized that there had been no concrete evidence of reform by the end of 2005. It had been eighteen months since the TAHOA meeting, and every time I had asked about inspecting trophies to estimate the lions' ages, I kept getting brushed off.

They would start next year; they would do it themselves.

But who exactly, and how? Wouldn't they need training? Severre kept talking about synergies between our two organizations, but where was Savannas Forever in this synergy?

I went to TAWIRI in Arusha, asked about CITES-related exports, and was told that, by international treaty, TAWIRI was supposed to verify trophies from every CITES-listed species, but they never saw anything because it was all done in Dar at the Ivory Room by WD staff.

Rolf Baldus was just as skeptical as George Hartley when I asked about hunting certification, and he insisted the WD would just politely agree with me but never follow up.

But I had been to CITES with Severre. The Tanzanian Hunting Operators Association had agreed to the age minimum.

"They got what they wanted out of you. Have you actually seen what they are shooting?"

No, no one did.

"They'll just wear you down."

So I went to Jo-burg in January 2006 prepared to play hardball, and Miriam's public endorsement of Savannas Forever had been the first order of business.

But then I needed to pay for this fabulous new organization. Inspecting lion trophies would be relatively cheap and easy. We could have someone in Arusha inspect the lions shot around Serengeti and Tarangire and station a second person in Dar to inspect the rest.

Measuring antipoaching efforts could be organized by the hunters themselves; they would only need to buy some of those cheap new GPS units, then we'd provide the software to download their rangers' movement patterns, and they could just e-mail us the data.

The hard part was going to be the community development projects. We would have to send people out to all the hunting blocks, then visit the nearby villages and take measurements and interview members of the community. That would require cars, staff, and expertise.

So if Savannas Forever was going to become a reality, we needed cash and, more fundamentally, we needed to figure out what we should measure in the villages.

A few days after the Jo-burg meeting, Susan and I went together to Reno for the 2006 SCI convention. We had two main quests. First was to meet the fabulously wealthy Friedkin family, who were concerned about the future of the industry—and their own chances of retaining their hunting blocks—in such a difficult country as Tanzania. Tanzania Game Tracker's booth displayed large photos of clients barely able to lift up their leopard trophies, kneeling beside fine male lions, and posing beside large antelope. Their booth was a mock-up of their hunting camp, with large canvas tents furnished with bookcases and luxurious safari gear. Dan Friedkin was only in Reno for a few hours, so we quickly went over ideas for improving the harvest practices for lions, leopards, and other key species, as well as improving his return on investment from TGT's community projects in the rural villages. Dan gave the green light and said that he had friends who could help match his donation.

We only had one more hurdle to clear. Gerard Pasanisi's booth wasn't that far from TGT's, and his photos of trophies and clients weren't that different from TGT's, but there was something in the air over there. It was like leaving the warmth of family in TGT's comfortable canvas campground for another country that was even richer yet a lot colder.

Pasanisi himself was the selling point for his safari company. He was about as haughty as an old-school Frenchman could be. But he was also a very old man, he was more than a little bit deaf, and his grasp of English had largely leaked away after a full day of promoting himself to prospective American clients.

There was only one man in Reno who could help us sell Savannas Forever to Pasanisi, and that was Charles Williams from TGT. Charles radiated charm and sophistication; his voice rang out like Sean Connery as the original James Bond. Charles had hunted over most of Africa, and he was a member of Shikar, the most exclusive hunting organization in the world—a sort of Free Masons of the trophy hunters. There were only three members of Shikar who worked in Tanzania: Charles, Pasanisi, and Pasanisi Jr.

So we stood in the harsh fluorescent lights of the Reno convention hall being passed like fine china from Charles to this gnarly old guy who only dimly seemed to comprehend what we were saying. In his occasional befuddlement, he would look to Charles for guidance, and Charles would reassure him and remind him of my fine work on the lions at CITES and also that Danny and Michel Mantheakis and half a dozen other companies had all signed on. Pasanisi shrugged and said, of course, why not?

We had the permission from Severre, the commitment from Miriam, the money from Dan Friedkin and the approval from Pasanisi. Now we just needed to figure out what to do.

I met Susan three years after the end of my first marriage. We had kids of roughly the same age enrolled in the same Montessori school in south Minneapolis. I had come to watch my eleven-year-old son Jonathan run cross-country, and I spotted this six-foot blonde about a hundred yards away. I had seen her at a few other school functions over the years, but I'd never had the courage to introduce myself.

That day I was determined to overcome my shyness, but by the time I reached her, she was in conversation with another dad, and I nearly lost heart. But then the dad turned to introduce me as "Mr. Menopause."

"Mr. Menopause?" she said with a twinkling laugh.

"He just found out that monkeys and lions have menopause, there was a story about him in the papers."

And what better introduction to a beautiful forty-something single mom at a cross-country meet?

Susan had an MBA and a business career. She was working as a freelance business management consultant when we met, and much of her work consisted of research—market segmentation, market analysis, customer satisfaction, and so forth.

So when we came up with Savannas Forever, Susan started studying the mechanics of successful conservation certification programs, and I recruited Monique Borgerhoff-Mulder, who had written a textbook on community conservation, and number of Tanzanians, including Simon Mduma, Bernard Kissui, and Benson Kibonde, who was the equivalent of chief park warden of the Selous Game Reserve.

Savannas Forever would cover the whole country—around all of the hunting areas—and by covering so many areas we would form the foundation for a unique new research program. We wouldn't just survey a bunch of villages and leave—we would work with each village so that it truly benefited from the hunting companies. We would come back over and over again for decades. It would be a long-term study of unprecedented scale, and the hunting companies had the money to pay for it—hunting certification was going to improve their business, and we had already raised over a quarter million dollars. . . .

We held our first planning meeting in Minneapolis in February 2006, a few weeks after Reno. Monique and most of the Tanzanians came, and within three days we had a plan. Our mission was to "provide practical solutions for conserving African wildlife populations while reducing human-animal conflict and promoting livelihoods in rural communities." The objective of hunting certification was the "conservation and protection of Savanna ecosystems in Tanzania through environmentally appropriate, socially beneficial and economically viable management."

We started drafting a set of household and village questionnaires and agreed to weigh and measure all the kids under five years old in each village—Monique explained that childhood nutrition was the best possible quantitative measure of poverty. We designed a rough sampling program of villages located within five kilometers of a protected area and returning every other year to measure any changes from a successful conservation/development intervention.

We would meet again in Dar es Salaam in June to finalize the questionnaires and hold a workshop with TAHOA before the start of the hunting season. Savannas Forever would be ready to go on the first of July 2006.

*

In March 2006, we were rocked by the news that Severre had been sacked by the new minister of natural resources, Anthony Diallo, but Severre made headlines around Africa by refusing to leave office. The Tanzanian press ate it up. Cartoons in the Kenyan and Zimbabwean press lampooned the impotence of the Tanzanian ministries. Severre had become the government's biggest embarrassment in a society that thrived on consensus and obedience to authority.

The impasse dragged on for weeks. I came back to Tanzania at the beginning of April to hire staff, buy cars, find a house in Arusha, and follow-up with the northern hunting companies. I recruited a staff of six young Tanzanians, three women and three men. Some had worked for the Danish Aid Agency conducting village surveys, some had just graduated in wildlife ecology at the University of Dar es Salaam. The fifth was a botanist and the sixth had worked on the dog-vaccination program around the Serengeti. They were all part of a new generation of educated Tanzanians that were barely distinguishable in their worldview from recent graduates anywhere in the developed world.

I went to see Severre and clarify whether his situation might affect Savannas Forever. "Sir," as I always addressed him, "you are being wasted in this office. You really ought to consider moving up to a better job. Maybe something with the CITES Secretariat or the UN."

Severre knew every detail of national and international wildlife legislation like the back of his hand, and he almost always used the letter of the law to expand the amount of wildlife area in Tanzania. His motives might only have been to expand his personal empire, but the outcome was almost always an extension of the protected areas.

He was a mere shell of his former self during that first meeting, but he noticeably brightened at the thought of being promoted to the International Union for Conservation of Nature or the United Nations Environmental Programme or some other international conservation organization, and he was obviously pleased that I came to see him while the rest of the country regarded him as a pariah. He reassured me that Savannas Forever would go forward, regardless of the outcome of his position.

I went back to see him every week or two, and his confidence grew the longer he managed to stay in office. Then the unexpected resolution: Severre's dismissal was ruled illegal by the chief secretary of the civil service. The director of wildlife was a presidential appointee, and Severre could only be removed by the president—not by Minister Diallo.

Severre is a well-built man of average height for a Tanzanian; he dresses well, wears gold-framed glasses, and his fine features are only slightly marred by the dark brown stains on his teeth—the outcome of too much fluoride in the drinking water in his hometown near Arusha. He is never afraid to smile; the gleam in his eyes more than compensates for his brackish teeth.

Somewhere in all this, I told him about our upcoming workshop and asked his permission to introduce Savannas Forever to all of the Tanzanian hunting companies at the next TAHOA meeting in June; we could either have half a day or perhaps schedule a second day of meetings once TAHOA had finished their usual business. He said he would work out something, and then as I started to leave the room with his usual henchmen, he shook my hand warmly, and said, "Welcome to the fraternity of hunters."

I felt like I had been initiated into the Mafia.

At the end of April 2006, I flew down to Maputo to help drum up support for a parallel version of Savannas Forever in Mozambique to be headed by Colleen Begg. I had met Colleen at the Jo-burg meeting; she had studied honey badgers in Botswana before moving to Mozambique to work in the Niassa National Reserve.

Niassa is nearly as big as the Selous, with enormous wild spaces and giant rocky outcrops the size of mountains. Niassa follows the Tanzanian border for hundreds of kilometers, and thousands of elephants swim across the Ruvuma River each year as they roam from the Selous to Mozambique. About twenty thousand people live inside Niassa Reserve and endure the most frightening existence imaginable: eaten by lions, trampled by elephants, and bitten by snakes, spiders, and everything else with fangs and teeth.

I walked the streets of Maputo one afternoon. The city market was an anthropological cornucopia: kiosks selling used plastic bottles and empty tin cans, stalls of hardware—new and used nuts and bolts—next to fresh vegetables of every shape and color, and the "medicine market." In Tanzania, the traditional medicines are mostly herbal, like tree bark, powdered roots, ground leaves, and flecks of peculiar plants. Maputo market featured the mummified hands and skulls of monkeys, dried snakes, and the moldy skins of pangolins, civets, and mongooses.

Mozambique had been through a hell that Tanzania has never known. In the city center, I admired the five hundred-year-old Portuguese fort and the Mediterranean architecture, but there were still rubble-filled city blocks that had been bombed out during the civil war.

Maybe there is an advantage from going through the wringer and starting over completely from scratch. Unlike the arrogance and self-righteousness of our Tanzanian delegation to CITES in 2004, the Mozambicans had sought advice on how to build the Niassa hunting industry from scratch.

Maputo, Mozambique

And the Maputo hunting operators' meeting was a revelation. I gave my Dallas talk, first thanking my hosts for graciously allowing me to speak. I warmly congratulated the regional government for their openness and wished out loud that the Tanzanians were even half as open, briefly listing a few examples of corruption in the Tanzanian industry. Then I described how to age lions and outlined why hunting certification would benefit their businesses.

An expatriate hunting operator from Tanzania, "Manic Mike," had attended the meeting in hopes of finding a Mozambican hunting block to adjoin his Tanzanian block on the opposite side of the Ruvuma River. He came up to me during the coffee break, talking like a clockwork toy that had been wound up too tight.

"You can't say that about Tanzania," he practically barked.

"But it was all true, you told me so yourself!"

"It may well be true, but we don't say any of that out loud. I came here looking for a new hunting block, and now no one will talk to me—just because I work in Tanzania."

He looked me straight in the eye and said, "Thanks to you."

But I had already heard why no one would talk to Manic Mike. At a recent SCI convention in Reno, he had nearly gotten into a fistfight with several of the Mozambican operators after accusing them of harboring poachers who crossed the border into his block and shot his elephants. He wanted a Mozambican block so he could patrol both sides of the Ruvuma.

The next day, I met a representative from one of the richest families in Europe who had also come to secure a hunting block. They had no intention of using it for hunting; they wanted to set up a high-end tourist destination. The very rich wanted extreme privacy, and price meant nothing, as long as they could spend their holidays as far from ordinary people as possible. The rep also dropped the name of a fabulously wealthy Englishman who wanted to do the same thing.

I asked him about obtaining hunting blocks in Tanzania, and he said, "No, the problem with Tanzania is that all the best blocks are already taken. The R—— family would snap up a block in the Selous in an instant, but Gerard Pasanisi has all the best blocks, and he will never let them go. But," he handed me his card, "if you ever hear of a block becoming available in the Selous, please let us know."

On the morning of the fifth of May 2006, I was driving out to the Wildlife Division when I stopped for a red light, and as the news hawkers crowded around, I saw the headline "Foreigners colonizing Tanzanian hunting industry." The article was about Pasanisi, how he had an illegally large number of hunting blocks, how he had all the best blocks in the Selous, how difficult it was for a Tanzanian citizen to get even a single hunting block.

I met with Pasanisi that afternoon at the Royal Palm Hotel. He asked me if I had seen the story in today's newspaper.

Yes, I said, it was the sort of thing with which Savannas Forever might be able to help. There was too much secrecy in the hunting industry; no one realized how well managed the best companies were or what the benefits were for conservation and for local communities. The weaker companies might suffer from a certification system, but that would help open up new blocks for the Tanzanians who wanted to start new and better companies.

What did I think of the WMAs—those new wildlife management areas run by local villagers rather than by the WD?

I said I was a scientist, and I had no data yet from the WMAs. If the WMAs worked, it would be obvious from the data. If not, the data would show how they had failed.

"I don't like these WMAs," Pasanisi spat. "These villagers are stupid; they're drunkards; they don't understand business."

I agreed that there might be problems with a lack of trained managers in remote rural areas, and if so, we would provide neutral and objective information both to the hunting companies and to the WD. A lot of hopes were riding on the WMAs at the international aid agencies, but the WMAs might not succeed as well as everyone hoped. We could help find out if they were worth the effort. . . .

But Pasanisi was not reassured, and he kept pressing, "Are you for the WMAs or against them?"

I tried to maintain my neutrality, but I needed his blessing. I tried to change the subject to larger issues facing the future of the hunting industry, but I got the idea he thought I was too naive to be useful to him.

In the following days, we tried to win him over by drafting a resolution in support of Savannas Forever that could be endorsed by TAHOA in the middle of June. We emphasized the advantages of raising public awareness about the conservation value of trophy hunting, the benefits from working with aid agencies that specialized in rural development, and the companies' improved return on investment from their antipoaching and community development projects.

We circulated the draft to our friendly contacts within the industry and tried to talk to Pasanisi, but he never replied.

We weren't too worried but felt we should cover ourselves in case everything went sour. So we started the process of registering ourselves in Dar es Salaam as a nongovernmental organization (NGO), and we invited a series of important allies to serve on our board.

High-rise construction in Dar es Salaam

Charles Mlingwa was an obvious choice. I had known him as director general of TAWIRI, the host institution for my lion research, and we had been to CITES together.

Then there was Rafael Mwalyosi. We were students together at Gombe in the early 1970s, and he went on to become the head of a powerful research unit at the University of Dar es Salaam before becoming an MP.

Finally, there was Mizengo Pinda, MP from one of the most remote districts in Tanzania, and a personal friend of Monique Borgerhoff-Mulder's. Pinda was the minister for local government, and Monique was impressed by his genuine commitment to rural development.

I confess to being a bit awestruck around the rich and famous, but Charles and Rafael were old friends and approaching them to help build Savannas Forever had been easy. But the first time I shook hands with Pinda, I was completely taken aback. We met in Dar and talked for maybe an hour. He was dressed entirely in black, and his exceptionally dark skin was creased and pockmarked. He looked like he had come straight from central casting, but as the conversation continued, it became clear that he was uncannily insightful. Not just because he knew so much about the corruption in the hunting industry, but also because he was perfectly suited to join our board.

The rural communities next to the hunting blocks had been deliberately neglected for decades. Natural resources were crucial to the economic development of the country as a whole and to the rural villagers themselves, but they were being excluded from the system. Instead of being engaged as partners in conservation, they were being treated as pests.

Mizengo Pinda is also a brilliant student of human behavior—a useful trait for a politician and especially valuable in the minefield of the Tanzanian hunting industry. I watched him weave his way through the hotel lobby: his bearing conveyed a profound gravitas, but he played the room like a magician, shaking hands in just the right way, knowing when to nod in sympathy, knowing when to move on to the next in line.

We still had money in those days, and we stayed at the Hotel Peacock. The Peacock is maybe a hundred yards away from the Econo Lodge, but it is on a major thoroughfare with a large dining room, rooftop bar, and glass elevators; the rooms all have televisions, bedspreads, and bathtubs. Susan's daughters, Mimi and Carrie, came out with my son Jonathan that summer. Three new Minnesota graduate students were with us for the launch of Savannas Forever, and Laurence Frank came down from Kenya, Monique from California, and Bakari Mbano, the former director of wildlife, took time off from his new job with USAID.

I explained to the group how we had signed up a dozen hunting companies to join us the day after TAHOA's annual meeting, and Susan assigned

Dar es Salaam. *Above,* office entrance to Game Frontiers. *Opposite, top,* safety slogans at the port authority. *Opposite, bottom,* vendor selling street art and used license plates.

our tasks: I would outline the rationale for establishing a hunting certification system, the scope of our activities, and some of the practical details for working together. Susan would then break us into smaller working groups to learn the hunters' needs and expectations; she would address the whole assembly on how to use certification to market the "gold-star companies" at Dallas and Reno.

And with about forty-eight hours to go, we received the following e-mail:

TANZANIA HUNTING OPERATORS ASSOCIATION

Ethics and Conservation

June 16, 2006

Mr. Craig Packer
Savannas Forever

TANZANIA PORTS AUTHORITY
Thank You
ACCIDENTS
NEVER GIVE
A DAY OFF
SAFETY
IS NO ACCIDENT
OUR AIM
NO ACCIDENTS
YOU ARE THE KEY TO

T 680 ACF
SXA~89422
SXA~89422
T 379 AQS

Dear Prof. Packer,

I thank you for your letter inviting us to attend the workshop you have organized to kick-start Savannas Forever Consortium . . . I have the following observations regarding Savannas Forever (SF) as an institution that is working towards soliciting cooperation and collaboration with many partners, including the hunting fraternity in Tanzania:

1. During our meeting between you and the undersigned at the Royal Palm, we advised you that in order for TAHOA to be involved with the Savanna Forever Project you needed to get a written authority from the Government that your Project is officially accepted. So far, you have not produced such a document. To us, this means that S.F is yet to prove that it is a registered entity in Tanzania, and that it has legal capacity to operate any business. For this reason we take it that your planned meeting as a private one, during which you have planned to discuss issues of personalities and operations, without linkage with the Government.

2. Reading from your message on the list of invitees it is clear, but very sad that proprietors of the hunting companies you have asked to participate in your meeting are "foreign owned." You know very well that . . . the hunting operators in Tanzania constitute local Tanzanians. There are also local companies owned in partnership between local Tanzanians and foreigners. Those companies that are "foreign owned" constitute of only 25% of hunting operators.

3. You know very well that all operators who have invested in this country are credible. The selection of the companies you invited some how shows discrepancy, where most of the locally owned companies have been left out . . . It will never be acceptable to a credible personality to work in a country loaded with ideas, by and large scientific in nature, but fails to uphold to the principles of good government, human justice with no prejudice to colour, and yet expect recognition from companies they would want to work with.

4. It appears that what you want to do is already being done by the Government and its institutions, and we feel that our participation to your workshop, which the government is not formally represented, also concerning an entity that has no legal backing will spoil the good relationship we are enjoying with the government.

5. We have no confidence at all with your undertakings in particular on the face of the international community in whose countries our noble markets rest now and in the future. It is noted that for your Project to succeed the Wildlife Conservation Act which is currently used has to be declared null and void as most of what you are suggesting in the S.F. pro-

posal contradicts the . . . [Wildlife Conservation] Act. TAHOA is not prepared to be part of this venture.

6. Reading from the agenda of the workshop you seem to be deeply involved in many areas in the conservation field and social well being. Your work, of this nature, may have direct serious caution on us. Its results and implications may be blurred for now but devastating some time in the near future. This is the case with international NGO's, which champion conservation in search of money to pay for their travel, up keep and salaries in the pretext of good management [of the NGO's]. Such cases, under these circumstances, invariably focus to ban trade in many wildlife species. For this, and other reasons, we are worried and have no confidence in your work at this point.

7. May I conclude by saying that we see no added value to our participation in this type of a workshop.

Yours Sincerely,

TANZANIA HUNTING OPERATORS ASSOCIATION

SIGNED	SIGNED
___________________	___________________
Gerard Pasanisi	Adulkadir Mohamed
Chairman	Secretary General

Cc. Permanent Secretary, Ministry of Natural Resources and Tourism
Cc. Director of Wildlife, Division of Wildlife

At least that is what Pasanisi said in writing. Verbally, he threatened to expel any attendee from membership in TAHOA, and he warned that the government would arrest anyone who came to talk to us.

We replied within an hour: our survey staff consisted entirely of Tanzanian college graduates; we had developed Savannas Forever in close consultation with Severre and other Tanzanian wildlife managers. The government was keen to have an auditing system for the hunting industry.

We phoned our allies in the hunting industry, and they promised to work on Pasanisi. But he was unyielding. Unless we received written permission from the government to meet with TAHOA no one would be allowed to attend our workshop.

I went to Severre's office, and he seemed genuinely puzzled by TAHOA's decision. I asked if he would write a letter assuring the companies that the government had no objections to our meeting. He laughed and said that the government had no power to prevent private citizens from meeting in the first place, so it was impossible to say that they permitted our meeting.

I said that Pasanisi seemed to be afraid of hunting certification. Severre asked why. I said I didn't know, but that he seemed to be running the industry like it was his personal possession.

I went to see the permanent secretary, Saleh Pamba, and he agreed with Severre—there was no way the government could annul a power they didn't have. Pamba sympathized, though, saying that the hunting industry was a dinosaur whose time was coming to an end. I complained about Pasanisi, and Pamba sighed, saying that Pasanisi was part of a much larger group that was extremely powerful, and there was nothing anyone could do about him.

I had agreed to advise a new cohort of Minnesota graduate students. We had cars, we had staff and a board of MPs, and we even had a house. We were all dressed up with nowhere to go.

So what else could we do?

Every year, Pasanisi held a gala dinner at the Royal Palm Hotel where the current president of Tanzania, Jakaya Kikwete, and the likes of George H. W. Bush, Giscard D'Estaing, and the archduke of the Austro-Hungarian Empire might appear as honored guests, and Susan and I needed a good party, so, hey, why not?

We weren't surprised to discover that our table was at the farthest corner of the hall, but we nearly laughed out loud when we found we were seated with Sheni Lalji.

Sheni had the worst reputation of any Tanzanian hunting operator. He had merited an entire chapter in a high-level investigation into grand corruption in Tanzania. He had gotten himself appointed to the board of directors of Tanzanian national parks, where he had granted himself permission to build new tourist lodges inside Tarangire and Gombe National Parks—during a ban on new construction in the national parks.

I had gone to see Sheni a few days before TAHOA's rejection slip, asking him to join Savannas Forever—but he dismissed the idea out of hand, saying that our Tanzanian staff would never work without constant supervision.

"They'll get drunk whenever you're not watching. They'll steal from you and lie about our business. You can never trust them."

"But our survey crews are all college graduates."

"Doesn't matter. They're all thieves."

Sheni might have been one of the Tanzanians that Pasanisi claimed we hadn't contacted about joining Savannas Forever, but Pasanisi hated Sheni with a passion—Sheni wasn't as easy to control as the other operators, and he had committed the unforgivable sin of forgetting that Pasanisi was

God. So Pasanisi spent two days at the SCI convention in January castigating Sheni during a closed session of TAHOA, and Sheni's insubordination had earned him a seat with us at the table of shame at the Royal Palm.

The former presidents of the United States and France failed to appear, and the crown prince of Lower Slobovia only gave a few lame comments from the dais, but Tanzania's new president, Jakaya Kikwete was on hand, and as he stood to address the banquet hall, we expectantly awaited his first public comments on trophy hunting since taking office in December.

Kikwete was ordinarily charming and disarming, an attractive, upbeat speaker who inspired hopes and focused on the greater good. But after starting off with the usual political banalities, his tone changed, and he seemed exasperated, "We are killing too many of our animals, and we are not doing enough for our people." He paused and drew a line in the sand, "This must change." The room went cold. He gave no specifics, and a few minutes later we all stood by our dinner tables as his entourage filed out from the hotel.

The evening was over, and Pasanisi was shell-shocked. I came up to say good night, and he only recognized me after we had started to shake hands. He swiveled his head away from me so fast, I thought his neck might snap.

Savannas Forever had reached a fork in the road. Our staff and students were ready to go, and we figured if we could just get started, the hunters would realize we could help their business—and our allies told us that if we could get written support from Severre, TAHOA would be forced to allow us back the following year.

So we sent Jonathan and Carrie down to the land of the man-eaters with Dennis Ikanda and my new student. Hadas Kushnir. They crossed the Rufiji River every day in dugout canoes, dodging hippos and crocodiles in the slow moving waters that flowed out from the Selous Game Reserve, the largest protected area in Tanzania.

Cell phone coverage in Rufiji was sparse, so Jonathan only sent the occasional message:

> A woman and a child were attacked last night in a village we had visited two days ago. They're both recovering in hospital. Some crazy old guy blamed us for the attack but others seemed to laugh at the notion.
>
> 26-June-2006 18:34: 01

Dennis and Bernard Kissui had recruited a couple of WD staff from the Ivory Room in Dar es Salaam for the first big man-eating survey in September 2004. As soon as they reached the outskirts of Dar, they were already in man-eating country.

No one ever forgot a lion attack. Survivors relived their experiences as if they'd only been attacked the night before. Relatives of victims still felt the loss as keenly as if their loved ones might miraculously walk in through the front door.

We asked the circumstances of each attack, the location, time of day, time of year. We also asked which prey species were available in the vicinity. Were there still any zebra or buffalo? Anything at all?

Most lion attacks took place in the wet season, at harvest time, and most of the attacks were in agricultural districts where there was nothing else for the lions to eat—no wildebeest or buffalo, no zebra or impala.

Most attacks occurred when lions pulled someone out from a temporary hut or *dungu*—usually just a simple A-framed roof above a raised platform a meter or two off the ground. These people were subsistence farmers with a single crop per year; their lives depended on a good harvest, and crop pests could quickly destroy an unguarded field—especially at night.

The worst man-eating areas lacked the lion's usual prey species, but there was an abundance of one particular crop pest. Like magnets, bushpigs drew lions to the farm fields, right into the areas where people slept in their flimsy huts. The worst hit areas were mostly Muslim, so people were unwilling to touch a pig—and there was no local market for bush pork.

The hunting season of 2006 officially started on the first of July, but one week later the guillotine fell. I received a phone call informing me that the new minister of natural resources and Tourism, Anthony Diallo, had announced an immediate ban on hunting lions, ostriches, and wild dogs. Wild dogs were "critically endangered" so they had been off quota anyway; ostriches were hardly an important target for macho trophy hunters. But the lion was the golden nougat at the heart of the trophy industry, and I was the Lion Guy, so Pasanisi was convinced I had gotten lion hunting banned to get even with TAHOA.

I was stunned by the ban and crestfallen by the rumor mill.

"Why," I asked the caller, "would anyone believe that I had anything at all to do with this? I went to Bangkok to save lion hunting. The whole point of Savannas Forever is to try to help the industry. How could it be my fault?"

"Nobody trusts you; they don't think you know what you're doing, and you spend all your time talking to politicians. You're much too close to the government," he said and hung up.

I called the permanent secretary, Saleh Pamba, and he told me that the hunting ban had already been reversed; they would issue a statement in the morning restoring lions to quota. I asked why they had issued the ban in the first place. He told me they wanted to set new lion quotas in each hunting block after completing the survey.

What survey?

"We wanted to know how many lions are in the country."

I went to the headquarters of the WD on the thirtieth of August 2006 to meet with Severre and his inner circle, including Miriam Zacharia and Erasmus Tarimo. Around the table were Simon Mduma from TAWIRI, Inyasi Lejora from TANAPA, and a representative from the Ngorongoro Conservation Area.

The room was tense. With great formality, Severre explained that he had asked representatives of the major wildlife agencies to hear my presentation. Their input would help him decide whether to accept a collaborative relationship with Savannas Forever.

And so I began.

We had developed Savannas Forever as a certification program for trophy hunting, habitat conservation, and village development by the private sector.

Tarimo was the first to jump in, "Why is village development any concern of the private sector?"

Severre agreed, "That is the duty of the government."

Politely, I replied: "But the companies are required to devote resources to neighboring—"

Tarimo stopped me again, "A proportion of the hunting fees goes back to the communities; some of the companies add a top-up."

Severre concurred, "Which is purely voluntary."

"OK, but even if it is voluntary, the companies could benefit by making their community programs more effective." And I rushed to the second slide.

"We want to work with your office to help address concerns by the government. First, in terms of the risk of overconsumption in the hunting—"

Tarimo interrupted again, "Why do you say that?"

"I don't know if there has been overconsumption, but there is certainly the *concern* that at least some of the companies are overhunting." And I'm thinking about the universal perception that the subleasers were cleaning out their blocks. They only worked an area for a year or so, and they had to earn back the $70,000–100,000 they paid the leaseholders—so they had every incentive to overhunt.

"There is also the concern that trophy hunting might not be the best way to generate revenue from wildlife."

"Who says this? We are the government, and the government recognizes the value of trophy hunting for generating revenue. This whole department," and Severre spread his hands to indicate the walls, "is supported by trophy hunting."

"I agree completely! But these are just *concerns* that have been raised by other people in the government."

"Who are these other people in the government?"

"Well, a fifth of the Selous has already been converted from hunting to photo tourism."

"That was our decision."

"Yes! And it was clearly very successful because you knew what you were doing! But there are a lot of people who think that even more land should be converted to photo tourism."

"Who?"

"Well, Minister Diallo and his permanent secretary prefer photo tourism to—"

Tarimo, Severre, and Miriam all started to laugh.

"We'll see about them," Severre said.

"Okay, I'm sure you're right, but there is also the concern that the private sector should do more to reduce poverty in rural communities. President Kikwete made that very clear at the Gala—"

Now it was Miriam: "We already provide 15 percent of hunting revenues to local governments."

True, the WD purportedly distributed 15 percent to the district governments. But a district office might be a hundred miles from the nearest hunting block—with no roads and no interest in helping anyone other than themselves—so there was no connection with the communities that actually lost the crops to elephants, the livestock to lions.

I was beginning to feel dizzy, but my next point was already on the screen: "The block allocation system is not transparent."

Severre seemed serene. "We have an allocation committee that works openly and fairly."

The block allocation committee was a rubber stamp; Severre made all the decisions with one or two members of his inner circle. The block allocation system was at the very heart of darkness in the WD.

"OK, again, I'm just repeating things I've heard." I wasn't feeling long for this world. "As we go along here, I hope you'll see how hunting certification can help address these issues with members of the government outside the WD."

And I flashed up the next slide before they could stop me.

"There has been a lot of concern among the hunters themselves because of all the rumors of impending reforms—the minister and permanent secretary want to start an auditing system, for example, and there has long been talk of a public auction for the hunting blocks. So the companies worry that the government might fail to recognize their investments and improvements."

That one only elicited a few scoffs, so onto the next point.

"They are also ill-equipped to take on the responsibilities for community—"

Miriam practically shouted. "Why should they? The hunting companies are businessmen. Their job is making money. Why should they waste their time helping communities?"

I wondered if she would go ahead and call the villagers drunkards and thieves, but she returned her attention to the screen.

I continued. "There is also the concern, by the companies themselves, that the hunting industry operates in a culture of privacy that fosters excessive secrecy."

Tarimo and Miriam said something about the need for the industry to protect trade secrets.

Next slide and we're all the way up to number five out of fifty-nine.

But number five was about the concerns of rural villages, about their loss of access to natural resources, their risks from dangerous animals, the poor communications with the hunting companies, and the villagers' general dislike of top-down policies.

No one seemed particularly interested.

So onto the next. "We are fortunate to be starting this discussion this year—there are a lot of exciting new opportunities in the country.

"First, within the government itself, President Kikwete has publicly committed his administration to conservation, poverty reduction, and transparency.

"Second, within the rural communities, the World Bank has established a huge new project, TASAF [the Tanzanian Social Action Fund], that focuses on public works and livelihoods; the money is allocated at the village level. The hunting companies could partner with the communities to bring more development funds to the villages near their hunting blocks.

"There is a new emphasis on personal responsibility in international aid programs. No more handouts. People have to be accountable. Instead of complaining about losses to wildlife, they should learn how to improve their livestock-husbandry practices.

"And there are new tools for reducing human-animal conflict and promoting alternative livelihoods."

They were listening, and I was starting to catch my stride. "For the hunting industry, there are new models for certification that enforce co-operation between stakeholders and improve—"

Then Severre exploded. "You keep talking about secrecy and distrust when everyone knows we have an excellent relationship with the hunting companies and the best reputation for trophy hunting in the world." He glared at me. Venomously, he spat, "This meeting is a waste of time." He pushed his chair back from the table and shook his head in disgust. "A complete waste of government money."

The room started to spin, my temper rose. I felt my head swell and heard ringing in my ears. By then I caught my breath, closed my eyes for half a second, and calmed down enough to move onto the next slide.

The next part of the presentation went more normally. I started with the importance of considering the demographic impacts of shooting male lions; the need for setting an age minimum; the numerous techniques we had developed for estimating ages (now including patterns of tooth wear, X-rays of pulp cavities, and coloration of the hind legs).

Then around the twenty-fifth slide, I showed a series of underaged lions that had been shot in Tanzania in 2004 and 2005—the first few years after TAHOA's pledge to shoot older males.

Next came X-rays of lion teeth from an operator whose clients shot twenty-seven males in 2003–4—nearly half had been under four years of age. Next, another two operators shot a total of nineteen males in 2005; about half were under four years.

"Who are these companies?" Severre demanded

"I can't tell you," I stood up as straight as I could. "I promised confidentiality if they would let us X-ray their lions' teeth. They wanted us to help them learn how to age lions more accurately." They actually *cared* about how many lions they might have ten years from now.

Next came photographs of grinning clients crouched over egregiously underage males. The first lion had no trace of a mane, and even in the hot climate of Selous he couldn't have been more than three. The other male was maybe three and a half, but he had been promoted in the *Hunting Gazette* as being "close to twenty years old."

And now almost gloating, I said, "I *can* tell you who shot *these* males. These photos came from the Internet, so there are no issues with confidentiality. This maneless male was shot by Zuka Safaris."

Severre scowled. "Zuka Safaris? Who is this? There is no company named Zuka in Tanzania. I have never heard of them."

"Well, sir, if you'd like to meet them, they have a big office out at Sea Cliff." Sea Cliff is the ritziest shopping mall in Dar. "We could drive out there this afternoon. . . ."

And I started kicking myself for trying to be such a wiseass, but Severre was preoccupied by Tarimo, who was leaning over to tell him something.

Severre waved off Tarimo. "Oh, Zuka is just one of those fly-by-night subleasers. There are so many in the country, we can't keep track of them all."

All those untracked companies that are illegal . . .

I merely flashed up the graph from our simulation models, illustrating how killing too many three- to four-year-olds can be catastrophic. Companies and clients who were happy to shoot outsized cubs were a clear hazard to the future of Tanzania's lions.

Next came a slide on this year's efforts in Mozambique. The Niassa branch of Savannas Forever was successfully running a program of trophy certification. If a company shot underaged males, their quotas were reduced; if they only shot mature males, their quotas increased.

Next came Botswana, which had reopened lion hunting after our meeting in Kasane. The Botswana government had limited quotas to one lion per block, and all trophies were now subject to inspection. But nearly half of the recent trophy males were less than four years of age, and there were no consequences—a worry since the antis were still so influential down there.

There were few fireworks through the rest of the talk. I raised the point that leopards might be just as sensitive to underaged offtakes as lions—leopards were also territorial and infanticidal. It might also be wise to keep tabs on the quality of Cape buffalo trophies, since they were such an important mainstay to the hunting industry.

Then I proffered some suggestions about antipoaching, a summary of risk factors for man-eating, and ideas from Dennis Ikanda and Bernard Kissui on reducing livestock losses.

And then I got to slide fifty-one: an organigram of colored boxes on a blue background illustrating what a hunting-certification system might look like.

A large white box showed the Conservation Stewardship Council that would be appointed by members from three separate chambers (smaller white boxes), representing rural communities, conservation groups, and the hunting companies. The council would set the principles, criteria, and indicators for the certification system, requiring approval by two-thirds of the stakeholder groups.

Savannas Forever (green box) would collect data that measured each company's performance on each indicator. We would pass the data to an independent auditing agency (red circle), which would decide whether the company merited certification. We would also work closely with each hunting company so as to help improve their performance in plenty of time before the next survey.

The entire process would be open and public, enabling the government (large yellow rectangle) to be sure that the criteria and indicators obeyed the law and were consistent with wildlife policy.

I had requested the meeting today to ask for three things, all of which we had specified in our pending application for research clearance at TAWIRI. First, we wanted to inspect all lion, leopard, and buffalo trophies before export—we already knew how to estimate ages of lions, and we were working on similar techniques for leopards. Second, we needed the hunting companies' data on poaching arrests and the number of confiscated guns and snares.

Finally, we were requesting written authorization from the WD so that the hunting companies could provide us with their data.

None of the initial tension in the room had dissipated, but I had at least managed to reach slide number fifty-nine, and I was done. Spent.

Severre thanked me coldly and asked his colleagues for comments.

In Tanzania, the entire audience is allowed to react before the speaker can finally reply—an awkward custom that complicates any back and forth that might resolve an issue. And in my current state, I was worried that I might forget most of their points before I could finally respond.

So I rummaged around in my computer case, looking for a pen and paper and trying to register the comments from around the table.

Inyasi Lejora spoke out. "I don't understand why you are so worried about Savannas Forever." He turned to face Severre. "They only want to help you."

Lejora was from TANAPA, and comparing TANAPA to the WD was like comparing Mother Teresa to Typhoid Mary.

Then Simon Mduma spoke quietly and cautiously about how he had been working with me to craft a memorandum of understanding between TAWIRI and Savannas Forever. There might be advantages to his organization from working with us; Savannas Forever could help build the research capacity of his staff and provide data on wildlife in rural areas. Simon's support for Savannas Forever was rather more tepid than I had hoped for, but in combination with Lejora's endorsement, his comments seemed to turn the tide.

Severre softened his tone slightly and asked me to clarify the relationship between Savannas Forever and the Tanzanian Wildlife Research Institute.

Almost breathing again, I returned to the organigram, and there was the big yellow box indicating TAWIRI as the governmental organization to which Savannas Forever would report. Everyone started discussing where TAWIRI should be placed. I had originally positioned TAWIRI within the green box for Savannas Forever, but the honest hunting companies had worried about our autonomy if we were connected too closely to a governmental agency.

Simon said it would be alright for TAWIRI to stay in the yellow box; Lejora wondered about putting TAWIRI in the green box. Tarimo said one thing and Miriam another.

Severre then pronounced that TAWIRI should be shifted from the yellow box of the government to the green box with Savannas Forever. The Tanzanian Wildlife Research Institute would work closely with Savannas Forever and would report our results directly to the WD.

I was glad to oblige, since Simon had been a postdoc on the biocomplexity project, and TAWIRI had been neglected for decades.

Severre declared the discussion to be finished; they would continue the meeting without me. The Tanzanian Wildlife Research Institute would contact me about next steps.

I packed my laptop and left the room, but I had failed to appreciate the significance of that little green box.

3 Luke Sidewalker

DAR ES SALAAM, 11 AUGUST 2008

The Sunday newspapers ran stories yesterday about the tenth anniversary of the American embassy bombings in Dar es Salaam and Nairobi. I went by the new U.S. embassy in Dar this afternoon; a large wreath of red roses lay wilting against the memorial.

Karyl Whitman was in Nairobi the day of the bombing, staying with Barbie Allen and preparing for her trip home. Karyl had planned to reach the embassy by ten in the morning, but she was delayed in another part of town when the bomb went off. Two hundred people died. Thousands more were hurt—mostly from shattered glass falling from neighboring buildings. Barbie was one of the many volunteers who went into town that evening to donate blood.

I was in Johannesburg with fourteen-year-old Catherine and eleven-year-old Jonathan. We were at the tail end of an adventure; we had listened to the World Cup final among the elephants of Amboseli, encountered a giant puff adder on the road to Tanzania, and ridden at dawn in a hot air balloon above long-shadowed giraffe and chased baboons around a friend's house in the Serengeti. Then we had flown to South Africa and toured a number of private game reserves. Jonathan started feeling feverish. We tested him for malaria, and his tests were negative; the clinic tentatively diagnosed him with tick fever. If the diagnosis was correct, he must have caught it somewhere in Kenya or Tanzania.

We started him on a course of antibiotics, but he was miserable by the time we reached our destination for the night. Makalali Private Reserve features a high-end private game lodge with ornate art nouveau Afro-Italian architecture and separate cottages that were made to look like mud huts from Mali in the style of someone like Salvador Dali. Our bedroom had a fireplace, our veranda had the feel of a tree house, and our shower looked out at the stars.

Lorry on the rim of Ngorongoro Crater

Jonathan was too sick for dinner, so I tucked him in under the gauzy white mosquito net enshrouding his four-poster bed, turned out the lights, and waited until he was safely asleep before joining Catherine at the dining area. We ate fine food and barbecue beside an open pit fire and tried to ignore the spoiled rich kid dancing the Macarena at the next table—we were guests of the manager, but Makalali cost normal people something like $1,000 per night and catered to guests like Michael Jackson and other mega-celebrities.

We walked back to our cottage, escorted by an armed guard with a flashlight, and the sound of our entrance caused Jonathan to half leap out of bed, delirious, asking, "Why is everything times six?"

Jonathan had mostly recovered the next day, and we went out to look for lions. I had started visiting South Africa to learn about the reintroduced lions in these restored ecosystems. Makalali, like many other private reserves in South Africa, had previously been a cattle ranch, but ecotourism had become far more profitable since Nelson Mandela's election had ended South Africa's status as an outlaw state. I wasn't that hopeful about the quality of science on these tiny populations in their fenced reserves, but I had reached a point in Tanzania where I needed a backup plan in case the then head of TAWIRI, Costa Mlay, forced me to quit the Serengeti.

Mlay had been director of wildlife and was transferred from the WD to TAWIRI as a punishment for various offenses, including complying with my plea for a temporary ban on lion hunting in Loliondo Game Reserve in 1990. The Tanzanian National Parks was worried about overhunting of lions in Loliondo, so I conducted a lion survey with forty students from the College of African Wildlife Management and found that the local hunting company had already exceeded their annual lion quota by the second week of a six-month season.

Mlay agreed to the ban, but then the matter came up in Parliament two years later, with MPs asking why the WD had stopped collecting revenues from lion hunting in Loliondo. Mlay asked for help, and Markus Borner arranged for a helicopter survey of the game controlled area, calibrated against the lion population in our Serengeti study area. Fly low over a thicket or river in a helicopter, and the noise drives the lions out into the open—the survey crew found every single one of our study lions, but when they scoured Loliondo, they found no lions at all.

Mlay's decision had been vindicated, but whatever else he had done to piss off the higher powers led to his immediate dismissal at the exact moment when the Loliondo operator lost his block to the defense minister of the U.A.E.—who represented some big shot in the Emirates—in one of the most corrupt hunting-block deals in the history of the WD.

Mlay went into a funk after becoming the director general of the Tanzanian Wildlife Research Institute. Markus wanted Mlay to engage his political savvy so that TAWIRI might actually become something more than an office building in Arusha with a couple of dispirited scientists with nothing to do except show up at their offices each day.

Markus brought Mlay to the United States to recruit someone to head FZS's new ecological monitoring program in the Serengeti. They visited Minnesota to interview one of my graduate students, we went to see *Jurassic Park*, and I took them to the Mall of America with its indoor amusement park.

So Costa Mlay and I started out well. He was charming, and he seemed genuinely interested in making TAWIRI into an international research organization.

But his mood started swinging from day to day. Some days he was delightful, some days he was short-tempered and mean. He became vicious to his staff.

He also started to steal.

It wasn't too bad at first, just a few research fees went missing; new standards of creative accounting were developed. But then he started embezzling on a larger scale, stealing whole grants and siphoning large sums of money.

He no longer pretended to be interested in anything; he only put himself in the middle so he could take away from everyone else.

Markus blocked him from stealing Frankfurt funds, so Costa started to hate Markus with a passion, and since I was Markus's friend, I was suspect, too.

But I gave Mlay an even better reason to hate me. I became so upset with his total disregard for research that I traveled to Dar one day and complained to various high-ranking government officials. Mlay lacked any scientific training; he lacked any interest in wildlife. The permanent secretary encouraged me to write a formal complaint; I circulated the letter to various prominent scientists. Once it reached the permanent secretary, he handed it to Mlay, and no one else. I had been set up.

I had too high a profile for Mlay to attack me directly, so my research assistants became the targets. Their research clearance was delayed, their visa renewal requests denied. Our research fees were raised.

Then on the day in 1998 when we tested Jonathan for malaria, I learned the Tanzanian Wildlife Research Institute had just denied research clearance for my latest field assistant. That left Peyton without help hauling around the lion dummies. I needed to get the kids back to the United States; we were supposed to depart in another few days.

Jonathan still wasn't himself. His delirium had passed, but he was still feeling poorly.

And his sister, ahem, his teenaged sister was, at fourteen, not the most tolerant of traveling companions. The thought of her being his sole escort halfway round the world was enough to send Jonathan and me both into a state of significant anxiety.

We stayed at another fancy lodge in Pilanesberg National Park. This one was high on a hill. Our cottage protruded from between two giant boulders. On the way to breakfast, Jonathan stepped on a small but deadly snake. His close encounter brightened his youthful spirits, but it didn't do much for my peace of mind.

By the time we reached the Johannesburg airport, we had agreed to go our separate ways. Catherine had vowed to remain on her best behavior so they could be the cool kids who had flown unaccompanied all the way from South Africa to Frankfurt, Washington, DC, and Minneapolis. But poor Jonathan had almost suffered an embarrassing intestinal accident on the long drive back to Johannesburg, and now he faced the prospect of being at the total mercy of his big sister for the next thirty-two hours.

I felt like the worst father in the world.

Standing in line at the Lufthansa counter with Jonathan sitting pensively on a suitcase and Catherine hovering reassuringly, we looked up at a TV screen and saw footage of bleeding bodies in Nairobi and shots of a smoldering gap in the U.S. embassy in Dar.

The next evening I was in Nairobi, riding in a taxi along the Uhuru Highway, six blocks away from the ruined embassy. Pieces of glass were still scattered across the road. There was almost no traffic all the way to Barbie Allen's house. Everything seemed broken.

I flew down to Dar the next afternoon. My taxi driver was a Muslim, and he spoke angrily about the terrorists as we drove past the damaged embassy. Tanzania was a peaceful country; Islam was a peaceful religion. These terrorists were foreigners; they weren't real Muslims. We drove in silence until his mood lifted, and he asked if it was OK to tell a joke about Bill Clinton and Monica Lewinsky.

His joke was pretty good ("Have you heard about the new laptop named after President Clinton? It has a six-inch hard drive and no memory"), but I had come to vanquish the Evil Costa Mlay, and I could only manage half a smile.

*

I was in the Serengeti again in 1999, watching lions with Peyton and Grant when the TAWIRI board came to the park for their annual meeting. My efforts to neutralize Mlay had so far failed, and he had become even more tyrannical over the past year, harassing my Tanzanian students and threatening to expel me for good. He had reached retirement age, but he wanted his appointment extended for another five years, and no one thought he could lose.

I had warned Peyton that we might have to abandon the Tanzanian lion projects and move everything down to South Africa. In the worst-case scenario, we could drive through some of the parks in southern Tanzania, then through Zambia and Botswana, expanding her sample size for dummy tests by looking for lions as we went. The odds were remote of seeing even a single lion on such an ad hoc ramble, so she was not amused.

Then the TAWIRI board met in closed session in the Serengeti for two days, and on the second day they voted to replace Costa with Charles Mlingwa, an ornithologist who had just completed his PhD in Germany.

The board invited me to join them for dinner at the hostel about half a mile away from the lion house. I arrived and saw Costa Mlay sitting in the middle of the dining room with officials of the TAWIRI board arranged in a wide circle around him. The room was dimly lit by a couple of kerosene lamps; everyone sat in grim silence. As I walked in, Costa took one look at me, stood up, and stepped outside.

Everyone greeted me warmly, and the board chairman directed me to sit in Costa's empty chair. Mlay remained outside for the rest of the evening, mingling with the creatures of darkness.

I only saw Costa Mlay one more time—at the TAHOA meeting in 2004 where everyone had hailed me as a savior. He had metamorphosed into an ebony skeleton from an advanced case of AIDS. Speaking to the assembled hunting companies, he congratulated me for helping to protect the lions for trophy hunting, but after leaving the stage he avoided making eye contact and left without saying good-bye to anyone.

He died four months later.

I had always wondered about his sudden mood swings, puzzling over whether he was addicted to heroin or speed. But maybe he was just suffering from the side effects of some early form of antiretroviral therapy, or maybe he was just railing against his impending doom.

Maybe he felt an urgent need to provide for his family, or maybe he was just a crook with no fear of the consequences.

I might only have played a minor role in his departure from TAWIRI, but it was enough.

I was ready for the next challenge with authority.

DAR ES SALAAM, 12 AUGUST 2008

I've stayed in a dozen different rooms in the Econo Lodge by now, so a climb upstairs to the fifth floor is a bit like passing through the strata of the past two years. On the third floor, there's room 302 where Jonathan stayed after getting back from Rufiji in 2006. Down at the end of the hall is triangular room 304, with the triangular bathroom (sans hot water, sans tub) where I listened to the first iTunes songs I had downloaded on the free Internet over at the Kilimanjaro-Kempinsky. On the landing of the fourth floor and room 406 is where Susan and I wrote our President's Emergency Plan for AIDS Relief grant proposals; over on the left, room 402, is where Luke Sidewalker nursed his wounds after the assault, and, down in the shadows near the end of the hall on the right, room 405 where Joanna went mad.

I always run up the stairs, three, four steps at a time—it is my only form of exercise—and I always feel my legs start to weaken as I reach the fifth-floor landing. There is room 508 where I stayed with Susan when Luke was trying to keep Joanna from self-destructing. Room 510 is where the shower head kept popping off and the waterline to the toilet leaked all over the floor; and then finally back home to room 512 where I've been staying for the past five weeks.

Hotel breakfast is down on the ground floor and consists of two small slices of white bread with butter and jam, a slice of papaya, and a choice of tea or coffee. But I keep a spoon and bowl in my room and serve myself muesli or granola fortified with bran flakes. The milk comes in boxes, and it keeps for a day or so after opening. I brew Rooibos tea by soaking four to six bags overnight in a liter of bottled water, and I also make sugarless lime juice from the fruit sold on the sidewalk near the Hong Kong Restaurant.

Down on the street this morning, there is the general bustle of pedestrians and traffic, sounds of construction from all the new buildings (a few weeks ago a new ten-story building behind the Hong Kong collapsed under its own weight; the newspapers claim that at least a quarter of these new high-rises are dangerously illegal), and, rising above the din, the high lyrical whining call of a boy selling cold water as he walks the streets.

At half past five this morning, the muezzins were appealingly musical, especially the first one. His tenor voice was so warm and soothing; it dipped and soared and carried me along to another world entirely. The next two voices overlapped with each other, and they seemed to have been inspired by that opening solo. Or maybe today has some special significance on the Muslim calendar.

Walking the streets of Dar at night, I pass by numerous mosques. This is partly because of the sheer number of mosques in town and partly because no muggers would dare linger in plain view of the Muslim worshippers.

And since it is always too hot to celebrate anything here during the daytime, the nights come alive whenever the Aga Khan comes to Dar es Salaam to open a new hospital—with colored lights draped along the sinuous contours of the mosques and strung above the streets—or when the Shiites observe the Day of Ashura—which commemorates the martyrdom of Hussein, the grandson of Muhammad, and marks the origins of the schism between Shiite and Sunni. Parades of aggrieved men and boys march the streets dressed in black and carrying banners, while their women watch mournfully from the sidewalks.

But the Islamic lunar calendar only has about 354 days, so Ashura—like Ramadan—occurs on a different date each year of the Gregorian calendar, keeping me in a state of perpetual cultural jet lag.

Dar is also home to enough Hindus to populate a dozen temples and to ensure large fireworks displays during the nights of Diwali. The Econo Lodge is an Indian-run hotel, and a huge party came here to celebrate Mahalakshmi Vrata last Friday with small boys in Nehru suits and little girls in saris, women dressed like swirling dreams of bright colors, and bored men in Western suits standing off to one side.

Dar es Salaam is Arabic for "dwelling of peace." The slave trade ran six hundred miles from Lake Tanganyika across to Bagamoyo, fifty miles north of Dar, across the water to the sultanate of Zanzibar. A thousand years of Arabs trading ivory and slaves were followed by the Germans employing the Arabs to manage their cotton and sisal plantations. The Arabs dictated conditions for the village laborers and earned cash salaries from the colonial masters. The British came next, bringing Indians to serve as clerks, engineers, and managers, effectively giving them the reins of power in every village and town.

By independence from Britain in the 1960s, the Indians and Arabs effectively owned and managed the country. Nyerere's socialist government soon set out to empower the vast majority of citizens by Africanizing businesses and educational opportunities. Similar resentments led to Idi Amin's expulsion of all the Indians from Uganda in 1970, and Tanzania made things sufficiently uncomfortable that maybe a quarter of Tanzania's Indians left for Canada and the United Kingdom.

So the remainder feels no great love for this country. They are frequently reminded that Tanzania isn't their country at all. But they have accumulated enormous wealth and untold power.

Their power remains deliberately untold because it is a lot easier to run the country that way. And it is a lot harder to figure out what is really going on.

The TAWIRI board turned down Savannas Forever's request for research clearance on the seventeenth of September 2006. The research committee at TAWIRI had approved our proposal, but Erasmus Tarimo had attended on behalf of Severre, and he complained that Savannas Forever was trying to do things that could only be done by government. The government was the only agency that could inspect trophies; only the government could perform antipoaching activities. The government already had mechanisms for monitoring these tasks. If there was to be any form of hunting certification, it could only be performed by the Wildlife Division.

We started hearing reports from the hunting companies that Severre had destroyed me during that meeting in August; Savannas Forever was finished. I had been banned from Tanzania.

Reports of my death proved to be premature, however, and I returned to Tanzania in the middle of December 2006 and went straight to the Ivory Room. I ran into the CITES officer, Julius Kibebe, and he nervously asked how long I'd be back.

I smiled and said, "As long as it takes."

I tried to meet Severre the same day, but he kept putting me off by one or two days at a time, keeping me in Dar twiddling my thumbs over Christmas, then beyond New Year's into the middle of January. I spent entire days at an Internet café next door to the Ministry of Natural Resources.

The Internet connection was barely adequate—"www" stood for the "world-wide wait"—but the air-conditioning fought off the dripping heat of the streets. The manager kept playing the same CD of Gospel music over and over until I could praise Jesus in my sleep.

Desperate for something to pass the time, I started organizing my iTunes collection in chronological order, looking up the date of composition of every classical piece and the date of performance for all my kids' recent favorites. Those were easy, thanks to Wikipedia, but the flamenco and sitar music proved a lot more challenging. I even started deleting tracks I couldn't date. More than a little crazy, sure, but I learned a lot about the web.

What can anyone say about boredom and frustration in a place like Dar? It isn't that different from boredom at home, just hotter, sweatier, smellier, and with nowhere near as many distractions.

Minister Diallo's reputation—after failing to expel Severre, backtracking on the lion-hunting ban, and a blundering attempt to reassign ownership of tourist lodges in the Selous—had reached the point where he was transferred to the Ministry of Livestock.

The new minister of natural resources was Professor Jumanne Maghembe, a scientist with a background in forestry. I kept trying to meet him, but he was on an extended tour of the United States and Europe, promoting Tanzania's tourist industry.

Another issue had emerged. All of the Southern African Development Community countries had banded together with a proposal to down-list their elephant populations from appendix 1 at the next CITES convention. Namibia and Botswana have well-regulated wildlife departments, their elephants were bursting out from the reserves, destroying crops, and trampling people, and the elephant's conservation status was making it difficult to manage their numbers. Zimbabwe, Zambia, and Tanzania all wanted to hop on board the bandwagon, so the WD had commissioned TAWIRI to conduct a countrywide elephant census. The preliminary analysis suggested Tanzania had masses of elephants, but the survey had been fatally flawed by the flight crew's refusal to follow the necessary elephant-counting protocol: they were eager to finish the job and go home as quickly as possible.

Stories had been circulating for months that the hunting companies had witnessed a recent surge in ivory poaching throughout Tanzania. A Chinese construction crew had just completed a new road from Lindi to Mozambique. It was called the Ivory Highway. With China's rising wealth, millions of middle-class families could afford to buy ivory signature stamps. The demand was insatiable, and since a signature stamp is about the size of a lipstick case, the tusks could be chopped into sections and packed in fifty-kilogram sacks to look like cassava or charcoal.

Severre threatened to shut down immediately any hunting company that revealed the true extent of ivory poaching on their blocks. He could only sell off Tanzania's huge stockpile of ivory if the country was seen as successfully controlling the elephant slaughter. He even hailed TAWIRI's survey as evidence that Tanzania should immediately cull its excessively large elephant herds.

The Environmental Investigation Agency, a nongovernmental organization based in the United Kingdom, came to Dar in late 2006 and found ivory on sale in the finest jewelry stores. The stalls at the Mwenge tourist market kept wooden carvings, beaded trinkets, and clay pots on open display, but the false ceilings concealed whole tusks of ivory.

*

The headquarters of the WD are called the Ivory Room because of the large warehouse that stores all the confiscated wildlife products from around Tanzania, including the seventy-ton stockpile of ivory that has accumulated since the CITES ban on the ivory trade in 1989. International law requires that ivory remain under the strictest possible control: nothing is allowed to pass out the steel doors of the storehouse.

Sara Bambara, head of a British animal welfare organization, sent a Chinese undercover agent to the WD to buy tusks from the Ivory Room. The warehouse manager readily accommodated his request. The undercover agent recorded the entire transaction on videotape.

The Environmental Investigation Agency made a second video showing the warehouse manager apologizing that he couldn't sell ivory today because Severre was out of his office.

In February, Sara Bambara met Minister Jumanne Maghembe in a London hotel. Maghembe is a round man with a gruff voice and an unpleasant face. He had assured her that he wanted to clean up the wildlife sector, but he insisted on talking to her in his hotel room, asking her to show him the evidence while making her sit beside him on the edge of his bed. She calmly showed him the video of the ivory sale at the WD and threatened to release it to the media unless Tanzania withdrew their request to CITES to reopen the legal ivory trade.

In the following weeks, Maghembe reassured her that Severre was under strict instructions, but he seemed more interested in pursuing Sara than conserving elephants.

The Ivory Room was a rather ramshackle series of simple one-floor buildings with central hallways and offices on either side. Many of the Wildlife Division staff wore continuous tough-guy scowls, while others seemed in a constant state of anxiety—nervous and suspicious. White hunters strolled around in shorts and khaki and waited their turn to meet the director.

Then one day in March 2007, the Ivory Room received another kind of visitor. No one is quite sure where Luke Sidewalker came from. Some said he was British, others thought he was Swiss. He might even have come from North America.

Luke met in a closed office with Source A. The window air conditioner was on high; they kept their voices low. Luke asked about Maghembe.

Source A. [*making a face*] He's a crook.

LUKE. How so?

A. When Maghembe first came to the ministry, he tried to stay clear of *him* [*points toward Severre's office*]. But it didn't last. The Big Man chased after him.

LUKE. Does Severre tell you these things?

A. He never includes me in anything. I'm not part of his circle.

LUKE. Who is in his circle? Miriam Zachariah?

A. Of course.

LUKE. Erasmus Tarimo?

A. Yes!

LUKE. [*after listing a few more names*] "Anyone else?"

A. Serakike [*whispering now*], treasurer for TWPF.

The Tanzanian Wildlife Protection Fund was financed by a 10 percent levy on all trophy fees; it was reputed to hold millions of dollars.

LUKE. What is the money really used for?

A. The Big Man uses it to pay off people.

LUKE. Who?

A. We had a big celebration here to congratulate Mama Meghji when she left to become finance minister. The boss gave her a Toyota, gifts, cash—it must have cost two or three hundred thousand dollars.

LUKE. But she's finance minister now; she can steal as much as she wants from the treasury.

A. No, they have safeguards over there. Not like here.

LUKE. Is Severre fond of her?

A. No. The Finance Ministry controls cash flows in and out of the country. He still needs her.

LUKE. Who else has he bought?

A. He gave Kayera [*the recently retired assistant director*] a hundred thousand dollars to finance his campaign for Parliament.

LUKE. Did he win?

A. No, he ran against CCM [*the ruling party in Tanzania*] and lost two-to-one.

LUKE. Why run against the ruling party?

A. Kayera was stupid. He couldn't win the CCM nomination, so he tried to win the election against the same man he couldn't beat the first time.

LUKE. What about Maghembe?

A. The Big Man bought him. He was cheap.

LUKE. How so?

A. He gave him a satellite phone [*worth maybe $500 or so*]

LUKE. Maghembe says he wants to withdraw the elephant proposal at CITES.

A. You can't trust him.

LUKE. [*leaning forward, barely audible*] He has to withdraw the proposal.

A. Why?

LUKE. They got caught selling ivory from the Ivory Room.

A. How do you know?

LUKE. There was an undercover operation—a sting. The warehouse supervisor was filmed with a hidden camera. If Tanzania doesn't withdraw the proposal, the tape will go public. Maghembe himself has seen it.

A. Don't believe anything Maghembe tells you.

LUKE. How do you know so much about him? About the Sat-phone?

A. My driver knows his driver. The big shots drive around town, and they all act as if they are alone, as if their drivers don't have ears. The drivers all talk to each other, bragging about the big deals their bosses make in their cars.

LUKE. What about Pasanisi?

A. They're business partners [*points again at Severre's office*]. Pasanisi lives in the boss's office whenever he's in Tanzania. He also has a fund that Severre can use whenever he's running low at the Wildlife Protection Fund.

Thus was the conservation money spent from Pasanisi's gala dinner.

LUKE. What about Sheni Lalji?

A. He built his office as close to here as he could so he could see the Big Man whenever he needed. They're all part of the same mafia.

LUKE. What about G——?

A. Now you are really trying to get me killed!

LUKE. Why?

A. If anyone found out I told you . . . [*slides a pointed finger across a nervous throat*].

LUKE. What is it about?

A. [*barely audible again*] They put him on the payroll. The checks arrive here every month.

LUKE. Checks for Severre and his circle?

A. Yes. Four of them.

A few days later, Luke Sidewalker talked to Source B, a subdued middle-aged man whose suit seemed about two sizes too large. B had once seemed likely to rise to the top of the WD, but he had been sidelined after failing to conform to the ethos of Severre's inner circle.

B's English was good, but he spoke so quietly and so nervously that he was hard to understand. He explained that the hunting quotas were not limits on how many animals could be killed in each block—they were production targets. So if a company managed to shoot more than their initial quota, then the WD would raise quotas retrospectively. Wardens complained about overhunting of lions, eland, gerenuk, or kudu, but Severre raised the quotas anyway.

The quotas had been fiddled with so often that no one actually knew how many animals were being shot. There was no attempt to monitor anything. Numbers were kept on small scraps of paper in locked drawers in a room where only Severre's inner circle could enter.

Sources A and B wanted to blow the whistle on the WD because they both resented the way that the game reserves were being treated as strip mines rather than conservation areas. But they wouldn't say why they were so frightened of Severre. They didn't seem particularly worried about the prospect of losing their jobs; their fear was much more visceral and immediate.

Luke sought out people who had worked in Tanzania for decades, people who had seen the full arc of Severre's rise. Most of the dirt seemed to involve the usual fabric of financial corruption that had permeated the WD since Ndolanga's era.

But then he heard a name.

Matthew Maige had been a rising star in the WD as one of the few people who made any difference in conserving black rhino. Tanzania had lost over 90 percent of its rhino population in the 1970s when poachers sold rhino horns to Arabs for dagger handles and to the Chinese as traditional medicine. Maige represented Tanzania at international meetings and became the treasurer of the Wildlife Protection Fund, but he wasn't part of Severre's gang.

And he wasn't compliant.

A band of armed men broke into Maige's house one night in 2003 and hacked him to pieces. Nothing was stolen. There was no investigation.

Everyone suspected that Severre had ordered the hit.

Chama cha Mapinduzi (CCM), or the Party of the Revolution, is the ruling party of Tanzania. Elections follow a five-year schedule, and they are largely free and fair, but the emergence of multiparty politics in 1995 has led to ever more expensive campaigns. In 2005, the owner of the Abu Dhabi airport offered CCM a substantial campaign contribution in exchange for the promise of a hunting block. One MP volunteered an area belonging to the Hadza tribe—Tanzania's equivalent of the Bushmen. The Hadza are hunter-gatherers, and they require healthy animal populations to survive. Unlike trophy hunters from Europe or North America, Arabs hunt for the pot: the UAE has no import restrictions on meat from East Africa. Thus, a couple of heavy hunting seasons might destroy the last vestiges of Hadza culture—one of the oldest on the planet.

Some of the local Hadza accepted the modest government payment for their land, but others resisted. Protestors were jailed; people died. After CCM's successful reelection, the helpful MP was appointed minister of governance.

Collectively, TAHOA and Severre's claque were known as the hunting mafia; a hunting company manager even confessed to a code of silence (omertà). They were untouchable despite Parliamentary probe teams and frequent newspaper exposés. Source A and Source B only knew the names of the figureheads—by law most companies had to be owned by Tanzanian citizens, and they were mostly just window-dressing.

Source C was the most outspoken of Luke's contacts and the least fearful. In a just world, C probably would have been director of wildlife instead of Severre, but even a just world might find it difficult to deal with

C's blunt evaluations. C seemed starved of educated conversation and he talked with an excited eloquence that must have been common in, say, eighteenth-century France, but it could be a challenge sometimes to pick out the nuggets from his meandering circumlocutions.

But then C handed Luke the official list of all fifty-four hunting companies operating in 2007.

Luke took the list to the Business Registration and Licensing Agency at the height of the wet season in Dar es Salaam. The days were stiflingly hot; the rain was warm. Cloudbursts left filthy rivulets flowing along the steamy streets. The air in the Business Registration office was musty from all the damp, crumbling paperwork. He handed over the names of the fifty-four companies and asked for their files. His eyes itched and his head throbbed as he thumbed through each page. Sweat poured off his face, his sides, but he kept digging and digging, copying down the names of every shareholder and every partner and where they came from.

Two of the biggest companies had Tanzanian partners, but they were just window-dressing. Pasanisi's partner was a man without much influence or renown. One of the Arusha-based companies was partnered by a semieducated mediocrity.

The block allocation system was so ossified that a new owner could buy the name of a long-defunct company, register it retrospectively for thirteen consecutive years on a single day, forging the signatures of people who were either dead or gone.

Several companies reported their profits and losses when seeking new licenses; the rich Arabs in Loliondo claimed a loss every year. Another company claimed large losses but then had the whistle blown by an employee—the arguments of the suit and countersuit went on for years.

Most of the really big companies were registered in tax havens like Panama and the island of Jersey.

Luke next went to the Tanzanian Revenue Authority, posing as a business consultant from Botswana, asking if he could review the tax returns of the hunting companies so that his firm could get a better idea of the size and scale of the Tanzanian industry; they were deciding whether to establish an office in this country. Tax returns are confidential, he was told, and could only be reviewed with written authorization from the government. He would have to obtain a letter from the Ministry of Natural Resources or perhaps the director of wildlife.

He went on the Internet and Googled "Sheni Lalji," along with his real name, Mohsin Abdallah. He found newspaper reports of tax evasion and fraud.

He Googled the name of every partner and shareholder, linking more and more companies to politics and corruption.

*

I went down to Lindi with Dennis Ikanda and Hadas Kushnir during all this, finally visiting the sites of lion attacks in person. We went to a village that had suffered three attacks in the past few years, walking for maybe two miles through high yellow-green grass and negotiating crude foot bridges before reaching a cluster of thatched huts with walls made of dried mud and wooden poles and containing absolutely no possessions except for the clothes on their backs.

A middle-aged woman told us how her grandson went outside to use the latrine one night. She heard a suspicious noise and ran out making as much noise as she could, and the lion ran away. She was disappointed that we had merely wanted to interview her. She thought we had come to give her a gun.

We tried to explain that there wasn't enough money in the world to buy everyone in Tanzania a gun and bullets, and, besides, most gunshot wounds were the result of friendly fire. Our logic made no sense to her, and she seemed to forget about us as soon as we started to leave.

We stopped, hot and thirsty, at a neighbor's hut maybe two hundred meters away and eagerly accepted an offer of fresh coconuts. A young man scaled the tree and tossed them down to his father. The old man opened one with his panga (machete), and I drank down the purest ambrosia—cool and fizzy and a fresh-feeling taste I had never encountered before.

We walked back to the car and drove on to the next site. There, the victim's mother appeared with a toddler strapped on her back. She reenacted the night she walked the thirty meters home from her neighbor's hut. She had left her five-year-old son on the trail while she ducked into the grass for a pee. A female lion ran up and bit him on the chest, just below his throat. The woman ran screaming toward them; the lion dropped the child and ran off into the night. The woman took her boy to the hospital, but his chest had been crushed, and he died the next day.

The woman recounted all this matter-of-factly until she placed her hand to her chest, just below her throat, and her face took on all the pain of any mother in the world.

This was the first time anyone had ever come to talk to her about her loss.

Every lion in Tanzania belongs to the government. Sometimes the government gives the victim's family a *pole* (POE-lay)—a consolation amounting to about $40.

No one ever counsels anyone.

Coming back from war is said to be tough. Soldiers suffer flashbacks and nightmares. But the soldiers at least know they've left the war behind. This mother walked past the spot of her child's death every day of her life.

*

In the meantime, Savannas Forever obtained temporary permission to survey twenty-six villages spanning from Mount Kilimanjaro to the shores of Lake Victoria. The village surveys revealed almost no benefits to local communities from ecotourism, trophy hunting, or anything else. Less than 1 percent of villagers around the Serengeti gained any sort of employment from anything to do with conservation. The only real benefit from wildlife was that people could eat it.

And bushmeat consumption was far higher than anyone had realized. One of our graduate students, Dennis Rentsch, found that bushmeat was one of the most important sources of animal protein in the diets of the six hundred thousand or so villagers who lived within easy reach of the park. To top it off, there was a phenomenal rate of migration into the areas adjacent to the Serengeti—mostly by the Sukuma, an agro-pastoralist tribe that sought out uninhabited areas and converted wildlife buffer zones to cotton fields.

By now I had discovered the free Internet at the Kilimanjaro-Kempinsky coffee shop and went there for lunch every day. The connection was the fastest in Dar, and I could do more work on the net than just look up dates of musical compositions.

If you go to the CITES website, you can download all the trade data for every listed species in every country in the world. With an additional two years since the Jo-burg meeting, the Tanzanian lion data showed a strikingly clear trend: the number of lion trophies had peaked at over three hundred in the year 2000 but had fallen to only about two hundred in each of the most recent years.

Botswana showed a similar peak in 1990–93, then never recovered to its former level. Zambia peaked in 1986–92. Zimbabwe peaked in 1992–94.

In every African country that had ever allowed large-scale offtakes of lions, the pattern was the same. A clear peak, followed by precipitous decline, despite the fact that the demand for lion trophies had continued to grow.

Damn.

Lions were in trouble and the hunters were the first to know—except that they didn't.

I went out to the WD a few days later and showed my graphs to the CITES officer, Julius Kibebe, and asked him to arrange a meeting with Severre. "You could be the heroes of conservation!" I assured him. "If Tanzania were to write the next proposal to up-list lions to appendix 1, everyone would see how much Tanzania cares about conservation. We could all be together again, just like in Bangkok!"

Severre could ignore Savannas Forever, but he couldn't ignore the collapse of his lion population.

We all met in his office, just like in years gone by. I showed them two sets of graphs, first for lions, then the same analysis for leopards. Leopards seemed to be doing just fine in all the same countries where the lions were in danger, which certainly made sense, since leopards were famously able to live in cities like Nairobi and Bombay, eating guard dogs and hiding in the smallest patch of forest. Then I showed two graphs that related lion offtakes to the quotas in Botswana and to Baldus's short-lived trophy-hunting database in the Selous: the recent declines were due to lower success rates, not to fewer hunters.

"We need to know if this is generally true—if the declines are due to a lower supply or a lower demand. If you could give me the size of your overall quota each year, it would be quite easy to sort this out for the whole country." And then, pretending to think of it for the first time, I said, "You know, it would be best if you could give me all the data for every block—we could see if the overall decline was mostly due to habitat loss in areas with too many Sukuma or too much farming or if it was because of retaliation for man-eating and livestock depredation."

Oh, what pretty words, and I halfway meant them, too.

And I'm not sure why I bothered other than to satisfy my curiosity.

The Tanzanian Hunting Operators Association was having a preseason meeting in Dar around this time, and I went out to have dinner with Charles Williams, Danny MacCollum and his wife, and a couple of other hunters. They were embarrassed to see me. Not only had they all ended up dropping Savannas Forever, but they also knew perfectly well how badly lions were doing throughout the country—and Pasanisi had refused to allow me to raise the issue in their formal meeting today. Severre had told him that I was planning to ban lion hunting altogether.

Luke Sidewalker met Joanna at the Protea Courtyard Hotel in late March 2007. The place felt like a throwback to the colonial era with a leafy courtyard, flowers everywhere, and a buffet luncheon beside the pool. Joanna's room had a TV, heavy curtains, and a bedspread the size of Rhode Island. She had come to Dar to follow up on the info that Luke had collected in the WD and at the Business Registration and Licensing Agency. Joanna had worked as a producer for an investigative TV series at the BBC, and she had just finished a grueling undercover story in another African country. She had brought out a hidden video camera and microphone, and she showed Luke how she kept the lens between her ample breasts—it was behind the top button of her low-cut blouse—so she could be sure that her subjects would stare straight into the lens.

The first two days she met with several of the honest hunters in order to gain credibility. She told them she was doing an article for *British Heritage Magazine*, a total fiction that was just the sort of title that would appeal to the hunters' rather conservative mind-set and the perceived gentility of their wealthiest clientele.

She wanted to sell the magazine on a big story about the importance of trophy hunting for wildlife conservation. The whole industry was misunderstood.

After the first couple of meetings, her interviewees introduced her to the rest of the industry. Everyone loved the aims of her story. Between appointments with the hunters, she met the reporter for *This Day* who had written the big story on Pasanisi the previous year. She also met with several Tanzanian organizations that were trying to nationalize the hunting blocks.

She was on a tight budget, and she felt she might need to stay an extra few days. To save money, she moved from the Courtyard to the Econo Lodge. She debriefed Luke every night.

Then she went to see Manic Mike.

Manic Mike went through an eccentric performance of hand kissing and clumsy flirtation as he stared fixedly at her cleavage.

She gave him her spiel about how hunting was misunderstood and underappreciated. She wanted her article to set the story straight.

He winced and grimaced agreeably enough through it all, winding himself up as he sat next to her.

Then she apologized, saying that while she herself firmly believed in the value of trophy hunting, she needed to cover the accusations of corruption that haunted the Tanzanian industry. She was sure the stories had been exaggerated, but it would be best to discuss the issue openly. . . .

Manic Mike screeched the proceedings to a halt, and in a voice as cold as death, he said, "If I ever find out that you have anything to do with the antis, I don't care where you go or where you live, I will come, and I will get you. There will be no escape."

Luke Sidewalker went to see Joanna for the usual catch-up session that same night. She let him into her room, and she seemed calm enough to begin with. She told him about the material she had collected from the newspaper reporter—he had survived an attempted poisoning shortly after publishing his story on Pasanisi. Then she started to talk about Manic

Mike. She seemed distracted; she wanted to review her tapes. They agreed to meet again later; tomorrow was going to be a big day. It was time they focused on Sheni and Pasanisi.

Luke came back to her room maybe an hour later. She was pale as a ghost. She told Luke that Manic Mike was going to kill her. If she said anything critical about the hunting industry, she would die.

Luke commiserated. This was going to be a tough assignment; it was why no one had tackled the Tanzanian hunting industry before, but she was must surely encounter this sort of thing all the time in her line of undercover work.

She exploded. "How dare you suggest I would ever do anything unethical." Then, spitting with rage, she said, "I am *disgusted* with you."

Her investigative agency airlifted her out of the country the next day.

DAR ES SALAAM, 18 AUGUST 2008

"I see you coming, and my legs are shivering," Benson Kibonde, project manager at Selous, says to Henry Brink, grinning broadly, bending forward and clutching his knees, "I remember I owe you money."

We shake hands, and I tell him that the new director of wildlife, Erasmus Tarimo, has agreed to let us collect the data on lion and leopard offtakes, but we wanted to move quickly before something went wrong. Benson smiles broadly again and says, "Take care to strike while the iron is very hot." He raises his arm and swings it down. "Hammer it as hard as you can." Spreading his hands, he says, "Make sure to flatten it completely."

We all laugh and shake hands again; we stand in the hot sun outside Benson's old office at the Ivory Room. The occasional car rolls along the drive, and we step in and out of the shade to let each one pass.

Benson continues: "The problem with this industry is that it is pegged in blood—there is no revenue unless the blood is oozing. We should sell chances to shoot the one big trophy; let the best man win." He shakes his head, as he always does about this subject. "I may not be thinking correctly, a man with my limited understanding, but that's where we went wrong at the very beginning."

He warns us that even if we inspected the original hunting licenses, the companies typically lie about the blocks where they have taken their lions. We talk about using GPS, but he says they make up their coordinates; he laughs and tells us about one Tanzanian lion where the hunters' coordinates indicated that it had been "shot deep within the Congo." I suggest that the hunters be required to photograph the reading from their GPS unit amid the sequence of photos of the lion's nose, face, mane, and so forth. He says they would get around that, too. In the future there should be on-site inspections, but the inspector would have to get there quickly, since the company would load the lion carcass onto a lorry and drive it to wherever they wanted to claim they had shot it.

"If the government tries to raise more revenue from the hunting companies, the companies have two exit doors. One exit door is by cheating the government and not paying anything. The other exit door is to charge more to the client. That's why the companies don't mind if we raise their trophy fees; they just hide the trophies or pass costs onto the clients. But block fees have to be paid by the companies themselves, and they always scream when we raise the price of their blocks."

Benson is in fine form this afternoon. "Why waste time in Parliament talking about someone being a thief? Everybody knows politicians are corrupt. Professor Packer is a thief, and I am corrupt. Fifty percent of the people in this country are corrupt. The question is why? Why? Why?"

Then in reference to nothing in particular, he says, "Our new director is a rudderless boat in high seas."

On the twelfth of April 2007, Luke Sidewalker met Sara Bambara in her room on the sixth floor of the Kempinsky Hotel. She had just flown in from London and was only staying the night. Their man was due to arrive in an hour, and they were worried.

"He's scared to death," Luke said as he came into the room. "But he's coming."

"Bloody hell," Sara said. "I only flew here for this one meeting.

"Are you OK?" he asked, looking for scrapes or bruises.

"I'm ex*haust*ed. I worried they might arrest me when I got off the plane."

Sara booted up her laptop, and they started to review Luke's PowerPoint, *A Beginner's Guide to the Tanzanian Hunting Industry*.

The first slide showed a box diagram of the key figures in the Ministry of Natural Resources: Minister Maghembe, Permanent Secretary Saleh Pamba, the directors of TAWIRI and national parks, then Severre and his inner circle highlighted in a red box.

Next came a diagram showing Severre's sources of ready cash: the Wildlife Conservation Fund of Tanzania (Pasanisi's charitable fund, which publicly sponsored the lavish banquets at the Royal Palm Hotel), monthly checks to four members of Severre's inner circle from a wealthy hunting company, direct bribes from the smaller companies whenever they wanted to renew their leases, and the Wildlife Protection Fund, financed by a tax on hunting trophies and presumed to promote conservation but which mostly served as Severre's second slush fund.

The chart was decorated with quotes from Severre, who proudly called himself the chief mafioso and bragged how he could bribe any MP he wanted, how he had managed to remove Diallo from the Ministry of Natural Resources.

The next slide listed the names of the government insiders who were known collectively as the Maasai mafia and who protected Severre from disclosure, then came a real rat's nest of a slide with a complicated set of interconnecting boxes naming the four ex-ministers, two ex–prime ministers, two ex-presidents, several current MPs, and the current prime minister, Edward Lowassa, all of whom had hunting blocks—and showing how everything linked back to Severre, as well as highlights of Severre's business partnerships with Pasanisi and Sheni Lalji.

Next was the Hadza scandal in Yaeda Chini, then the murder of Matthew Maige.

Then came the hole in Sara and Luke's hearts: the outline describing how Severre sold ivory to Chinese businessmen from the government stores in the Ivory Room.

But exhibit A was missing.

Sara had taken the Tube into central London on April 8. It was Easter Monday, so the station was nearly empty. She came up the escalator and walked maybe twenty meters along Oxford Street when a BMW rolled to a stop beside her. Two black men leapt out of the car, pushed her down, and grabbed her briefcase. They tore off down the street in the BMW and vanished around a corner.

She had come into town to copy the video of Severre's ivory sale onto a CD. There was no backup file.

Luke waited in the lobby, next to the elevators. Saleh Pamba had grown up with President Kikwete; they were more than just childhood friends. Kikwete's father had died when he was young, and he practically lived with the Pamba family. When Luke asked Source A how best the Severre dossier could be used to shake up the WD, the answer was immediate: "Show it to Pamba." Pamba had worked with Minister Diallo to try to sack Severre; Severre had publicly humiliated Pamba dozens of times. Pamba had an inside track with the president. Severre had only managed to cling to office because he was a presidential appointee.

Luke had gone to see Pamba a few weeks earlier, and Pamba had practically begged for the dossier. "Severre wants to see me sacked—or worse. My life is in danger."

"Should we show it to Maghembe?"

"No, no," pleaded Pamba. "He's working with Severre now. They both want me out."

"And can you get it to Kikwete?"

"I can see the president anytime I like."

"My friend will be calling you," Luke told him. "Her name is Sara—can she call you in your office or would it be better to call your mobile?"

"My office phone will be okay."

"But doesn't Severre have people who can tap your phone?"

"My mobile phone may be tapped but not my office. The line belongs to the government."

But Luke wasn't so sure. He had been told about the Tanzanian government's secret surveillance team. During the socialist days of President Nyerere, Tanzania had been a leader in the nonaligned movement. Tanzania, though not officially Communist, had close diplomatic relations with Mainland China and East Germany. The red Chinese built a railway from Dar es Salaam to Zambia in the 1970s; the East Germans had piloted all the aircraft for Air Tanzania.

Tanzania's global orientation had shifted toward the West by the time the Berlin Wall fell in 1989, but the demise of East Germany left their secret police with nothing to do—and not much reason to stay at home. So the Tanzanians supposedly imported them to run their security networks.

The sun still bore down on the Kempinsky even though it was after six P.M. Luke sat reading a newspaper until the sliding glass doors parted, and Pamba walked uncertainly inside and looked around through his thick glasses. Luke looked up in mock surprise, waved at him with a public greeting, asked after his health and his work, and then lowered his voice as he gave him Sara's room number. Pamba barely said a word before he tottered over to the elevators. Luke sat back down and resumed his inspection of the local paper.

The news was mostly about grand corruption. The closest neighbor to the Hotel Kempinsky was the Bank of Tanzania with its twin towers—which cost more to build than if they had been constructed in downtown Tokyo. There had been power shortages all over the country because the generators had been bought for vastly inflated prices and broke down almost immediately. Newly constructed paved roads turned into swamps at the first rains. The most recent ex-president had started a private business while still in office and owned the largest coal mine in the country.

A half-hour went by, then forty-five minutes. At last the elevator door opened, and out stepped Pamba. He glanced nervously toward Luke, waddled across the lobby, and was gone.

Luke waited a few minutes, then went up to Sara's room.

"Ach, he was useless," she grumbled. "I finished everything, and he just said, 'I already knew all that. Everyone knows all that. What am I supposed to do about it?'

"I said, 'Show it to the president, please. Just let Kikwete know that everyone knows—it's not a government secret anymore.'"

"Do you think he'll take it to the president?"

She sighed, "He promised he would."

They went down to the fancy Asian restaurant on the mezzanine floor of the Kempinsky and sat about ten meters from a giant aquarium near the cashier's desk. Sara's militant vegetarianism got the best of her, and whenever a waiter came by, she complained about the cruelty of keeping animals in a restaurant—about the evils of eating meat. Luke tried to lighten her Teutonic gloom, but she wasn't to be deflected.

They ate vegetables and mushrooms, and Sara knocked back several glasses of red wine. Luke learned that Sara rode motorcycles through bucolic England and that she had a boyfriend who was about fifteen years her junior. Back when she worked in the Serengeti, a Tanzanian scientist had thought she was a witch because of her constant proximity to animals. He had asked her to remove a barn owl from a storeroom. Owls were magical animals; the Tanzanian avoided any possibility of a hex by averting his eyes as she carried it outside.

Sara had met with the Dalai Lama and inspired him to condemn the use of animal skins in traditional Tibetan robes, she was trying to stop tiger farming in China and to up-list the potto (a small primate that lives in African rain forests) to appendix 1 at CITES. Her staff regularly danced with death: China was particularly dangerous for undercover work.

Sara's dinner performance sometimes verged on the operatic, but Luke kept turning the conversation back to the WD, to what to do next in case Pamba didn't follow up.

Luke had spoken with several ambassadors and high commissioners, and they had all agreed that if the Tanzanian government didn't sort out Severre on their own, the next step would be to publicize the dossier. Open up Pandora's box and let Kikwete deal with the consequences publicly.

Sara wasn't thrilled, but there was no other choice.

They said their goodbyes, and Luke slipped outside into the dark.

I have walked from the Kempinsky to the Econo Lodge maybe fifty or seventy times. I sometimes get so caught up in my e-mail that I lose track of the time, and suddenly it is ten or eleven o'clock at night. I can't afford a taxi, and the heat and humidity is so oppressive during the day that any movement is a struggle, but by late evening, the air is usually tolerable.

Walking at night in any African city is foolish, I know, but I follow a route that is reasonably well lit all the way from one hotel to the other. And there are night watchmen on almost every corner.

Luke knew this route well, and he strode purposefully past the taxi rank and guards in front of the Kempinsky Hotel then out onto Kivukoni Front. He went past the Ministry of Justice then drew up to the Lutheran Church on the waterfront—the only unlit segment on the route home. In the darkness, he felt a sudden snap of his computer strap, a shove from behind.

He sprawled awkwardly forward and crashed down hard on his knees and hands, the tops of his bare feet scraped along the asphalt.

There was no pain; only the dim vision of a thin man in a white shirt running across the black street, and there was no way on god's earth that Luke was going to let him escape.

Luke knew the man had a knife or razor, but he didn't care. His guts could always be sewn up again, but if they got his computer, he was dead. Pamba might already be dead.

Luke rose up like an acrobat. Every muscle was like steel. He didn't run—he flew.

The man paused in the middle of the road, startled by Luke's fury. Luke grabbed him and lifted him up in the air, yelling and cursing in a low hoarse voice. "Motherfucker, give it back! GIVE IT BACK! NOW! YOU FUCKING ASSHOLE."

The man dangled at least a foot off the ground. He flung Luke's computer about five meters away and slashed Luke's right hand with his razor blade.

Luke tossed him aside, leapt to his computer, and realized that about five passenger cars had stopped to watch his little drama. He looked over his shoulder to see the thief glaring back at him from the shadows of the night before slipping through a chain-link fence down toward the waterside.

A driver came up just after Luke had checked that his laptop was intact. In the glare of the headlights, Luke collected all the papers that had fallen from the case: hand-written lists of hunting companies, their owners and patrons, and the sketches that outlined the pattern of corruption, every page now covered with the dirt of the street and the blood from his hand.

"Need any help?" the driver asked.

"Not anymore."

4 Limbo

DAR ES SALAAM, 16 JANUARY 2009

The new director of wildlife, Erasmus Tarimo, has a disconcerting habit of closing his eyes when he is thinking, and my PowerPoint appears to be putting him to sleep.

I'm trying to show him graphs of population trends. Lions are doing fine in the middle of the Serengeti (I speak up as he starts to snooze) and the photo tourism area of the Selous (he snaps awake to say, "Because there is no trophy hunting in the photographic area"), but the Crater lions are fading because of persistent inbreeding (his eyes reshut), while Tarangire and Katavi are in steep decline (he stirs and says, "Because of too much trophy hunting").

Then I show him that lion offtakes have declined across the country as a whole. We suspect the trends have been worst in any area where the Sukuma have recently settled.

The director nods in agreement. "These are the most destructive practices in our country. Agro-pastoralism. The Sukuma move in and cut down everything."

I show him a human population map that illustrates the growth rates in each ward (the rough equivalent of a county) in Tanzania. "The most rapid growth has been in the west and near Tarangire. Even in the areas that don't look too bad now, the human growth rate is about 3.5 percent. Populations will double again in the next twenty years. Twenty years ago, we had twenty million people—now we have forty million. If the population hadn't doubled, we wouldn't be sitting here worrying about lions."

"Yes, and we had nine million people at independence," he says and then, pointing to his two associates, "We need this map. We must show it to Parliament, so they can understand what we have to deal with."

I cautiously suggest that if I were permitted to see the annual lion offtakes from all the different hunting blocks in the country, we might find that lion harvests fell farthest in Sukuma land or in districts with the fastest human population growth.

"Or in the areas with the worst hunting companies," he laughs. "Why not? We must get everything out in the open."

Mission accomplished.

Revving up again, I show him data from Savannas Forever's surveys around the Serengeti in 2006, a series of pie charts on a map indicating survey villages right next to the park boundary. "A lot of human population growth is through migration. Look at how many people have immigrated into these villages."

"And they moved there to be poachers," says the director.

"Perhaps on the northwest side of the Serengeti. But on the southwest side it's Sukuma again, moving to unoccupied areas that can be cleared for farming. . . ."

"And cattle." He finishes my sentence. "All the cattle in these areas come from Sukuma."

Then I show him a map with the number of tree species that have been lost to firewood in each village over the past few years. Next I pull out a graph with the overall loss of forest and woodlands in Tanzania compared to seven neighboring countries—and Tanzania is the worst. Even worse than Zimbabwe, and we don't have the excuse of an aging despot, economic collapse, and mass starvation.

"Get these graphs, too, we most show these to the MPs," says the director to his assistant, Obed.

I produce, for his viewing, more Savannas Forever data showing how often people around the Serengeti eat bushmeat. "Just taking the inner ring of wards that border the park boundary to the west, half a million people live here, and we estimate they eat 9.5 million servings of bushmeat per year." This translates into a lot of wildebeest, not even counting however much meat makes it to Dar es Salaam and Nairobi.

"And even to Darfur," the director says. "Part of the problem is that people think of wild meat as exotic. They pay a lot more for it."

Then with the loss of habitat and the reduction in prey populations, we get to the surge in man-eating lions in the 1990s and 2000s. "The districts that have the least lion prey suffer the greatest number of attacks on people. Those with the most attacks have the most bushpigs."

"Why bushpigs?"

I explain, and the director practically jumps out of his chair, "I nearly got eaten by lions because of bushpigs!" He had been warden in Mahale Mountains when it was still a game reserve, and the pigs living next to park headquarters got so tame, they would come into the storeroom of his house. The lions followed them inside.

So our discussion is happily rolling along, but Director Tarimo seems to be getting even more tired, and he shuts his eyes whenever I click to the next slide. I'm trying to maintain the momentum, trying to keep everything on the same high plane of excitement. And he is resting his elbow on his knee, chin in hand, and, I swear to god, he drifts off to sleep every five to ten seconds. I expostulate to keep things moving, raise my voice in excitement, set up my denouement with a dramatic pause, and bark out my conclusion. The contrast wakes him up, and he is still on topic, he turns again to his staff, tells them to get copies of my presentation.

So we have all these people and all this conflict: lions are eating livestock, elephants are destroying crops. There have been calls from Parliament to relieve the suffering of the people. There is a widespread clamor for the Wildlife Division to start paying villagers compensation for wildlife damages.

I describe the wolf reintroduction into Yellowstone. Wolves were only accepted by state officials in Montana and Idaho because of the promise of compensation to cattle ranchers. The wolf population soon grew into the hundreds, but it ranged over such an enormous area that the program was economically tractable—but only for the first few years.

All around the world, compensation runs the ubiquitous risk of fraud (no, honest, these cows weren't sick—they were eaten by lions!), and cash payouts for wildlife damages can entice people to move into wildlife areas in the first place and even provide a disincentive for proper husbandry (why mend my fence? I'll get paid for my losses!). And compensation would impose an unbelievable expense for the government of an impoverished country with ten thousand lions and a hundred thousand elephants. . . .

Tarimo dozes occasionally, but he always bounces back after a few seconds. I suppose I should be worried about the effects of these micronaps on his short-term memory, but everything seems so intuitive to him that he will surely remember that some white guy came into his office once upon a time with a bunch of information that might prove useful somehow someday.

"Instead of compensation, a lot of countries are considering some sort of insurance scheme, which has two major advantages. First, by paying even a small premium, people are encouraged to show personal responsibility, but, second, and most important, people's premiums could vary according to their behavior.

"We all drive, but if I have more accidents this year, I will have to pay a higher insurance rate next year. And if I live in an area with a lot of traffic accidents, I'll also have to pay more.

"Do this with wildlife damage, and people would be more motivated to fix their fences. You could also discourage people from living near wildlife: just set a higher premium close to the protected areas. People living in the centers of agriculture—Morogoro and Dodoma—would hardly have to pay anything."

"And we could encourage people to move away from wildlife," Tarimo says with delight. "This is just what we need. Obed, I want you to follow up straightaway. Start work on this before Parliament sits again."

The director wanted to meet with me on this particular day because Parliament is currently deliberating the new Wildlife Act. "Can you come to Dodoma to present this same talk to Parliament?"

A flattering request, but within a few seconds, reality is restored, and we agree that Bernard Kissui should make the presentation, since it wouldn't be as effective coming from an *mzungu* (white person).

Now the main event: "OK, so the ultimate source of the lion problem has been human population growth, leading to habitat loss and human-wildlife conflicts. But given all the pressures on lions, we have to consider that trophy hunting might make things even worse."

The temperature in the room hasn't dropped, exactly. They are still on my side, but they aren't looking forward to whatever might come next.

"This graph shows the lion harvests each year in all seven of the major lion-hunting countries, and you can see that virtually every country had a peak in lion offtakes, but then there was always an obvious decline—the decline in Tanzania took place over the past ten years, Zimbabwe's offtakes collapsed back in the mid-1990s and never recovered, likewise in Zambia and Botswana, CAR [Central African Republic], and Namibia.

"We can compare countries by calculating how these raw numbers translate into the number of lions harvested per thousand square kilometers. Tanzania shoots a lot of lions, but we have the most lion habitat, so our offtakes are less intense than they've been in Zimbabwe. However, Botswana and Namibia are desert countries, so their harvests were much too intense."

The director is resting his chin again with his eyes closed, he is weaving slightly, his eyes seem to be darting around as in sleep, so I go back and forth between slides, reiterating the obvious, modulating my voice until he snaps to attention again.

"And in those countries with the most intense lion harvests, their subsequent offtakes fell most steeply." All eyes are open, so I go ahead with it: "This really means that trophy hunting itself must have helped drive down the lion populations. Tanzania is not the worst offender, but you can see that we fall right on the overall trend line, and our offtakes have been too high."

No one raises any objections. Resignation may even be verging on open agreement.

Now to my hobbyhorse: "A lion male can't compete successfully against other males until he's four years old. He's a devoted father and he needs two years to help rear his cubs. If he is killed too young, his successor will kill his offspring. But if you wait until he is six, you can harvest him without harming his family."

Then I show Karyl's graph illustrating how restricting offtakes to six-year-old lions can never lead to overharvest, but including younger males can be disastrous.

And now 2009's version of the Hall of Shame: photos of a half-dozen cubs and subadults lying dead in front of a collection of grinning idiots who think they're Real Men because they just shot a lion, each photo with the year when they were shot: 2004, 2006, 2007, 2008. None of the lions were more than two-and-a-half years old.

"We got all these pictures off the Internet. The operators use them as advertisements; they act like they're proud to be shooting babies."

Much shaking of heads, then Tarimo says, "We are getting a lot of pressure from the government to do something about the declining quality of our trophies."

I show my last slide, and say, "There are only seven countries with significant lion hunting. But last month, Zambia and Mozambique both lowered their lion quotas. Botswana banned lion hunting again in 2007. Zimbabwe is a disaster, and CAR and Namibia have tiny markets for lion hunting.

"Tanzania is the premier destination for lion hunting in the whole world. You have nothing to lose by raising your standards so that Tanzania is known for the highest quality of lion trophies in Africa."

The director is all over it. "We have to do this." He points to one of his deputies, and says, "Obed, work with Dr. Craig to get this done."

We agree that Tanzania can leave their lion quotas unchanged, but they must initiate a system of inspecting each lion trophy, fining the operators and PHs for shooting underaged males, and preventing the export of underaged males to keep the client in line, too.

"From now on, the hunters will have to wait for the right lion," says the director, and I pipe in with some cliché about emphasizing quality rather than quantity.

I ask: "You're going to the SCI convention in Reno next week?"

"The government is sending a delegation; we leave day after tomorrow."

"You've been before, right? You've seen how so many of the hunting companies promote themselves by practically guaranteeing a kill?"

The director doesn't hesitate. "We have to change that." He pauses, and sits up very straight. "Tanzania shouldn't be selling dead animals. We should be selling excitement."

ARUSHA, 5 JULY 2009

Some of our students refer to the Savannas Forever house as the Pink Palace. I prefer to think of the exterior as being salmon colored, but I suppose most people would call it salmon pink. The house is only two years old, the walls were made from uncured cinder block, and a white crust has already formed beneath the paint close to the floor. Large swatches of latex have peeled off inside and out.

We bought our furniture from roadside carpenters along the main highway in Arusha. The uncured lumber has since warped so much that the tables, chairs, and beds all wobble on the concrete floors. Little holes have appeared in the shelves as tiny wood-boring insects emerged through the varnish.

Our night watchman is barely five feet tall and so mild mannered that he doesn't seem like much of a deterrent in the event of armed robbery. Other parts of town have had their share of home invasions, but Arusha is nowhere near as dangerous as Nairobi—at least not this close to the housing complex built for the judges and lawyers of the United Nations Tribunal on War Crimes in Rwanda. Perpetrators of the Rwandan genocide have been on trial here for the past ten years, and the housing demands of the well-paid United Nations staff have jacked up rents to Western prices.

Every house in the neighborhood has a watchman, and we all have panic buttons to alert the private security forces. We have a reliable supply of running water and our own gas cylinders for cooking, but electricity is intermittent; we only have power for a few hours a day. The Savannas Forever staff work here during the day; my lion project staff stay here while in town for car repairs and supplies.

The floor plan is rather awkward (Susan and I have to go down the hall to the nearest bathroom), but the rest of the house is reasonably cozy, and it seemed like the safest spot for Susan to live by herself while I teach in Minneapolis each fall. We're also close to a volcanic cinder cone called Themi Hill, and we walk the two kilometers to the top every afternoon—the altitude in Arusha is around five thousand feet, so it's like hiking around near Denver. On the walk home, we can stop for dinner at a cheap Indian restaurant or a game of pool at the bar in the half-abandoned shopping center that was built for the United Nations staff. Evenings are cool this time of year, and the neighborhood is so quiet that I usually sleep like a lamb.

At midnight, there is a loud crash at the front gate. Our dogs start barking. Two or three men shout in Swahili. The watchman opens the gate, and a car rolls up the drive. Car doors bang, people climb the steps to the front porch, someone unlocks the iron-barred outer door and fumbles with the wooden inner door.

The lights go on in the living room. Two African men guide a third, who has a bandaged head, like a mummy.

"What's going on?"

The mummy's head is barely audible. "It's me, Kissui."

Back during my battles with Costa Mlay, TAWIRI presented an ultimatum. I would only be permitted to continue the Serengeti lion project if I agreed to hire two Tanzanian field assistants. In earlier years, I had supervised two Tanzanian Master's students, one of whom counterfeited all his lion-sighting data, and the other was reassigned to the WD before he could finish his thesis. The bosses at TAWIRI and TANAPA always seemed apologetic about the way things had turned out, so Mlay's new directive was clearly intended as a punishment.

I agreed readily enough, but I also made it clear that I would retain the right to turn down anyone that I deemed unfit for the project.

Two candidates came to the lion house for a weeklong audition in the Serengeti. Julius Nyahongo was the nephew of some high official in the government in Dar. He was reasonably intelligent, but he soon alienated Karyl and Peyton. After an afternoon driving around looking for lions, he came home, put his feet up, and ordered the two women to make him a cup of tea.

Fortunately, the Norwegian government had just started a new wildlife-related program in Tanzania, and they were looking for PhD students to study at Trondheim University. Julius ended up going to Norway, but he knew we didn't much care for him.

The second candidate was Bernard ("Benny") Kissui, who had been recommended by the head of the Zoology Department at the University of Dar es Salaam as "a once-in-a-generation student." Both of his parents had been grade-school teachers in a rural district in central Tanzania; he went to ordinary state-run schools until being selected to attend Tanzania's national high school and eventually winning a scholarship to the University of Dar es Salaam.

Reinforced livestock boma near Tarangire

Kissui was modest, enthusiastic, and sunny. His English could be a bit difficult to understand at times, but he loved the Serengeti, he loved the work, and he loved to think about lions as populations as well as individuals.

I didn't have enough Land Rovers for everyone, but I did own an obsolete Toyota Land Cruiser that no one could keep roadworthy. Benny had an uncle who ran a second-hand car dealership in Dar, and the old wreck was soon transformed into the most reliable car in the fleet.

I had been reluctant for Benny to conduct his fieldwork in the Serengeti because of the geographical isolation and the fact that most of the researchers were expatriates. But the Ngorongoro Conservation Area is considerably closer to Arusha, he had gone to school with several of the NCA staff, and the Ngorongoro Crater attracts so many Tanzanian tour drivers that there was more of a middle-class social scene than in the middle of Serengeti National Park. So Kissui worked in the Crater until just before the CDV-babesia outbreak in January 2001, and his eventual paper on the vulnerabilities of the Crater lions to infectious disease was a major contribution, but it wasn't enough for a PhD thesis.

*

The research coordinator at TAWIRI asked me to evaluate a lion project in Tarangire National Park that had been running for several years. TANAPA was worried about the status of the Tarangire lions. Tarangire is at a far lower altitude than either the Serengeti or Arusha. The weather is hotter and damper; the park is a famous herpetological hot spot, with more species of snake, lizard, and amphibian than anywhere else in Africa. During the wet season, every rain-filled puddle is crawling with frogs and terrapins, and large pythons drape themselves from umbrella acacia trees in the middle of the waterlogged plains.

The park boundaries extend far enough to enclose the only permanent water in a huge semiarid area, thus safeguarding the dry-season refuge of a much more extensive migratory ecosystem. The wildebeest and zebra spend over six months of the year grazing in far-flung Maasai pasturelands every wet season. The lions presumably followed the herds outside as well, but no one knew for sure.

Alessandra Soresina had worked in Tarangire for five years by the time I visited in December 2002. She had no radio collars, so she searched for lions by driving along a network of improved roads. She had encountered the lions often enough to assemble ID files for a hundred individuals, and she was convinced that the population had been dropping. She would see a lion two or three times in a year then never see it again; whole prides would disappear.

Alessandra blamed the sport hunters. Tarangire National Park was bordered by hunting blocks. She didn't know how many lions were shot each year, but all the trophy males must have come from the park.

But overhunting should mostly have affected the survival of males and their cubs. Looking at her data, it seemed the population lost as many adult females as males.

Dennis Ikanda had learned enough in the NCA by 2002 to make me suspect the neighboring Maasai were contributing to the population drop. The Tarangire lions might well leave the park just far enough to subsist on livestock while the migratory wildebeest roamed a safe distance away.

Tarangire seemed the perfect place for Kissui to come into his own. The work would have practical applications, and the park is barely a two-hour drive from Arusha, making Tarangire a potential showcase for a Tanzanian-run lion project. He was excited to move on from Ngorongoro to Tarangire, Alessandra was happy to have help, and TANAPA gave us permission to attach five radio collars in the heart of her study area.

Alessandra gave Benny all the lion-identification files. The following year, Benny attached two GPS collars, which use satellites to record the lions' locations no matter where they go. He discovered that both animals spent most of their time in Maasai pasturelands well outside the park—one had traveled at least a third of the way to Arusha.

Nowadays he has enough collars to monitor ten prides on a regular basis. His assistants can drive up to each pride several times a week each dry season, but he still relies on GPS collars to locate the lions in the wet season when black cotton soils are too sticky for tracking from a Land Rover.

It turns out that Alessandra had been right: the Tarangire lion population really was in decline. But she was wrong about the reasons. Several of Benny's collared lions have died at the end of a Maasai spear; others died from poison.

So Bernard Kissui's research has indeed become practical and applied, and, beyond just finding out where and when the lions attack livestock, he tries to prevent it. Whereas the vegetation inside Tarangire Park is quite scrubby, a lot of the surrounding pasturelands are virtually treeless—so the Maasai don't always have access to enough thorn bushes to make lion-proof bomas. Some families rely on thin straight strips of wood—the sort of thing you might use to build a visual barrier around a privy, but hardly robust enough to confine a panicked herd of livestock. Lions charge straight toward these flimsy bomas, provoking the cattle or goats to break out—at which point they can easily be captured. Other places have so few thorn branches that the lions have little problem broaching the rickety defenses.

Benny reckoned that if the bomas were reinforced somehow, the lions wouldn't be able to enter and the livestock wouldn't be able to escape. Chain-link fencing would be plenty strong for the job, and bushmeat hunters wouldn't be able to refashion the stiff wire of a chain-linked fence into snares.

But how to get the Maasai to invest in preventive measures? Their cultural wealth is measured in the here and now by the number of cows they possess.

Benny hired enumerators in a dozen villages to keep track of lion attacks, the number of livestock lost, and the number of lions killed. He followed up once a week. He was a friendly face who was just as concerned about the pastoralists' losses as he was with the wildlife fatalities. But he was stern when it came to husbandry practices. Why didn't they do more to protect their herds? Why did they leave themselves so vulnerable?

The force of Benny's good will was enough to encourage a few Maasai families to do the unthinkable: each sold a cow, handing Benny the money to buy fencing in Arusha, which he delivered for free.

The reinforced bomas worked like a charm—no lion attacks, no livestock losses. Benny hoped that, with such positive outcomes, the project would start to take off. But the rains failed the next year, and the Maasai lost so many cattle that no one would part with any of their surviving stock.

The African Wildlife Foundation had been looking for a Tanzanian scientist who could improve relations between conservationists and pastoralists in the Maasai Steppe. Conservation organizations were worried about the future of Tarangire, because much of the pastureland around the park was being converted to agriculture as city people moved down from Arusha to start their own farms. The Maasai hadn't fully comprehended the consequences of selling their land, and they didn't trust conservationists who might one day expand the national park and take away even more of their grazing lands.

The African Wildlife Foundation hired Benny to try to ease the tensions, and they agreed to cover half the costs for the chain-link fencing—a far more palatable cost-sharing system to the Maasai—and Benny quickly reinforced about sixty bomas.

The first time I visited one of his chain-linked bomas, the Maasai family greeted Benny with a quiet sense of pride. They had suffered no more livestock attacks. The only problem seemed to be the rising floor of accumulated animal dung that would soon require them to relocate to a new patch of bare earth.

"Benny, what do they think? Do they like it?"

"They think it's good. They like it. Especially the wife."

"Why's that?"

"Before, when there was a chance that a lion might attack, someone had to sleep in the boma with the animals. That was usually her son. He's twelve. But now there's no risk of attack, so he can sleep in the house with the family." He paused for a moment. "Now the wife can sleep through the night, because she doesn't have to worry about her son."

Benny now keeps track of the lions inside the park while keeping an eye on the lion conflicts outside the park. His wife is studying for a Master's degree, and Benny still displays the same positive vibe that first won our affections in the Serengeti. He goes to the United States every year to visit the African Wildlife Foundation in Washington, and he has visited Peyton in Manhattan and Grant Hopcraft in Vancouver. Hardly a day goes by that I don't thank the ghost of Costa Mlay for aligning the stars in our favor.

But on the night of 5 July 2009, Benny hadn't been so lucky. After a long day with the lions and the Maasai, he drove back to his research camp—a temporary wooden structure comprising two rondavels connected by a narrow kitchen, a nearby wooden outhouse, and a flimsy wooden shower stall. It hadn't rained in months; there was no escape from the dust. His camp staff helped him raise a bucket full of warm water fitted with a showerhead at the bottom. The dim light of a kerosene lamp glowed feebly on the ground. He stepped into the darkness of the shower stall, turned on the water, and reached down for the soap.

He didn't see the spitting cobra coiled in the corner until it reared up and spat directly into his eyes.

Blinded, he stumbled outside, shouting for help. His assistants poured water into his eyes, covered his face with towels and eased him into the car. They bounced along the deeply rutted washboard road for twenty minutes, then turned right onto the tarmac. After half an hour they reached the Mserani Snake Park, a roadside tourist attraction midway between Tarangire and Arusha.

The snake park staff had antivenin for spitting cobra. They rinsed out his eyes and sent him to Arusha.

He has arrived at the Pink Palace looking like an escapee from pharaoh's tomb. After identifying himself, he goes straight to his room, and we don't see him again until morning. He emerges without the mummy wrapping, and his eyes are so swollen he looks like he's been through eight rounds with Mike Tyson.

But his good humor is intact, and he laughs that he's had to come all the way to town just to finish his shower.

ARUSHA, 11 JULY 2009

Our studies of man-eating lions in southern Tanzania attracted a lot of attention. *National Geographic* sent out a film crew in 2006, as did the BBC, and another team from German TV. The consultancy fees supported most of Dennis Ikanda's fieldwork, and I raised additional funds from the United States to bring out a professional trapper to attach radio collars to the lions in some of the worst affected areas. The idea was for Dennis to track lions in the agricultural areas that hadn't yet become man-eaters—and learn enough about their basic ecology to better predict the circumstances when they might turn to human flesh.

Dairen Simpson, the trapper, wasn't exceptionally tall, but he was five times larger than life, sporting a handlebar mustache, a nineteenth-century hairstyle, a Southern accent, and the demeanor of a mountain man. He had taught most of the carnivore ecologists in southern Africa and Kenya how to catch lions, leopards, and hyenas by setting foot snares beside well-placed baits. He was the best trapper in the world.

Dairen had left the United States with four large trunks of equipment,

Mother and child in a *dungu* in Rufiji District

but airport security was suspicious of the contents and somehow lost his flight details after the inspection. So we spent the first ten days of Dairen's four-week expedition in Dar es Salaam waiting for his trapping paraphernalia to arrive. I could still afford the Hotel Peacock in those days, and Dairen and a journalist from *National Geographic Adventure* magazine had rooms just down the hall.

We ate breakfast together every morning in the windowless basement of the Peacock beside strings of flashing green Christmas lights that draped around a broken fountain. The buffet consisted of canned beans, frankfurters, chalky eggs, and greasy potatoes. Dairen recounted his adventures in the desert sands of the Kalahari, the jungles of the Amazon and Indonesia, and the mountains of the American West. He explained how to set a trap so that even the wariest cat would be drawn to the bait, given no place to step except in the middle of the snare. I worried that a snare might damage the animal's paw, but his setup included a "throw" that flung the snare up to the animal's armpit, binding the animal tight on a short leash and preventing a frantic response.

Once his gear finally arrived, Dairen and Dennis and the *National Geo-*

graphic guy drove south to the man-eating areas in late April 2006. They started trapping during the harvest season when the lions were likely to follow the bushpigs into the fields. Dairen set his traps each evening and checked them before dawn each morning. They walked for miles and miles day after day for two weeks—and only caught a leopard. There were signs of lions everywhere, but never around the traps. The rainy season was good for finding tracks in the mud but terrible for finding a specific site where the lions might return on another day.

Dairen had never failed before, and he charged me nearly a thousand dollars a day, so we were both unhappy, and Dennis was left with nothing. The fundamental problem seemed to be the wet weather, and so I brought Dairen back in the dry season of 2006 to try again when the lions might show up at a water hole. They also tried using loudspeakers to call up the lions with hyena noises—just as Peyton had done for her dummy tests. Their first night, a large male lion ran right up but then kept going past the bait and departing on the opposite side. The next night a lion again hurried toward the speaker, and Dairen managed to fire off a dart—but missed.

And that was it. In three weeks, they never saw another lion.

Now they turned against each other. Dennis hadn't arranged for enough darts, they never had any animals to use as bait—Dennis might disappear for half the day.

Dairen was difficult. Dairen was impatient. Dairen had missed when he had a clear shot.

I was teaching in Minnesota at the time, and I called Dennis one day while I was walking across the Washington Avenue Bridge above the Mississippi River. I agreed that Dairen could be gruff, but he was the best trapper in the world, he worked like a demon, et cetera, et cetera, and why on earth couldn't they find any bait? Dennis said the Wildlife Division wouldn't let them shoot any antelope, and none of the local people kept cattle or goats in that part of the country.

But couldn't he drive to Dar for the day or go somewhere else with a livestock market?

Not for any reason that I could ever understand.

While still a teenager, Dennis rode to Dar es Salaam with his father and older brother in their car, which crashed on the edge of town. No one offered to help, and Dennis was trapped inside the wreckage while bystanders stripped his father's and brother's dead bodies of watches and wallets.

Dennis went into a profound depression that lasted for years. His college grades suffered, but he completed his degree. He was working for a tour company in Arusha when he contacted me about the lion project. He started out like a house on fire—hard working and reliable. But then he fell in love, and his fiancée became the focus of his life.

Suddenly, his lion work fell down on his list of priorities.

He continued to perform his basic duties, but instead of working thirty days a month, he was in the field maybe only twelve. But the worst part was never knowing what he did on the other eighteen. No one seemed to know where he was. But as long as he fulfilled his obligations with the Ngorongoro Crater lions and the Maasai herders, I decided to leave him be. He was in love, and his new family seemed to be very high maintenance.

After he finished his Master's degree on Maasai-lion conflicts in the NCA, he wrote a first-rate scientific paper outlining the factors leading to retaliatory lion killings and confirming the continued existence of ritual killings. Shortly thereafter, he tackled the man-eating surveys in southern Tanzania with renewed energy, organizing the staff, the questionnaires, and the itineraries and ensuring the quality of the data. He was as brilliant as ever.

But somewhere in the following few years, his absences lengthened. He was always two hours—or even two days—late for an appointment. He seemed constantly distracted, worried.

I caught him, once, on a long drive, hauling the belongings of his parents-in-law to their new home. I read him the riot act about using a lion project car for private purposes, but he said he always paid a portion of the car maintenance out of his own pocket.

By the time Hadas Kushnir came back to Tanzania from Minnesota in early 2007, Dennis had become surly. He resented having to share any aspect of the man-eating work with anyone, especially Hadas.

I've always minimized competition between students by assigning a clear division in their respective research topics. The dividing line between Dennis and Hadas was simple at first. Dennis would collar lions and study lion ecology in the agricultural areas; Hadas would interview the victims and their families and compare villages with and without lion attacks to see how the villagers' behavior influenced their risks of being eaten.

But then Dennis and Dairen's lion-collaring expeditions failed, and Dennis couldn't focus exclusively on the lions anymore. Without collars, they would be impossible to track. I assured him that he could use interview data to study the lions. The pattern of attacks would reveal how man-eaters moved between villages, why certain lions became habitual man-eaters whereas others might never attack a second time, and so on.

But he no longer trusted me. I had antagonized Severre and the WD, and the hunting companies still blamed me for the temporary shut-down of lion hunting in 2006. Dennis was looking toward his future, and whereas I could always work in the United States if the WD decided to deport me, he was stranded out here. Besides, the WD had given him some cash for his fieldwork and might even be grooming him for a job someday.

Hadas was a nuisance, he said. She kept bugging him to get to places on time, to go that extra mile every day to get that one last interview.

But I thought there must have been something about Hadas being a *woman*. And a *foreign woman* to boot. My trip to Lindi in 2007 was a transformative experience in many ways—not only by meeting the mother who had lost her five-year-old outside her home but also by the incessant negativity that Dennis poured on Hadas.

On our last night together, I took everyone to dinner at a resort hotel in Mtwara, a border town with Mozambique, and Dennis started complaining about how Hadas had stolen his project and how I had given it to her, and how he had started it all, and it should all be his.

Hadas left the table in tears, and I confronted Dennis, asking what else was wrong besides the work. It couldn't just be the work—there was more than enough work for both of them to do. We were trying to save lives, and there was no room for this sort of *crap*.

He blamed Dairen and me and Hadas for all his failures. It wasn't even his project anymore.

"That's not enough," I persisted. "What else is making you so angry?"

But there was no reason that I could ever understand.

Over the next two years, our relationship was a roller coaster. Dennis got a job at TAWIRI and his newfound security seemed to help a bit at first, but then he started behaving like one of those horrible customs officers from your worst travel experience anywhere in the world. Julius Nyahongo, the failed applicant for the Serengeti lion project, had come back from Norway and now worked for TAWIRI, too. Shortly after Luke Sidewalker's dossier on Severre went public, Nyahongo circulated an "exposé," claiming that I funded my research by selling lion blood for a hundred thousand dollars per liter. Nyahongo had insulted Dennis during a wildlife conference a few years ago, but after Dennis enrolled in Norway, Nyahongo became Dennis's role model.

Now Dennis was a government employee, and he would tell me that the government always knew best. The government had the power, and so, by proxy, did he. He lectured me about how his status as a wildlife biologist made his work of putting radio collars on the man-eaters more important than working with villagers. He applied to the Tanzanian Wildlife Protection Fund for more money: he wanted to set up a system of loudspeakers to broadcast roars of adult males all night every night in order to repel the local lions.

After I met the new director of wildlife in January, I went to see the new research officer in the WD. She told me about Dennis's grant application, and I stupidly burst out laughing.

For the past few months, Hadas and I had been working with a new Tanzanian assistant, Harunnah Lyimo, designing a project to prevent bushpigs from eating people's crops. The man-eating area extends over fifty-nine thousand square kilometers of subsistence farmland—family-sized fields in an irregular patchwork across a landscape lined with rivers and punctuated with hills and small forests. Pigs sleep in the riverbanks and the forest patches; no one knew where the lions might be hiding.

Bushpigs could never be exterminated from southern Tanzania. It has been almost impossible to eradicate feral pigs from the Galapagos Islands at a fraction of the size. But if people learned how to keep pigs out of their fields, farmers wouldn't need to sleep in the open. And even if we couldn't eliminate the risks from man-eating lions, reducing crop losses would at least improve food security.

Farmers near Arusha exclude bushpigs by digging trenches around their plots: "Pigs have short legs, and they don't like to jump," as a local farmer once told me. People can dig their own trenches, and low fences can easily be made from short sticks and homemade twine.

But buying thousands of high-quality loudspeakers, in contrast, and broadcasting lion roars night after night in villages without electricity—I shouldn't have laughed out loud, but I did, and Dennis found out.

He is now in Norway, finishing his PhD at the University of Trondheim. I recently received a draft of his thesis and sent back a long list of suggested revisions and questioned some of his statistics. He e-mailed me today to say that I was just a typical *mzungu* (white person), acting like I still ran the country and that I only cared about writing papers for scientific journals. He, in contrast, was a man of the nation who was saving lives.

He could get me thrown out of Tanzania whenever he wished.

WASHINGTON, DC, 19 JANUARY 2010

Congress is increasingly skeptical about USAID. Voters are angry about government spending. There are growing demands for accountability of scarce tax dollars. Improving the effectiveness of American aid projects must surely be a bipartisan issue.

Tanzania is one of the biggest recipients of American funding. We've decided to call Savannas Forever the Whole Village Project to better capture our holistic surveys of village life. We want to work with the American government to help USAID gain a greater return on their investment in Tanzania's rural economies. Poverty alleviation can't be viewed in isolation. If people aren't healthy, they can't work. If children herd cattle, they can't attend school. If natural resources are lost, rural economies will worsen.

I am in Washington seeking three million dollars a year to convert Tanzania into a living laboratory with a grid of 240 villages that could be surveyed at two-year intervals to objectively measure rural well-being before, during, and after each development project. The U.S. Agency for International Development and the Millennium Challenge Corporation could then learn from their successes and failures. The senate aides I meet—Democrat and Republican—are all keen. Although development funding is an essential tool for foreign relations, it can be unpopular with voters, so Congress has been looking for ways to hold the U.S. Agency for International Development accountable.

But USAID projects are already tied up in red tape. Aid funding is tight and given that monitoring and evaluation are expensive, scientific measurements would subtract from funds that could save starving children. When explaining the goals of the Whole Village Project to the upper echelons of the agency, someone tells me flat out that if we think we can turn Tanzania into a laboratory for improving aid projects, we "must be living in la-la land."

Beltway bandits receive billions of dollars in government contracts. The status quo will surely hold. These great gray government buildings are huge, the vast public spaces around the Capitol are grand.

I am two inches tall.

SERENGETI, 12 MARCH 2010

"Sorry, that was too wordy," I sigh. "I can do better."

Collect yourself, slow down. This is easy; just keep it simple.

OK, we're back on our marks, the camera rolls, and we start walking again through the tall grass past a gray granite kopje and a scattered stand of umbrella acacias.

Jonathan Scott repeats his prompt, "Craig, you've worked in Africa for over thirty years, and you've set out to answer one of the most fundamental questions about lion society, namely, Why do lions live in groups?"

The scenery fades from my eyes as I step through the grass and start talking to Jonathan and all the rest of the world. "When I first came to the Serengeti in the 1970s, I thought, like everyone else, that lions were social because of the advantages from group hunting. But as you've seen for yourself with the Marsh Pride, lions don't always cooperate when they hunt. Often, just one female does all the work of catching the warthog or the wildebeest, then the rest of her pride just piles on."

He nods in recognition and smiles as if it is the first time he has ever heard me say any of this.

"The successful hunter gets *so* annoyed," I continue. "She's clamped onto its throat, throttling the thing, and her sisters and cousins have already started opening up the abdomen and eating all the best bits: the liver, the spleen, and heart. Meanwhile, the hunter's obsessing over whether the thing is dead, so she's snarling at everyone through clenched teeth."

"Yes," says Jonathan, "she can get quite upset, and the rest are careful to stay just out of reach, so she can't quite swat them."

"Lions do indeed hunt cooperatively in certain situations," I resume. "They always cooperate when trying to catch something as large and dangerous as a Cape buffalo. But if a singleton can catch it by herself, she doesn't wait for anyone's help, and the rest of the pride are just as happy to let her do all the work—and then join her at the table when supper's ready.

"The most difficult project I ever conducted was back in the 1980s with a grad student named David Scheel. We watched lions nonstop for four days at a time. We recorded everything they ate, no matter whether they hunted or scavenged.

"Everything depended on the migration, so when prey were plentiful—when the wildebeest were back in the Tanzanian side of the Serengeti—all the lions ate their fill, no matter what size group they were in. But in the dry season, there's a lot less prey down here than in the Mara, and life is especially bleak out on the Serengeti plains—just the occasional warthog, or maybe a gazelle they can steal from a cheetah. This is the time of year when feeding success is really critical to survival, and it turned out that, under these conditions, a lone female was able to catch a pig or scavenge a gazelle perfectly well on her own.

"The other essential dry-season prey is the Cape buffalo, and females in really large prides can and do team together to catch a buffalo every week or so, whereas smaller prides mostly survive on the occasional warthog. Solitaries get plenty to eat from a pig, but groups of two to four females have to divide a single pig into smaller chunks.

"And even though large groups can catch buffalo, there are so many mouths to feed that they only eat as much per capita as a solitary gets from a warthog. So there's no real advantage from group foraging in terms of full bellies. But at least by catching a prey as large as a buffalo, these big prides don't suffer from feeding competition with their companions. And when times get really awful a female can always split off to catch a warthog by herself."

"They have," Jonathan suggests, "a fission-fusion society that allows them to split up when necessary, but then they can recombine when they need to. But that still suggests they benefit somehow from living in groups . . ."

"Yes," I respond, "there must be a good reason for being social, but the advantages don't come from catching their very next meal.

"We next checked whether females live in prides to protect their cubs against infanticidal males. Incoming males refuse to be stepfathers and want to breed immediately. Any cubs in the pride serve as contraceptives until they're about two years old, and males generally only have about two years to produce their own cubs before being replaced by yet another coalition.

"Protection against infanticide is certainly a powerful force in lion society. Mothers with cubs of the same age pool them together to form a crèche. The crèche is the social core of the pride. The moms spend almost all their time together until the cubs are about a year and half old. Other female pride mates may come and go from one day to the next, but the mothers are together almost constantly.

"The crèche turns out to be an effective defense against infanticide. A female hasn't got a chance in protecting her cubs one-on-one against a male—a male lion is about 50 percent bigger than a female. But a group of females acting together can easily chase a male away. Sisterhood is powerful!

"But then again if this was the sole reason why lions lived in groups, we'd expect it in all the other cats, too. Leopards, tigers, cougars—even housecats—are infanticidal, yet none of these other species are social."

We have been filming about a hundred meters from the lion house. The cameraman kept zooming out as we approached his tripod; the soundman mostly recorded us with remote microphones, but he also duplicated the proceedings with a shotgun mike.

It took about three attempts before I finally managed to finish my spiel without too much sputtering or bad grammar. Colin Jackson, the producer, is happy, Jonathan is pleased, and we all drive back to the BBC's campsite a couple of miles away. We're doing a two-part program called *The Truth about Lions*. The first hour outlines all the fun topics on the basic biology of the beast, so I've been having a good time so far. The second hour will inevitably cover lion conservation, but we won't have to deal with that unhappy topic for another few days.

Jonathan Scott is a tall, charming Brit with a dark mustache and a good head of hair. He came to East Africa in the late 1970s to work as camp naturalist for a small lodge in Kenya's Maasai Mara Reserve at the northern limit of the Serengeti ecosystem. He soon became well known as a photographer and writer, focusing on Serengeti wildlife in general and the Marsh Pride in particular. We first met in 1985; he asked if he should enroll in a PhD program, and I assured him that with his skills he shouldn't waste three to five years on a thesis. He is handsome, articulate, and a natural for television; he subsequently became a regular on the BBC, recording the lives of the Marsh Pride lions as well as more exotic destinations around the world.

The Truth about Lions aims to merge Jonathan's perspectives on the daily lives of the Marsh Pride with our more global view from the Serengeti—we've followed a hundred prides over the past forty-five years, but we've spent far less time with any single individual. He has a nearly clinical understanding of the dozen members of his pride; we can report the statistical trends within a cumulative population of five thousand. The BBC has filmed his animals in minute detail; we only write down a few salient points every few days on our data sheets. So while my voice-over will mostly talk in generalities about lions, the documentary will be showing specific illustrations from archived footage filmed in the Mara.

I feel like we only ever get an occasional glimpse of the inner lives of our Serengeti lions. They sleep all day, they are hard to watch after dark, and we have twenty-four prides to monitor each week—so we all just flit from one pride to another, collecting a series of snapshots, while Jonathan's crew has assembled a more precise narrative on a single family.

The strength of the Serengeti study emerges from all those numbers consistently collected year after year. We know how long the lions can live, what happens to entire prides over the span of several decades, how many descendants they leave behind—this is the stuff of evolution. We can also say what happens during the periodic droughts and floods or when the woodlands regenerate or a disease strikes—this is the stuff of ecology. The Marsh Pride, in contrast, reveals a great deal about the personality clashes and sexual proclivities of particular individuals—this may be the stuff of soap operas, but it is a lot more engaging for the general public!

Soit le Motonyi is a fortress-like kopje at the boundary between the plains and woodlands in the eastern Serengeti. We are standing on the summit beneath the broad green canopy of an enormous fig tree that is surrounded by leopard droppings and scattered gazelle bones. Leopards have used the low branches as scratching posts and carried a dead wild cat up to a higher branch.

The cameraman has set his tripod on the brow of a huge pink-granite

Group of mothers chases away a male

Top, males on patrol in Ngorongoro Crater. *Bottom*, four females collectively attack a stranger in the Serengeti.

boulder, covered in orange and yellow lichens. We're ten meters above the surrounding grassy plain on top of a ridge that looks down on the crowns of scattered umbrella acacias and the tree-lined tributaries of the Ngare Nanyuki River.

As the sun emerges from behind a cloud, we start to roll, and Jonathan begins: "Craig, you've brought us here to see this magnificent landscape of the Serengeti, because it's the landscape itself that holds the key as to why lions are social . . ."

"I wanted to show you firsthand," I reply, "the hot spots that are such valuable real estate for the lions. If you look off to the north, you can see the main drainage line of the Ngare Nanyuki, and here and there you can see where little tributaries meet the main branch of the river. Those confluences are the key to lion sociality.

"My recent graduate student, Anna Mosser, took all of our long-term lion data to map out the lions' typical reproductive rates across the long-term study area—she literally produced a Darwinian real estate map of the Serengeti. Some spots were associated with high reproductive success; other spots were useless.

"And the results were clear: decade after decade, reproductive rates were highest for lions that occupied the confluences along the major rivers.

"The confluences have everything a lion needs to raise her cubs. As the prey move across the landscape, they're reluctant to cross any of the rivers—there might be lions lurking in those bushes! So when the wildebeest or zebra reach a major drainage line, they walk parallel to the stream until they reach a confluence—in effect they've trapped themselves in a funnel where they will finally have to cross—and the lions are waiting for them in the mouth of the funnel.

"In addition, the water flowing in the merging tributaries produces the greatest turbulence in the streambed and digs out the deepest water holes. So these convergences provide water longer into the dry season and promote the growth of woody vegetation. Thus confluences provide more shade and more shelter for the lions' cubs.

"So here we have the three rules of real estate: location, location, and location. In this case, the most desirable real estate provides food, water, and shelter. These are by far the most valuable spots on the map, and the lions compete most intensively when they're battling for territory. Unlike hunting, lions always cooperate when they're defending their home.

"So visualize a complex landscape where all these lions occupy every square inch, but only those females that control the confluences can reproduce successfully. Imagine you are a solitary female and everyone else is solitary, too. If you allow your daughter to stay with you, you now have a companion who can help you defend your favorite confluence, and the two of you can go chase away neighboring solitaries from their favorite spots—you can expand your territory to include multiple confluences, and you're on your way to forming a proper lion pride.

"Another graduate student, Margaret Kosmala, worked with Anna to construct a virtual Serengeti with virtual lions that competed over virtual confluences or "hot spots." Then they sorted out the rules for this virtual competition and modified the virtual landscape to discover the circumstances that favored the spread of a mutant social lion in a world where everyone else was as solitary as a tiger or jaguar.

"They found that if a landscape was dotted with hot spots—like we see in the Serengeti—and if population densities were sufficiently high, the sociality gene usually replaced the gene for solitariness."

"So lions," Jonathan interjects, "are the only social cats . . ."

"Because they live in such a complex environment," I elaborate, "and also because they are so abundant. Lions specialize on the plentiful prey: the wildebeest, zebra, and buffalo. So they live at higher densities than other cat species. With more competitors, lions have to be on the constant lookout for invaders, increasing their need to work together. Combine high population density with the opportunity for groups to monopolize the best river confluences against solitary neighbors, and sociality is almost inevitable."

"That makes sense for lions," Jonathan says, "but leopards live in this same habitat . . ."

"In comparison to lions," I say, "leopards are scarce in the Serengeti, so there's less pressure for a leopard family to remain together to compete against neighboring families. And if the leopards were to defend the river confluences, they'd run into a lot of lions—lions kill leopards whenever they can catch them."

I have worked with Colin for nearly a week now. I'm usually pretty offhand with film teams, but if they want to focus primarily on our research, I treasure the opportunity to spread the word about lions and to try to excite the electorate about the scientific enterprise. We get most of our funding from American tax dollars, and ideally these film appearances help encourage broad public support for basic research.

My first experience with prime time was back in the 1980s, when Alan Root made a film about our research for CBS. At the time, we felt pretty confident that group hunting and infanticide weren't the keys to sociality. But we didn't yet understand the origins of group territoriality, so the show lacked a big finish.

However, Alan's cameraman did manage to film a spectacular sequence in which an incoming male killed three small cubs one after another on short green grass in golden sunshine.

But by far the best part of filming was spending more time with Alan, the true father of high-end nature filmmaking. I'm still amazed by his keen insights into the behavior of the weirdest animals (snakes in mortal combat with monitor lizards, hawks that hunted their prey in the cracks of large rocks), and the man had the quickest wit of anyone I've ever known.

When *Out of Africa* was filmed in Kenya, Barbie Allen met Meryl Streep and rented out her horse and buggy for a street scene of nineteenth-century life. Alan was hired to shoot scenics and wildlife footage, and one day he found Robert Redford hanging out on the set with nothing to do. Alan asked if Redford would like to see the Serengeti, and they took off around midday in Alan's four-seat Cessna. The border between Kenya and Tanzania wasn't officially open, so Alan gave a bogus flight plan to the Nairobi tower, and they flew off above the Rift Valley and turned south to the wide-open spaces of the Serengeti.

It was the middle of the dry season, the plains were empty, and Alan decided to land on the short-grass plains at the Gol Kopjes, large gray boulders with flat-topped acacias, whose austere beauty resemble a Japanese garden. The ceaseless winds off the Ngorongoro Crater highlands are like the chinook or the mistral, and Alan judged his landing accordingly.

But just before he touched down, the wind switched directions, and the plane did a sudden one-eighty. Time froze and a terrified Alan Root visualized tomorrow's world headlines:

"Robert Redford dies in plane crash! Pilot error blamed."

But, in fact, the plane had stopped safely on the ground.

A calm voice from the back seat said, "Tallman turn."

"What?"

Redford repeated, "Tallman turn."

Allan looked back, puzzled.

"I did a film once called *The Great Waldo Pepper*. It was about stunt pilots, but the best of them all was a guy named Frank Tallman. He was the only one who could land a plane like that."

"Well," said Alan, "we don't have a name for it out here . . ."

For every Alan Root, there are also filmmakers that don't leave the best impressions. For the first few years after the big canine distemper outbreak in the 1990s, we became a hot spot for celebrity cameos. Anthony Hopkins came to the Serengeti one year, followed by Cameron Diaz and Justin Timberlake—and they only stayed long enough for photo ops with darted lions.

Then there was the day when Andy Brandy Casagrande the 4th (ABC^4) called from National Geographic Television. He had read a press report from the *Arusha Times* newspaper, describing how one of our study prides had just broken the record as "largest pride ever": forty-one lions in the Transect Pride. He wanted to come film this Super Pride, assuming that if five females could catch a buffalo, the Super Pride must catch really big prey like hippos or elephants—maybe even dinosaurs!

I explained that while the Transect Pride had indeed set a new record, it had only been temporary. Sometimes large cohorts of subadults hang around for a while after the birth of their moms' next set of cubs. At their peak, the Transects had contained a dozen adult females, but they had never moved around together as a pack—there was never enough food for everyone. And, no, they hadn't eaten any elephants. Furthermore, the report in the *Arusha Times* was over a year old; the Transects had already split into three separate prides.

Not to be deterred, ABC^4 arrived in the Serengeti the following month and promptly requested an introduction to the Super Pride. My field assistants explained that the Transect Pride had since divided into three.

He then asked if there was another lion study in the Serengeti—maybe someone else was studying the Super Pride. . . .

The filmmakers I enjoy most are those who can promise airtime for products that would make simply fabulous research equipment such as, say, life-size toy lions.

Peyton West was sitting in my office in Minneapolis one day in 1998. She had started writing her thesis proposal and was ambivalent about her research topic. On the one hand, solving the mystery of the lion's mane would be a spectacular achievement for a graduate student, but it would require a ton of money, and student research funds were scarce.

Behavioral biologists had been using wooden models since the 1950s to test how birds and fish respond to variations in the size and color of male ornamentation. But no one had used similar techniques to study mammals. Mammals are smart; surely the models would have to be lifelike and elaborate. Peyton had just been to FAO Schwartz and discovered that a half-sized toy lion was priced at about ten thousand dollars. She would need two of them to provide the lions with a choice—four would be ideal since the lions might eventually rip them apart.

Peyton was frustrated by her poverty, and I was at a loss about how to launch her research project, when the phone rang, and it was someone named Brian Leith. He was a filmmaker from the United Kingdom, and he wanted to make a TV documentary about the Serengeti lions. He ex-

Male lions are the most proficient buffalo killers in the Serengeti (*above*). Note the behavior of the female (*above, top and center*).

pressed interest in some of the research we had finished a few years earlier. I wasn't keen to revisit the same old stories about distemper and group hunting, but then he switched gears and asked if we were planning anything new.

The penny dropped. I glanced at Peyton and said, "Well, yes, actually, there is."

We could never have gotten anyone to donate forty thousand dollars to buy toys in order to play with lions, but a filmmaker could offer a perfect marketing opportunity. For the price of four life-size lion dummies, some lucky company would get star billing in an hour-long documentary in the United Kingdom and the United States. Brian initially contacted Steiff, the world-renowned maker of high-quality plush toys. Steiff was excited at first but then backed out a few weeks later. Then Brian found International Bon Ton Toys in Leiderdorp, Holland, which may have been less famous than Steiff, but they were more adventurous. They made unusual plush animals like snow leopards and lifelike lion cubs for the World Wildlife Fund. Could we please send them some photographs and hair samples from the Serengeti?

For the previous ten years, my field staff had routinely snipped off long strands of mane hair whenever they darted a male; so we had an archive of the full range of mane color from blond to black. We had also collected the black hair of the lions' tail tips, and discovered, as well, that many of the blond mane hairs were the same color as body fur.

We also had photographs of two aberrant lions that had been born into one of our Serengeti study prides. As cubs, they both looked like females, but by the time they were two and half years old, they were so enormous that when we saw them beside their same-aged sisters, they looked like mothers with cubs. One of these animals had been named Wosu as a cub, but after reaching full size, we changed its name to Woppin'Sue.

Woppin'Sue eventually took over a new pride with its brothers, chased out the subadults, then came back and tried to mate with one of the mothers—a peculiar sight since Woppin'Sue had no mane and, without external genitalia, couldn't connect with its new bride. At its initial approach, the pride female reacted as if Woppin'Sue was a neighboring female that had invaded her territory, but as Woppin'Sue drew near and adopted the posture typical of a courting male, she ran up and solicited a mating. After the mating attempt, Woppin'Sue tried to spray mark a low bush, but the arc of urine fell well below target.

Woppin'Sue turned out to be a genetic male who suffered from a disorder that prevented a normal response to testosterone. He had normal *levels* of testosterone, but he never grew a penis or a mane. So his mutation provided the designers at International Bon Ton Toys with an unobstructed view of the underlying contours of the head and shoulders of a fully grown male.

I visited the International Bon Ton Toys factory in Leiderdorp and tried to control myself when I first saw their mock-ups, which were a curious combination of cute and macho. Their faces lacked any sense of menace and danger, and their legs were much too thick. They shouldn't look like quarter horses, but they shouldn't look like Stonehenge, either.

But International Bon Ton Toys had neatly solved the problem of the manes. Plush material can only be made to have "hair" that is up to about two inches long, but they molded the material around a thick collar that matched the overall contour of the mane, making the mane "hair" look much longer. They had also attached Velcro so that each mane would stick tight to the neck and shoulders of each dummy. The design team had already completed the long dark mane and the short blond mane, and they were fabulous!

The toy designers sent the full set of patterns to their factory in Hong Kong and finished everything in less than a month. The dummies were immediately shipped from Hong Kong to Holland and then down to Nairobi. Brian Leith chartered a small plane in Kenya, crammed the dummies inside and flew them down to Tanzania. Peyton, Grant Hopcraft, and I arrived at the Seronera airstrip to find the four dummies still stuffed inside the plane. We helped unload, lined them up on the airstrip, and attached the manes.

We laughed so hard we nearly fell over.

The dummies certainly looked lion-like, and they looked like four different lions, with their unique mane combinations of color and length, but they were still too cute. Their faces were a bit too wide and short, their mouths too beatific and their legs too thick. We hauled them up onto the roofs of our Land Rovers and drove them carefully back to the lion house. When we brought them down onto the verandah, a baboon ran up a nearby tree and started giving alarm barks—"*wahoo, wahoo, wahoo*"—terrified by the sudden arrival of four male lions.

Whew! If these things could fool a baboon, we might be in luck.

Brian was eager to start filming, so that same afternoon, we found a group of three females that were resting in a brushy area about fifty meters off the tourist track. We drove another hundred meters away to a small clearing where the lions couldn't see us. We unloaded the two dummies from the top of the car, placed them about ten meters apart and positioned a Klipsch Heresy speaker at the midpoint between them. Brian's sound guy had fit our car with a hidden mike, and he was recording everything that Peyton and I had to say. The cameraman signaled that he was ready to roll.

Peyton didn't have a lot of confidence in the playback equipment. She had never used it before, and the sound was a bit quirky when she tried a practice run back at the lion house.

I didn't have much confidence in anything at that point. The sun was about to fade, the wiring between the amp and the speaker seemed a bit wonky, and Peyton was so nervous that she could barely keep herself together.

And my thoughts finally caught up with me: *you know, this might not work*. I had kept pushing everything forward between Minnesota and Britain and Holland and Hong Kong by confidently asserting that, yes, this will work just fine with lions. Lions are thick as a brick—barely any higher on the tree of life than an iguana. . . .

But now that Brian and International Bon Ton Toys have come through, it sure would be embarrassing if the lions didn't even react, wouldn't it? What a way to introduce Peyton's science to the general public. Whatever happens today will be viewed by millions of people, and it may be hard for her to live down.

Peyton fiddles with the tape recorder and the amp. The sun is getting lower and lower.

Nothing happens.

For the benefit of our future TV audience, I say in a stiff, stilted manner, "Gee, Peyton, what seems to be the problem?"

"I can't get it to turn on."

"Gosh, Peyton—"

"Wait a second; this should do it."

She tries again; a couple of seconds go by.

Nothing.

She plays with the equipment again.

"HAHAHAHAHAHAHAHA," comes blasting out of the speaker at a deafening volume. It is a recording of hyenas squabbling at a kill. But the hyenas sound like they are about fifty feet high.

She turns down the volume to a more tolerable level.

Now we wait.

"Gosh, Peyton, hope that didn't scare them off."

But then, from the direction opposite to the lions, a dozen real hyenas come running up to the speaker, pass the dummies, and continue over the horizon: they were sure they heard strangers around here somewhere. Must have been huge. . . .

Nothing yet from the direction of our lions.

"Gee, Peyton—"

Then a female lion peers from behind the bushes about thirty meters away, hesitates for a moment then walks cautiously forward.

We don't dare breathe.

The female is definitely looking at the dummies, but she is still about thirty meters away, and it isn't possible to tell which one she's looking at. We're off to the side; the film crew is in another car more directly behind the dummies, maybe they can see the direction of her gaze.

I start to say something obvious and inane for the sake of the TV audience, but before any of the words come out, the lion resumes her forward motion.

She walks slowly, slowly, closer and closer, and by the time she reaches the speaker, her choice is clear—she is going to Julio, the black-maned dummy. Blond-maned Fabio doesn't stand a chance.

Peyton is saying something like: "Oh, my god, oh, my god, oh, my god."

The female comes right up to Julio, and, a bit disappointed by his indifference, she ventures around to his rear and sniffs under his tail.

The remaining two females have arrived. The first female has moved about ten meters away and is now flopped down on the ground. Her two sisters go up to Julio, flank him on either side, go around and then sniff underneath his tail.

The first female gets up again, comes back to her sisters, they greet each other briefly by rubbing their heads together, then they all flop down in a semicircle in front of Julio.

It's as if they worshipped the ground he stood on.

The mane film came out the following year on ITV in the United Kingdom and the Discovery Channel in the United States, and it has been in circulation ever since—everyone loves the sequences with the lions and the dummies. A couple of years after its initial debut, International Bon Ton Toys showed an edited version in a kiosk at Amsterdam's Schiphol airport at Christmas time. What better endorsement than the sight of those three giddy lionesses with their schoolgirl crushes on an International Bon Ton Toys dummy?

The mane story was the most fun I've ever had with a film crew, but I have been just as excited to work with Colin Jackson's team this year. We've recorded voice-overs for the entire sociality saga, finally finishing the story started in Alan Root's *Queen of the Beasts*, and Colin has invited me to take advantage of his title *The Truth about Lions* to dispel some of the more stubborn myths about lion behavior.

I want to start with the myth that females do all of the hunting.

A *New Yorker* cartoonist once drew three female lions chasing after a herd of zebra, while four males relax nearby, sphinxlike, smoking cigars and sipping port, as one male says to the others, "Shall we join the ladies?"

While females do catch all the zebra and all the wildebeest and warthog, these are fleet-footed prey that can only be caught by speed and surprise. But it is the males that catch most of the buffalo. Buffalo aren't quick on their feet, so catching a buffalo doesn't require much stealth or strategy—just brute strength.

I once found a female lion striding along a tourist road beside the Seronera River in the Serengeti. She walked up to the four resident males in her pride, rubbed her forehead against each one of their heads and pestered them until they stood up. She then led them to a lone bull buffalo with a broken leg. One male, then two, leapt on the buffalo's back. It pivoted and stabbed at them with its horns as the male lions clung on for dear life.

Suddenly, a second buffalo came out of nowhere and thumped the males with its horns. The lions quickly released its friend, but the hobbled buffalo was too lame to flee. After about a minute, the healthy buffalo decided that the odds between 4.00 male lions versus 1.75 buffalo were too scary, so he left the scene. The abandoned bull was quickly covered by all four male lions and went down for good.

During this entire drama, the female lion stood stock still, watching comfortably from a safe distance. As soon as dinner was ready, of course, she partook of the feast.

Second myth: the lion makes a lousy mother.

The Serengeti lion project was started in 1966 by one of the true giants of wildlife ecology, George Schaller. Schaller was already famous before working on lions. He had been with the survey team that eventually set aside Arctic National Wildlife Refuge in Alaska, and he had studied tigers in India and gorillas in Rwanda. His book *The Serengeti Lion* remains a classic, and though my team has since written more than a hundred scientific papers on the beast, we've never needed to make any important clarifications. He got the most important points right the first time.

Except.

Except he also wrote a number of popular books and articles about his years in the Serengeti, including *Golden Shadows, Flying Hooves* and *Serengeti: Kingdom of Predators*. But somewhere between these less formal accounts and an article in a now defunct natural history magazine called *Animal Kingdom* titled, "The Lion Makes a Lousy Mother," he gave the distinct impression that the lion makes a lousy mother.

He largely based his argument on two anecdotes. First, he had seen a mother carry her one tiny cub out into an open area then abandon it, blind and helpless, to the elements.

This may not sound very wonderful, but a female lion can give birth to as many as four cubs in a litter. After birth, she will have to commit two full years to her new cubs before she can breed again. In the years since Schaller's study, the same behavior has been seen in grizzly bears and black bears, and we've seen several more lions abandon tiny cubs, but in all cases they abandoned singletons. By replacing a single cub with a larger litter, they can eventually rear more cubs in total, hence becoming better—or at least more productive—mothers overall.

Second, Schaller encountered a female lion feeding on a carcass that had been scavenged from a leopard; the leopard was watching safely from a nearby tree. After feeding for a while, the lioness went toward the Seronera River. While she was gone, the leopard came down from the tree and resumed feeding. A few minutes later, the lioness returned, carrying a small cub. The leopard zipped back up the tree, and the lioness fed briefly with her cub meowing at her side. But then she went off to fetch her other cub, leaving the first cub alone beneath the hungry leopard.

Leopards routinely eat small carnivores—domestic dogs, housecats, jackals, bat-eared foxes, lion cubs. . . .

By the time the lioness brought back her second cub her first cub was gone. Which certainly shows that lion mothers can make stupid mistakes.

But Schaller also described how male lions typically tolerated their cubs while feeding at kills, whereas the females swatted away the cubs and took most of the meat for themselves, and this seemed to be the behavior that bothered him the most. He painted a picture of a warm paterfamilias in contrast to a cold and unfeeling mother—the sort of mother that would leave you needing years of psychoanalysis.

My first wife, Anne Pusey, and I compared the tolerance of males and females at a hundred carcasses, recording how often each animal monopolized the best feeding sites. Although a prey carcass can be fairly large, there are usually only a few access points to the meat: with thick-skinned prey, someone must first tear open the abdomen. The entrails are consumed first, and the hide is eventually pulled back from the bones so that the lions can chew on the muscles. So if there are more lions at a kill than access points to the goodies, things can get tense, and there is a predictable pecking order, with the males being the bullies, routinely excluding females, subadults, yearlings, and cubs from the best bits. Females, in contrast, are far more civilized, almost never supplanting cubs and subadults from feeding sites.

Males have important work to do: they defend the territory against invading rivals, and they often only join their families at dinnertime, tired and hungry, needing to fortify themselves for another day at the office. They may even be two-timers who need to fortify themselves for the commute between families. Life is tough, and they eat like there is no tomorrow. If it is the dry season, and if the males have caught a buffalo, they may still act like louts, but their pride will be fed for a week. The mothers, conversely, keep close to their children at all times, more completely dedicated to the care of their current brood. They're in for the long haul.

Not at all cold and unfeeling.

Although most of Schaller's descriptions of lion behavior would be recognizable to anyone visiting the Serengeti today, there has been one dramatic change in the park's ecology that has likely transformed his mean moms into the mild-mannered moms of modern times.

In the 1960s, Serengeti was still managed by British wardens, who believed in active management of the national park. For example, the chief park warden routinely lit grass fires to clear the brush around the rivers, making it easier for tourists to find the lions.

These practices, which were still common during Schaller's era, drew large herds of Thomson's gazelle to the rivers during the dry season—attracted by the "green flush" of fresh grass shoots sprouting up from the ash and to the water holes that lined the rivers. So in the dry seasons of 1967, 1968, and 1969, Schaller watched large numbers of Thomson's gazelle stumbling toward the lion-lined water holes of the Seronera River each morning. The prides were full of cubs and subadults, and a frenzy followed whenever a mom caught a twenty-kilo gazelle. If a male was nearby, he would claim the carcass, and keep the hungry females at bay—and what was it to him if a few cubs sauntered up and fed on the crumbs of his majesty's feast?

So one small prey packet at a time became available to the hungry hordes, and if mom finally got a chance to feed from the partial remains, she might have cuffed a few cubs, sure, but, hey, she needed to keep her strength to provision the rest of the royal court again tomorrow.

By the time Anne and I came along, the parks management had adopted a much more laissez-faire attitude, and they had changed their fire policies so as to protect the woody vegetation from the ravages of repeated burning.

No longer attracted by the green flush along the Seronera River, the number of gazelle in the lion's diet dropped by 90 percent the year after Schaller left. Every year thereafter, the Serengeti lions have mostly fed on wildebeest and zebra in the wet season and warthog and buffalo in the dry season. Feeding competition can be a bit intense at a warthog carcass, but pigs are considerably larger than a gazelle—and the lions eat a lot more buffalo these days, so we rarely observed the desperate housewives that Schaller viewed as such lousy mothers.

Lionesses are wonderful mothers, and even more interesting is that they are remarkably egalitarian. Female lions do not form dominance hierarchies, and pride mates can all expect to have similar reproductive success over the long term. We might take this for granted in the suburbs, but it is an extraordinarily rare arrangement in social carnivores.

Think of a wolf pack. There is an alpha female, an alpha male, and several subordinates who tend to the alpha pair's offspring. The same is true in foxes, coyotes, wild dogs, and jackals. It is what dogs do—and so do a lot of other small carnivores like meerkats and dwarf mongooses. The dominant female in a clan of spotted hyenas is almost as obnoxious; she is a baby machine with high cub survival and short interbirth intervals, whereas her subordinates skulk around the community den, barely able to breed.

These top-dog females really are mean girls, who may harass the subordinate moms into submission or even kill their babies outright.

But lions couldn't be more different. Females pool their litters into a crèche and rear the cubs collectively over the next two years. Over each female's lifetime, she will raise as many offspring as any other female in her pride.

So why are lions such democrats?

First off, female lions are great proponents of self-reliance. In pack-living species like dogs and mongooses, the females are always together, so the dominant female can be as nasty as she wants, yet her subordinate sisters will slavishly stick by her side, and she exploits their insecurities by coercing them into becoming nonbreeding helpers.

But if a female lion loses her litter, she is perfectly content to go her own way until she can conceive again. Thus a hypothetical mean-girl lion would never be able to maintain her posse of adoring fans.

And as long as a female lion has cubs of her own, she needs the other moms to pool their maternal instincts with hers. Groups of mothers can better protect their cubs as a gang against the much larger males, but, if you've zapped your sister's cubs, she won't be there to join in the collective defense of *your* cubs.

Second, you are a lion. You have big teeth and four strong claws on each of your four strong paws. Worse, so do your sister and your daughter. If you tried to kill their cubs, they would fight back with the same lethal weaponry as yourself. Accidents happen; if you were to fight over breeding rights, someone could get hurt—and you don't just need your pride mates for cub defense, you have to maintain your territory against neighboring gangs of females.

So female fairness results from necessity and from mutually assured destruction. You need your sister, and she needs you.

Just don't mess with her.

There is just one more thing about the Serengeti lions, but this one isn't for television. Not because it isn't suitable for family viewing, but because some things change too slowly, too diffusely to be seen on film. Circumstances can inexorably alter, and conditions can eventually arise, that spur the sudden overthrow of the status quo.

This one begins with a virus.

When the Italians attempted to conquer Abyssinia in 1887, they provisioned their troops with livestock brought from India. One or two of the cattle were infected with rinderpest, a virus related to measles and CDV that causes fatal bouts of diarrhea in ruminants. By 1897, the disease had spread south from the Ethiopian plateau to the Cape of Good Hope and across to West Africa, killing 90 percent of domestic livestock across the entire continent. Control programs were initiated throughout Africa, and by the 1960s, rinderpest was restricted to only a few areas, and Serengeti was the last major reservoir in Tanzania. A cattle-vaccination program around the Serengeti finally eliminated the disease from the wildlife inside the park in 1963. Liberated from rinderpest, the wildebeest, buffalo, warthog, and other ruminant populations grew exponentially until they reached their current levels in 1979.

The lion population grew, too, but in a very different pattern.

The lions in the wooded habitat of our study area remained stable from 1966 until 1973, when the population suddenly leapt to a new equilibrium, then remained stable for another ten years before leaping again in 1983. The ruminant population had nearly tripled between 1966 and 1973—what held back the lions for so long? And what happened in 1973 and 1983 to cause those particular jumps in lion population?

Lions in our Serengeti study area endure an annual pattern of feast and famine: the migration brings the wildebeest and zebra within easy reach during the rainier months, but the herds move north to Kenya each dry season. In normal years, our lions manage to survive on warthog and buffalo, but this is really only enough food to sustain the adults—cubs often starve, leaving only enough survivors to replace their aging mothers.

But the dry season of 1973 was the rainiest in decades, and the unseasonably green grasses attracted the wildebeest and zebra to our woodlands study area more or less continuously until the normal rains returned in November. Without the usual dry-season famine, virtually every cub born in 1973 survived.

Here is the interesting part: these surviving cohorts were large enough to form entirely new prides that could compete successfully against the prevailing social order and redraw the map of pride territories. Tough new girl gangs squeezed their way into the neighborhood, finally allowing the population to rise to a higher post-rinderpest plateau.

The recovering herds not only provided more meat on the hoof, but the wildebeest's insatiable appetite for grass subsequently modified the habitat in the lions' favor. An awful lot of grass was left uneaten back when the wildebeest population was held down by rinderpest. Grass fires roared through the park each year, burning the young acacia trees to stumps. But the expanding wildebeest population became the world's largest lawn service, mowing the grass down to the nubs over thousands of square kilometers—creating fire breaks through much of the park. By the mid-1970s, less than a quarter of the Serengeti burned each year, and tree saplings were able to grow unhindered. Recovery rates in the tree populations reached a peak in 1980 and persisted for ten years.

Lions need cover to hunt more successfully—remember the Hopcraft effect? And 1983 was the first year with favorable dry-season rainfall in this new improved world for hunting lions. Once again, the woodland prides recruited large numbers of young—large enough to spawn an expansion of new prides and redraw the map, with yet more groups packed more tightly than ever before.

The woodlands population crashed with the canine distemper–babesia outbreak in 1994—lion numbers fell back to levels unseen since the late 1960s. But in 1999—the first post-outbreak year with favorable dry-season rainfall—the woodland population bounced all the way back up to the same level as in 1983–93.

On the plains, the population's initial post-rinderpest spurt occurred sometime after Schaller's departure in 1969, reaching a new plateau by 1974 when monitoring of the plains prides resumed. The plains population remained unchanged until November 1997, when El Niño brought the heaviest rains in forty years. The grasses on the plains had started growing taller during the early 1990s, and the El Niño floods kept the migration out on the plains for the longest period in decades. A single year with a more consistent food supply was enough to allow the plains lions to spawn whole new prides in the taller grasses.

As Lenin once said, "Sometimes decades pass and nothing happens; and then sometimes weeks pass and decades happen."

These were happy revolutions—good times leading to more and more lions—and our study population is currently in great shape—we have nearly three times as many lions in the Serengeti study area today as in the 1960s. But lion population trends across the rest of Africa couldn't be more different, and I wasn't eager to start filming part 2 of our series, the part about lion conservation.

UP IN THE AIR SOMEWHERE BETWEEN MINNEAPOLIS AND DC, 29 MARCH 2011

Peyton West's husband, Frank Lowenstein, is the foreign policy adviser for Senator John Kerry. One of my deans at the University of Minnesota, Brian Atwood, was the head of USAID during the Clinton administration. Frank and Brian have introduced me to the aides for numerous senators and congressmen and to upper-level staff at USAID and the Millennium Challenge Corporation. Along the way, I've met dozens of Beltway bandits in Bethesda, and I've met with top officials at the Department of Agriculture and at the Food and Agriculture Organization.

The Senate Committee slipped in an endorsement for the Whole Village Project in the report language for the 2010 Senate Appropriations Bill. But when Ted Kennedy died, and the Democrats lost their super majority in the Senate, the Obama administration placed all its capital on the passage of health care reform rather than on the 2010 budget. After the Tea Party insurgency in the fall elections, the Senate Appropriations Bill never even made it to the floor.

However, USAID is about to start a new food security program in Africa called Feed the Future. The project has been cut to just a handful of countries, but one of them is Tanzania. There will be a separate budget line for independent monitoring and evaluation. The provost at Minnesota has arranged for me to meet the assistant secretary of state for Africa today. The secretary, like Congress, is eager to reform the aid industry. Collecting outcome-based measurements of USAID projects in Tanzania would be a good start. He schedules me an appointment with the head of Feed the Future this afternoon.

Feed the Future will indeed require independent monitoring and evaluation; the Whole Village Project looks to be an ideal partner. The head suggests we meet again in Dar in ten days.

I feel like this ridiculous gamble has finally paid off. I've invested so much time and energy on the Whole Village Project the past few years that I nearly lost NSF funding for the lion project. Susan and I have lived so far out on the edge that we are on the verge of emotional and financial collapse. She has worked pro bono for five years; property taxes in Minneapolis have skyrocketed; the kids' tuition bills are overwhelming.

I have assembled a team of highly qualified economists and anthropologists, and Susan has trained a first-rate survey crew, but I've started to wonder whether we had merely been collecting bottle caps to build a rocket to Mars.

I call the team back in Minnesota and report on today's meetings at the State Department and at USAID—and that I've already got an appointment to see the head of Feed the Future again in Dar es Salaam next week.

What could possibly go wrong?

5 Milk Stains on White Trousers

EVANSVILLE, 23 SEPTEMBER 2011

I travel across town in Evansville, Indiana, to the headquarters of the American Patriot Group. The suburban office building is an oversized snow-white antebellum pastiche—part Monticello, part *Gone with the Wind*. The atrium is decorated with two large oil paintings, one of a female flamenco dancer, the other of a toreador; both are about eight feet tall. I'm told to wait for Mr. Chancellor in the board room, which is adorned along the inside wall with large oil paintings of the Capitol Building in DC, the White House, and the Lincoln Memorial.

I spend the next fifteen minutes piecing together a PowerPoint of lion population trends, impacts of trophy hunting, estimates of the true costs of conserving a lion in the wild for a year, the importance of fences. I can't imagine actually showing any of this to anyone today, but they paid my way from Minneapolis, and I don't want to appear disrespectful or unprofessional.

I am guided upstairs and pass another large oil painting—this time of a woman with long blond hair mounted imperiously atop a thoroughbred. We enter Mr. Chancellor's office and approach a massive black desk with silver-painted woodwork along the edges, a damasked silver Arab dagger and a silver-handled folding knife resting on a black blotter.

And there, sitting at his desk, Mr. Steve Chancellor, the multimillionaire, lion hunter extraordinaire, and top-tier donor to the Bush-Cheney tickets in 2000 and 2004. His fortune comes from coal, and he has always lived in Indiana.

Off to his left, another large oil painting of the blond woman, this time naked, her left breast discretely hidden by the tail of the male lion whose mane she strokes while gazing toward Chancellor's chair.

His handshake is firm; so is mine.

He starts talking, and it is clear that this is how the meeting will go.

Him talk, talk, talking and me occasionally getting a word in edgewise. But that is OK. I honestly don't know why he summoned me here today. He has consistently refused to talk to me over the past seven years, letting his displeasure with the age minimum on lion harvests be known to everyone and preventing numerous hunting organizations from collaborating with me.

He talks about Africa, about how the wildlife is disappearing—encroachment by local people is stripping the habitat from the animals, whereas trophy hunting brings value to wildlife and must be preserved. He has done most of his hunting in Botswana, and Botswana is a mess because they banned hunting on lions—now the prey populations are crashing, he says, because there are way too many lions.

I manage to get in a few words about Tanzania. Tanzania is losing its wildlife fast because it has forty million citizens and the pressures on the land are intense; antipoaching is far more expensive than people realize; revenues from hunting are inadequate to cover these costs—only a few well-capitalized companies can afford to manage their blocks properly.

This brings him to his point. "I have every respect for Paul Tudor Jones; he's very professional. He is the top of the hunting profession. But I've had serious problems with TGT. I first went with them to Moyowosi and shot two males. The flood plain was dried out completely; we saw so many lions dotted around everywhere. TGT had this six-year rule and they said my males were fine. They put their pictures on their brochure for the following year. Then they called me back and said those males were actually only five-and-a-half years, so I shouldn't have shot them at all. But they left my lions in their brochure!

"The following year, I was scheduled to go to Maswa for two lions, and they called me a week in advance to say that another party had cancelled and would I like to pick up a third lion?

"I agreed to take it, but when we went back to Tanzania, their PHs were terrified about losing their jobs because of that six-year rule. We went out and saw forty lions—that's males and females, too—but I would have shot eight of them. But each time, the PHs panicked and wouldn't let me take any of them."

Steve is in his early sixties, he bears an uncanny resemblance to Johnny Cash when he is animated. The creases on his face deepen while he talks. "I found out later that they hadn't shot a single lion the year before—and they didn't shoot any lions at all in Maswa that year either—and two years earlier they only shot two males. So, tell me, how could they have expected me to get three males in one year when they'd only shot two males in three years?"

Pause.

"I've talked to so many people with the same experience. If you ask me, they're crooks. They say they are doing such great things for lion conservation; but I'll tell you, they're doing it with other people's money.

"Everyone I've talked to says the same thing, there's no way you can be sure if a lion is six years old. You know, a big lion comes up to a hippo bait and after a day or two it's digging its way so deep inside the carcass that its mane is slicked back with grease, and you can't be sure of its age. It might be an impressive big male, but it will look like nothing with its hair slicked back like that.

"And now the U.S. government is talking about using the Lacey Act to regulate lion hunting. I met the new head of U.S. Fish and Wildlife last month, and he was complaining about his agency's 20 percent budget cut, and I said to him that he couldn't afford to take on the lion. Anyone who tries to stop importation of a lion because they think it is less than six years of age; USFWS will get tied up in court so long their budget will just vanish."

Finally another pause—long enough for me to say: "I'm disappointed to hear that about TGT."

Meanwhile I'm thinking: well, TGT is trying to run hunting as a business, and if they devote a portion of their profits to conservation, that is kind of like any other form of corporate giving, isn't it? Profits come from customers, don't they?

But why argue? Trust has to be earned, and an unscrupulous operator could certainly abuse any set of rules to the detriment of its clients. I've always held TGT in high esteem, but they were still casting about for a consistent set of aging criteria the year Steve came to Maswa for his three lions.

The bigger issue for the moment is the Lacey Act, which Luke Sidewalker has been pushing the past year in DC with the U.S. Fish and Wildlife Service (USFWS). The Lacey Act allows the U.S. government to ban the importation of wildlife products from any country that fails to enforce its own laws. The Tanzania Parliament passed a new wildlife act in 2010 that specifically forbids the shooting of lions less than six years of age. But the WD has so far made no serious effort to enforce their new law.

"Well," I say aloud, "I certainly agree that no one can reliably identify a six-year-old male compared to a five-year-old, so in my discussions with various people, I've always said that we should focus instead on the most conspicuous violations. We can certainly tell if the animal is less than four—if it's just a teenager. And if we only stop the shooting of lions under four years of age, we'd prevent the worst damage to the population."

He is surprised to find that he actually agrees with me. Somewhat startled, he says, "I've always been very discriminating. I mean, thirty years ago I was impatient and didn't know what I was doing, but now I'm much more careful. The problem is that most people go on a once-in-a-lifetime safari, and they don't know anything, and they'll believe anything their PH tells them."

I'm relieved, too, and say, "Yes, so if the rules only really affect the broad bottom of the market, it will leave the high-end pretty much alone, and lions will be better off if none of the juveniles are taken."

The ice broken, he starts to talk about specific hunting trips, particular trophy lions. Then he remembers to raise another objection to the age minimum: unless the young males are shot, they will gang together and cause problems for the older males, and there won't be any older males left. He talks as if young and old males are two different species. I explain that this year's younger males will eventually grow old themselves, and, the more young males that are out there now, the more old males there'll be in the future.

He starts rummaging around his desk drawers; he wants to show me photos of a particularly dark-maned lion but can't find them. Then he starts looking for pictures of a male that was deemed too young by the nervous PHs at TGT, but he can't find those either. He shuffles through more drawers of his desk and looks inside brown envelopes. Finally he finds pictures of a large male leopard shot by his daughter.

"This is what makes everything worthwhile," he says, showing his fourteen-year-old next to the carcass of a massive spotted cat.

We drive in his Bentley through the edge of town to a mansion the size of Arkansas. The floor of his garage is cleaner than most kitchens. We walk inside through a glass-walled living room, turn left, and he unlocks the door to his museum. He flicks on the lights, but they are so dim it takes a while to make out the menagerie—the stuffed hyena in a frozen lope, headed toward a stuffed lion that has grabbed hold of a stuffed zebra stallion. On the opposite side of the aisle, two more lions are taking down an impala. The walls are covered with mounted heads of African antelope. Steve stops to tell where he shot each lion and which species all those heads belonged to. We enter another corridor through stuffed forest animals, bongo, bushbuck, carnivores from Cameroon, go down a stairwell, past a gerenuk, to a longer corridor—head left and there are wolves and bears—including a polar bear that took sixteen days of hunting in below-zero temperatures; Steve describes how he nearly got swallowed when the ice floe suddenly broke up.

Steve says this is where he likes to bring school groups from town, so they can see animals in a different way than in a zoo—interacting with each other in a naturalistic manner.

Everything is superbly mounted; almost every animal is a true trophy—a spectacular example of its kind. No collection in any of the major museums anywhere in the world comes even close to the quality of these specimens.

Steve always walks in the middle of the corridor so I have to follow directly behind him. We retrace our path a dozen yards or so and step into a vast room, walls thirty feet high are covered with dozens of elk heads, deer heads, elephant tusks, a moose head, a couple of rhino heads on pedestals, stuffed bears, and in the middle, two male lions slapping each other while standing on their hind legs. High up on the far wall, those two elk heads, he tells me, were the world number one trophies, one for a nontypical (many more points on the antlers than normal) and the other for the southern Pacific. On the near wall, fifteen to twenty more elk heads that don't measure up to world-record standards, freak white-tailed deer with tangled antlers. Everything is bathed in soft light like in a giant aquarium.

The two lions were shot in either Botswana or Zambia; I've forgotten already. One of them charged Chancellor at some point—but then every stuffed lion has a story.

Sometimes, instead of telling me where that animal came from, he turns to tell me that I'm one of the few people on earth who can really understand what his life in the wild has been like. The time he has spent tracking animals on foot or waiting beside baits, the campfires in the evening, the months and months spent in the bush over the past thirty years.

We turn left into an even larger space with display cases filled with knives and guns and more heads: crocodiles, alligators, gazelle, monkeys, sheep, wildebeest, hartebeest, oryx, addax, ibex, musk ox. In the next room, there are more guns, more knives, bearskin rugs, wolf-skin rugs, dozens and dozens of bison heads, deer heads, sheep heads—and when I say sheep, I'm not doing these things justice. They have curled horns, twisted horns, thick horns, pointy horns. Everything is in triplicate at the minimum. Somewhere down here, he showed me three world-record Argali sheep from Kyrgyzstan—each newer specimen taking over the top spot in the record book from the trophy before it.

Before we came in, he told me he had 1,050 trophies of 385 species. I had no idea that they would all be on display. Besides the heads on the walls, chamois are skipping across plaster boulders, cougars are resting on trees, leopards are climbing up trees, another pair of lions is fighting in the middle of the floor. There are roan, sable, blackbuck, blesbok, klipspringer, dik-dik, red and blue duikers, suni, squirrels, possums—some animals are as big as a horse, others are only a few inches tall.

And then we go into the final basement room—also large, but this one feels cramped: two full-sized stuffed elephants, another ten leopards (world number one and two among them), six or eight hyenas, vervet monkeys, baboons, and fifteen lions (also world number one and two). He has a lot of stories to tell about the animals in here. He is in full flow, and he touches one of the leopards, strokes its fur. He musses up the mane of one of the male lions, which, he says, has lost its battle with gravity.

The lions are all busy in the pale light of the room. At least three pairs are working together to pull down various prey while a large group of males and females is stalking an eland, I think, or a greater kudu. The mixed group is standing, but the rest are all reared up on their hind legs—it is kind of like a workout room at a gym, but nothing moves.

Humility may not be the first word to come to mind down here, but Steve has stopped several times to tell me how his wife shot the larger elephant, his daughter shot the larger leopard when she was fourteen. At one sheep, he tells me how his daughter had shot the animal from 350 yards when she was twelve. She has gone hunting with dad every year since then, but she is now sixteen, so this phase of the father-daughter relationship may be fading away.

Her name is Hunter.

By the end of the tour, we have seen at least fifty lions, either mounted in various poses or displayed as skulls on shelves. "As you can see," he says, "I have a special love for lions."

We go back into the living room and encounter a high-strung golden doodle, which looks at me uneasily under the touch of Steve's calming hand.

We hear sounds from the kitchen.

Steve calls loudly, "Terry, I'm about to go out."

His wife is standing in the kitchen in a diaphanous black gown that touches the floor, the kind of dress you might expect to see in a Hollywood dream house. She is stockier now than in her portraits at the office, but her hair is still as long and blond. Her movements are uncannily precise, her smile tight. Steve gives her an awkward peck on the cheek.

Nothing seems real.

As we drive off, Steve tells me that he used to have three dogs, but their two German shepherds died over the past six months. The male of the pair was a splendid specimen, he says with such emphasis that I wonder if it is on display somewhere down in his basement.

We arrive at his local restaurant too late for lunch; Steve apologizes, saying that even though we had rushed through the museum so fast, it still took two hours. He reckons it takes three or four days to really absorb everything and that he frequently hosts fund-raisers down there with five hundred to seven hundred guests at a time. He has to hire a large crew to keep the place clean, a hundred fifty caterers at a time.

I'd had an accident at the airport early this morning. I got out my box of breakfast cereal, but when I opened the bottle of milk at Gate G19, the excess liquid splattered all over my pale khaki trousers. Milk stains on white trousers. There was no chance to clean up. I could only hope that they would be invisible for the next few hours.

As I left Steve's office at the end of the day, he said, "I want to help you out." Fiddling with the half-folded silver-handled jackknife at his desk, stabbing the point into his blotter, "I'll be sending you something to help with your work."

We shook hands, I walked downstairs, went into the restroom, and checked for any stains on my trousers, but by that point I wasn't sure if I even cared.

LAS VEGAS, 31 JANUARY 2012

Eric Pasanisi has been trying to ask me a question from the far end of a very long table at the Fleur Café. He has a heavy French accent, and he labors through a long rephrasing of his original question. I really want to catch his meaning, and I'm embarrassed by my incomprehension, but no one else at my end of the table can understand him either.

Almost everyone here has been polite to me, and I've nearly recovered from my initial petulance at the events manager's announcement that it would cost a thousand dollars to rent a PowerPoint projector for the next two hours. We had been promised a private room, but the Fleur Café only has a semiprivate dining area separated by a sheer cloth curtain. Although they have lowered the volume of the piped music, the white noise from the Mandalay Bay Casino strains communications.

The Safari Club International convention starts here tomorrow, and John Jackson, Charles Williams, and George Hartley are seated around the table with a dozen bigwigs from the hunting fraternity. But there is a pall on the proceedings. John can't look at me without snarling. A few years ago, I e-mailed him an analysis showing how lions had been overhunted in Botswana, Namibia, and Zimbabwe. He e-mailed back, first threatening to arrest me, then threatening to have me committed to an insane asylum. When I sent him the news about overhunting in Tanzania, he threatened to have me deported. He is worried that the lion will be up-listed to appendix 1 at CITES in 2013. He considers me to be public enemy no. 1.

John is sitting across the table from an American woman who could barely bring herself to shake my hand. She has been trying to establish an aged-based harvest system for lions in Zambia, but she has systematically overestimated the ages of the trophy males. Her husband is a Zambian PH, and their family income depends on keeping lions on quota.

Eric Pasanisi is spitting his heavily accented words through his teeth. He seems more interested in conveying his disgust for me than in making himself understood.

But at least a meeting is happening. And even though Eric obviously doesn't want to be here, his father, Gerard Pasanisi, didn't impose another boycott. This is the first time I've been allowed to talk to the Great White Fathers of African trophy hunting since the elder Pasanisi prohibited the Savannas Forever meeting in Dar es Salaam in 2006.

I spent the flight from Minneapolis in professorial mode, organizing my thoughts and preparing my latest PowerPoint presentation. Start off with a quick review of Karyl's original simulation model, show a selection of underage trophies that had been advertised on the Internet, follow up with evidence that lions had indeed been overhunted throughout most of Africa, then focus on Tanzania.

When Erasmus Tarimo gave us access to the block-by-block hunting data, he finally opened Pandora's box. Henry Brink spent weeks in the Utilization Office at the Ivory Room, poring over moldy files of hunting tags, tracking down the number of lions and leopards shot each year in each hunting block in the country.

Henry is a true cosmopolitan. Half Danish, half Indonesian, he spent his childhood in Africa and went to college in the United Kingdom. He worked as a field assistant on the Serengeti lion project for three years, before gaining permission to study lions in the Selous. We worried that he might be tarred by my conflicts with the WD, but he is so amiable (his smile reminds most people of the Buddha) that he quickly earned the trust of the hunting companies.

We already knew that most Tanzanian hunting companies had been shooting underage males for at least a decade. Severre had been pushing hard to raise "production" each year, and the operators had obliged by shooting every male lion they could find. As long as it was a lion and it had balls, it was fair game.

If our harvest model really did work, the resultant infanticide should have driven down subsequent offtakes—and the effect should have been strongest in the same blocks where lions had been the most heavily hunted.

The results from Henry's efforts confirmed our predictions: lion harvests had plummeted in almost every block between 1996 and 2008, and they fell farthest in the blocks that had been hammered the hardest by Severre's overeager hunters in the mid-1990s.

And, to my surprise, it turned out that nothing else mattered. When talking with Tarimo, I had presumed that lion offtakes would have fallen farthest in hunting blocks that suffered the most rapid human population growth or the greatest amount of agricultural activity. But, no, the only thing that predicted the drop in lion offtakes was the number of lions shot in 1996–98.

Males are keystone members of their prides: if they are killed, their offspring will be eliminated far more efficiently than if their habitat is disrupted. After all, trophy hunters pinpoint their quarry and selectively remove the very individuals that have the greatest impact on population stability. Habitat loss might eventually knock back the lion population, too, but the hunters got there first, and they hit the lions as hard as they could.

It was important to treat the bad old days like a scientific experiment. In most parts of the world, such an aggressive harvest strategy would have been terminated at the first signs of trouble. But since Severre had hidden the data so successfully, his "removal experiment" continued long enough to allow decent estimates for sustainable offtakes.

And those estimates were easy to calculate. Rolf Baldus had set up a careful monitoring system for trophy hunting in the Selous. The Selous is famous for being home to Africa's largest lion population, so hunting impacts could be measured on a block-by-block analysis. By plotting the harvest trend in each block of the Selous against the number of lions shot in the mid 1990s, Henry and I could see that offtakes always fell if initial harvests exceeded one male per thousand square kilometers. In blocks where initial offtakes had been lighter, subsequent harvests remained stable.

In the rest of Tanzania, the data were scrappier, and lions were less plentiful. We lumped adjacent blocks together and compared one part of the country to another. In these larger-scale comparisons, offtakes dropped precipitously in regions where more than one lion was shot per two thousand square kilometers.

When Tarimo first saw the results, he wasn't surprised to hear that lions had been so seriously overhunted during Severre's tenure, but he was disheartened that harvests should be kept so low. He called me one day in 2009 to discuss the relative merits of an age-based offtake system (which would allow the current quota system to continue but would restrict whether a given animal could be shot) or to adopt a sustainable quota based on the new analysis (which would require an immediate reduction in the operators' advertised quotas). Tarimo said the government could only adopt the age minimum. Any reduction in advertised quotas would put a lot of small operators out of business: if their one and only hunting block in Selous was smaller than a thousand square kilometers, they couldn't offer any lions at all.

And without lions on quota they could never make a profit.

So in July 2010, the Tanzanian Parliament passed a new law, mandating a six-year age minimum for lion trophies. But then the Tanzanian president sacked Tarimo in early 2011, presumably for being more interested in wildlife conservation than in maximizing short-term revenues for government officials. Dozens of well-connected Tanzanians were clamoring for blocks that they could sublease to foreign operators. A good hunting block was a reliable source of cash, and more and more politicians wanted blocks for themselves.

About a quarter of the country's blocks were no longer viable. They only held marginal wildlife populations to begin with, and the Tanzanian operators hadn't invested anything in conservation. They forced Tarimo to subdivide the better blocks so they could skim off the cream somewhere else. Tarimo fought back by supporting the foreign companies who had successfully protected the wildlife in their blocks, and he was subsequently sacked a month short of retirement—one of the most dramatic punishments ever meted out to a high-ranking government official.

By legislating an age minimum for lion trophies, the Tanzanian Parliament had acknowledged one of the most conspicuous failures of past management practices. But Tanzania has enacted a lot of well-meaning regulations that just sit quietly in their law books. I had been told that the hunting fraternity was ready to endorse a meaningful system for assessing the ages of Tanzania's lion trophies in Vegas today—they were fully aware of the power of the Lacey Act, and they also accepted the fact that lions had been seriously overhunted in the past. But lion aging was just one of the issues I intended to discuss.

Researchers from all across Africa had provided me with the results from their latest lion censuses and long-term population studies. Andrew Loveridge and his colleagues in Oxford had developed a computer model that could accurately estimate how many lions *ought* to exist within a given wildlife reserve, as predicted by the abundance of prey species or the extent of local rainfall combined with the richness of the soils. By knowing how many lions actually lived in a particular area compared to how many ought to be there, we could measure the conservation effectiveness of each reserve. If the latest lion count matched the predicted population size, we could say that lion conservation had been 100 percent successful. If there were only a quarter as many as expected, conservation was only 25 percent successful.

With data from forty-two sites across eleven African countries, we could see whether state-run parks did better or worse than privately run reserves, whether overall levels of corruption impaired management practices within a country's parks and reserves, and whether trophy hunting was inevitably damaging to lion populations.

The effects of corruption turned out to be more complicated than I would have guessed. On one hand, most of the management budget for Kenya's Maasai Mara National Reserve had been diverted to the bank accounts of a few well-connected individuals in Nairobi, and the Mara lion population had recently dropped by 40 percent. But on the other hand, while President Robert Mugabe had destroyed the Zimbabwean economy over the past dozen years, all of the privately owned Zimbabwean wildlife reserves had sustained healthy lion populations.

Except for a few national parks that were surrounded by overhunted blocks, trophy hunting didn't turn out to be particularly important in this overall analysis—we could only report the status inside the few hunting blocks across Africa where managers regularly monitored their lions, and few of these well-run reserves had raised harvests to Severre's devastating standards.

From Benin, Botswana, Cameroon, Ghana, Kenya, Mozambique, Namibia, South Africa, Tanzania, and Uganda to Zimbabwe two factors primarily determined the conservation success of African lions. First, lion populations that were bounded by lion-proof fences were all thriving; protected from the temptations of cattle killing and man-eating, these lions were never subjected to retaliatory killings. Their prey were buffered against poaching, and their habitats were protected against the nibbling plows of neighboring farmers.

Second, lion populations in unfenced reserves only reached half of their potential densities if management budgets exceeded two thousand dollars per square kilometer each year; any reserve managed on a budget below one hundred dollars per square kilometer had essentially lost all its lions. None of Tanzania's hunting blocks were fenced, so if the hunting industry wanted to maintain at least half of the lions in their hunting blocks, they would need to find $600 million a year to manage those areas. Current budgets were only a few percent of that total.

If unfenced reserves require two thousand dollars per square kilometer, the average hunting block is two thousand square kilometers, and the hunting operator has never invested a dime in antipoaching, community conservation or anything else, how could he possibly increase his revenue stream enough to protect his block and still stay in business? Increase the number of clients by several orders of magnitude without shooting any more lions? Keep the same number of clients but raise prices through the roof? Or find partners to share the costs?

Any of these options might be feasible but only in a few special areas: the romantic heart of the Selous, the magnificent fringes of the Serengeti. But most other hunting blocks would surely wither and die. In the absence of decent management budgets, a quarter of Tanzania's hunting blocks had been abandoned over the past five years—all the game was gone. There was nothing left to hunt. It was time someone started accepting financial responsibility for safeguarding these precious scraps of wild African wilderness.

In the absence of a projector, I've told myself to switch to a more conversational approach—less the professor and more the pregame coach. I tried to summarize my key points as clearly and briefly as possible, standing at the far end of a long table in the side room in a Vegas casino, waving my arms and practically barking my words in a herky-jerky fashion.

Eric Pasanisi has heard rumors of my economic viewpoint from his brethren, and he is accusing me of suggesting that hunting should be abandoned everywhere except Serengeti and Selous, that all other areas should be deliberately dropped.

I finally catch on and say, "Sure, some sort of triage is going to be inevitable between the remaining wildlife areas. Some places will be able to attract enough photo tourists or trophy hunters to cover the underlying costs. The rest will die a natural death from neglect and poaching. I can't possibly predict which areas will survive—but surely, there's no way all three hundred thousand square kilometers are going to make it to the next century."

"That's not what I'm asking," Eric shakes his head. "People say you're saying that hunting should be stopped in all but a few places."

"No, I don't know where hunting can survive. It's up to the people who lease each concession—they'll need to engage in whatever activity raises the most money. But it will have to be a lot of money. Millions—millions a year every year from now on."

The problem, of course, is that there are only two or three hunting companies that can afford this level of investment, and they don't have to make a profit to stay in business. Two of the most prominent companies are owned by American eco-philanthropists, Dan Friedkin of TGT and Paul Tudor Jones of Grumeti Reserves. Dan's father, Tom Friedkin, founded Trans World Airlines and sold out in the 1970s, and the family owns most of the Toyota dealerships between Houston and Miami. Tudor Jones is a hedge fund manager who had famously predicted the Black Monday market crash of October 1987.

The Friedkins run TGT as a high-end hunting company with exceptionally well-managed blocks. They advertise safaris at the hunting conventions, and their compound in Arusha employs a cadre of professional hunters, trackers, and caterers. They were the first company to realize that lions had been overhunted across the country as a whole. They were the first to impose the six-year minimum. They inspired—and funded—the formation of Savannas Forever.

Paul Tudor Jones owns properties in Zimbabwe and Zambia, and he holds two blocks along the northwestern boundary of Serengeti National Park, where he has built a five-star lodge for ultra-high-end photo tourism. He paid Severre all the hunting fees as if he had shot all of the animals listed on his quotas, but his friends and family only bow hunt a few impala and buffalo and never shoot any big cats. When he first took over the Grumeti-Ikorongo hunting blocks, he invested something like fourteen million dollars to eliminate poaching and manage the reserves, effectively enlarging the Serengeti National Park by 965 square kilometers.

But when he first set out to restore the area, he ordered his staff to erect a fence along his outer borders to prevent bushmeat poachers from commuting into the reserves and national park. But, oh, no! You can't build fences in Tanzania, Mr. Tudor Jones—that isn't how we do conservation in this country.

For the past few years, TGT has told me how poachers circumvent their well-guarded hunting blocks in the Maswa Game Reserve to go straight into the Serengeti National Park. Before TGT gained the Maswa blocks, cotton farmers had eaten away nearly 20 percent of the reserve by plowing a little farther into the protected area each year. Now there were so many people crowded up against the reserve boundary that even TGT's vast sums of money for antipoaching were being overwhelmed. Could I help push for permission to build a fence?

"So," I said to the folks in Vegas, "maybe it's time we started talking about fences."

ARUSHA, 21 FEBRUARY 2012

I'm in the kitchen of our modern three-bedroom house. A few years ago, we traded in the Pink Palace for a larger house with a clear view of Mount

Meru. We have three en suite bedrooms and a separate office space for the Whole Village Project accountant, two night watchmen, and three dogs, fluorescent lights, and various batiks and objets d'art scattered around the living room and halls. The night is warm, and I'm cooking on the gas stove. The aroma is soothingly exotic: to one cup of simmering brown rice I've added a bit of olive oil, a half-dozen whole cloves, a half-dozen whole cardamom, several cloves of garlic, and a half-cube of chicken bouillon. I have a song running through my head, but otherwise the neighborhood is quiet. We have electricity tonight, and Susan and I have been happier than we've been in months.

Then I hear a scream from the bedroom. Not just a cry for help, but a horrified banshee screech: prolonged, loud, panicked yet oddly muted—like a lost sound from far, far away.

I run fast as I can down the hall, turn into the bedroom half-expecting to see her bleeding beneath the hulking form of a marauding bandit. Then I catch sight of her lying on her back in the middle of the bathroom floor. Her arm is extended straight out from her shoulder. She is no longer screaming, just sobbing and trembling—unable to move.

"What happened?"

"I fell. The floor was wet, and I slipped."

"Are you OK? Did you break something?"

"Something's wrong with my arm."

Her arm has a funny crook in it, up near the top. But I'm convinced that she has merely dislocated her shoulder.

I try to take the situation lightly, saying, "I dislocated my shoulder a few years ago in the Serengeti. It hurt like crazy for a few moments, but then it popped back into place, and it was fine again." Susan has always had a low threshold of pain—surely she is exaggerating.

I call Dr. Mark Jacobson, who heads the Arusha Lutheran Medical Centre. I ask for advice, still convinced that she has a dislocated shoulder. He tells me to get her off the floor, lay her on her front on a table and pull down her forearm as hard as I can.

We have a houseguest, Marg, who is acutely sensitive to other people's pain. She shows considerable courage in helping me, despite the fact that every time we touch Susan, she screams and wails.

Susan is slender, but she is over six feet tall and can't contribute any power to her resurrection from the hard tile floor. We eventually manage to get her up on her feet and onto the table. The hard part is getting her to lie on her front as any movement to her arm is excruciating.

Finally, she is positioned with arm dangling down off the side.

I pull as hard as I can. I pull slowly at first then with as much of a yank as I can muster.

It doesn't hurt as much as she feared, but it doesn't improve things either.

And I think of all the pain she suffered the past few months, the emotional toll, and I want more than anything on earth to banish the hurt in her arm and in her heart.

We have been married for about twelve years now. We met cute, as they say, and fell in love. We had kids almost the same age that all went to the same school. We owned our own houses, and our lives were reasonably OK, but we both wanted to remarry. My kids and I were happy together, but I always felt a sharp sense of anxiety when they weren't around.

So here was this beautiful woman standing across a crowd of parents at a cross-country meet in the Twin Cities. We shared a goofy sense of humor, and she had a pent-up demand for adventure, so we decided to pool our resources and live happily ever after.

Our honeymoon was made possible by an invitation to France. I had published a paper in *Nature* on menopause a few months before the cross-country meet, and the French Fondation IPSEN had a long-standing interest in the biology of aging. So they asked me to a conference in Paris, and we scheduled a side trip to the south of France where I had been invited to view the cave paintings in Grotte Chauvet.

I had always wanted to test ideas about the evolutionary significance of menopause. When I first went to Gombe in the early 1970s, it was such a small park that it had lost most of its natural predators, so the baboons lived far longer here than in the open savannas. These doddery old codgers moved unsteadily along the ground with their scruffy fur and rats' tails. One particular old female (named Cauliflower) was particularly decrepit. She could only gum at her food and couldn't even climb up trees at night to sleep with the rest of her troop.

I was particularly intrigued that Cauliflower and all the other old ladies had stopped their menstrual cycles. Many of my scientific colleagues had speculated whether the menopause might allow grannies to focus their care on their daughters' offspring rather than to try to pop out a few more infants of their own. If the risks of giving birth at an advanced age were sufficiently high, it might pay grandma to forgo further reproduction and to instead spread her genetic legacy by assisting her youthful daughters.

This idea seemed rather fanciful in the baboons. Cauliflower had an adult daughter, Clover, with whom she seldom interacted. Cauliflower liked to hang around the garbage pit near the staff camp. She often stank of rotten bananas, and her face was streaked with flecks of cassava and whatever leftovers she could glean from the depths. Clover, meanwhile, spent her time in the forest eating real food with the rest of her troop. Cauliflower seemed like more of a liability than anything else.

But maybe something subtle was going on. After all, the baboon study had only been running for five or six years by then, and we had no idea how long these animals might live—how much of a long-term advantage grandmas might confer on their grandkids.

When I took over the lion project in 1978, I again saw some really old females. No animal kills an adult lion except another lion, and lionesses in large successful prides are well protected against neighboring enemies. I was particularly fond of an old girl named Shenzi (Swahili for "shabby") who had no teeth and a droopy lower lip—and who had stopped breeding during my first year in the Serengeti. The lion project had already been running for a dozen years at that point. Lions only live to about eighteen years. Thus, we had a better idea about overall life spans, but very few old gals had lived as long as Shenzi.

So in the back of mind I always wondered about aging and what sort of impact elderly female lions and baboons might ultimately have on their adult daughters.

Nearly twenty years later, I had collected enough data on both species to finally see what happened at the far end of their life spans. Female baboons could live to about twenty-seven years of age; the oldest female lion had died just short of her eighteenth birthday. (We have since followed one lucky lioness who almost reached twenty.) In each species, female reproductive rates plummeted at a particular age: twenty-one in baboons, fourteen in lions. They ceased breeding altogether a few years thereafter.

So, OK, if menopause is adaptive in these species, the old gals should make a measureable contribution to the reproductive rates of their daughters. So we compared adult females whose mothers were still alive with adult females who had been orphaned. Thus we could test if daughters with living mothers had earlier menarche in baboons (no), larger litters in lions (nope), or shorter interbirth intervals (nada, in both species).

There *was* an effect of grandmothers on the survival of grand-cubs—but only if grandma lion was still actively breeding. Female lions raise their cubs in a crèche, but granny is only part of the gang if she has cubs of her own. But if grandma lion has gone through menopause, she is off somewhere on her own. As far as her grand-cubs are concerned, she might as well be dead. In our species, we might think of grandma as someone who bakes us cakes and provides a gentle, guiding hand. Or maybe some of us think of grandma as that old woman who moved to Florida.

Once the biological clock turns off for a female lion, she's off to Florida.

We couldn't find a single positive effect of reproductive cessation on the number, survival, or reproduction of grandchildren. But if menopause serves no useful purpose, where did it come from?

Female reproduction is a bit liked planned obsolescence. Spare parts should only be built strong enough to last the life of the car—otherwise Toyota, Mercedes, and Ford would be wasting their money on something no one needs.

Similarly, there are so many slings and arrows of everyday life that most female baboons fall out of the trees, get munched by leopards, or struck by lightning, and most female lions are killed by their neighbors or gored by a buffalo, and thus very few individuals survive to reach old age. By then, so few animals are still alive that any mutation that harms their reproductive machinery can never be weeded out by natural selection.

Selection acts most strongly on the young, when we all experience the effects of our genes. But as our cohort thins out through time, fewer and fewer of us will be subjected to the tragedy of Alzheimer's or macular degeneration. So our children will also carry around genetic disorders that lie dormant until they, too, grow old.

But women's reproductive rates tumble around the age of forty, when they are otherwise in good physical shape. Perimenopause hits lions and baboons when they are obviously old, and hence much later in their relative life spans.

Here is the essential difference between humans and other mammals: a lion cub is doomed if it is orphaned in its first year of life, an infant baboon will die if orphaned before the age of two, but human children are dependent on their mothers for far longer than any other species. Thus a female lion leaves no progeny born in the final year of her life, a female baboon leaves no descendants from the final two years of her life, whereas a woman might not leave survivors born in the last ten to twelve years of her life.

In the late stages of life, selection cannot extend female reproduction if none of their late-born offspring ever survive to carry the trait into the next generation.

So whenever I go into any detail about the evolutionary implications of hot flashes, irritability, and the rest, I simply say: "Ladies, be grateful to be living in the twenty-first century. At any other time in our history, most of us would have been dead already."

Or as Maurice Chevalier said, "Old age isn't so bad when you consider the alternative."

*

Susan and I went to France on our honeymoon in 1999, and I gave my talk in a beautiful part of Paris. We had a lovely time. We dined on a barge along the Seine, took a day trip to Monet's gardens, and spent two days at the IPSEN meeting. We learned all about Madame Jeanne Louise Calment, France's national hero of longevity who lived to be 122 and who had sold pencils to Van Gogh as a child. She had a marvelous sparkle right up to the end. When she was about 115, she said, "I only ever had one wrinkle." Pause. "And I'm sitting on it"

But while my menopause research had been the official reason for the conference invitation, it turned out that my elegant French host from the Fondation IPSEN had a long-standing passion for lions, and he interviewed me for a local magazine about the Serengeti. But nothing I could tell him was anywhere near as exciting as the fact that Susan and I would soon be visiting Grotte Chauvet. He was amazed that we would actually be allowed to go inside—almost no one was granted permission.

We arrived at the camp of the French Ministry of Culture two days later, introduced ourselves, and met our host, Jean Clottes. Jean was the leader of the team that painstakingly cataloged every image, every artifact in Grotte Chauvet. Jean is an artist at heart, and cave paintings are his passion. He, like his entire country, is enormously proud that most of the great examples of Paleolithic art have been found in France. Jean's interpretations of the art sometimes tend toward the mystical, but he has always been careful to collaborate with field biologists to learn about the natural behaviors of ancient animals.

Before dinner, we met with students and scientists from around the world who had been digitizing and color coordinating their photographs of Chauvet's treasures. Everyone was proud to be part of the team. Spirits were high.

But when they saw that I had brought Susan with me, they tended to mumble to themselves and stare at the ground. We apparently posed a problem. I had been invited because I was a lion authority. One of the only other non-French visitors to the cave had been a bison expert. Jean Clottes had been seeking alternative interpretations of the artwork. Were the drawings symbolic or were they realistic? He had sent me a set of photographs, but they provided little sense of scale. He wanted me to see them on the curved surfaces of the cave.

I thought I had been clear that my new bride would accompany me to France. Everyone was perfectly graceful during the introductions, but how could she possibly be allowed inside?

My heart sank for the first time since our wedding. I knew I was being granted a rare privilege. It had never occurred to me that it was virtually unique.

That night we ate dinner with the rest of the team. The mood was spirited and lively. But my heart had sunk a few notches, and it wasn't until the next morning that Jean Clottes confirmed the bad news. I was to be only the fifth non-Frenchman to enter the cave since its discovery. Even his own wife had not been allowed inside.

I would later make light of my decision—saying that while I would have all the rest of my life to have fun adventures with Susan, this would be my only chance to go inside Grotte Chauvet—but in fact, I felt miserable about leaving her at the base of the cliff as Jean Clottes and I wound our way along the trail up to the cave entrance. Susan was left to wait among the olive trees in the oxbow valley across from Pont d'Arc—a natural arch in the winding stone canyon of the Ardèche River.

We climbed a narrow path until we reached a steel door set into the limestone cliff. Jean Clottes pressed his hand against a glass plate that read his palm. He was one of only three people in the world whose print was accepted. He then stepped in front of a camera lens and chatted with the security team in Paris who monitored the entrance twenty-four hours a day. Convinced of his identity, electrical lights flashed, and the door swung open.

We stepped into the semidarkness of a stone-lined chamber. One side was cluttered with piles of limestone that had collapsed a few thousand years ago, blocking access to the cave until a group of French spelunkers discovered a narrow entrance on top of the plateau in 1994.

Now we had to change into orange overalls and put on a pair of rubber boots that had been treated with fungicide to prevent the introduction of stray mold. My boots were a bit tight, and my overalls were too short, but I was more or less covered, and my toes weren't too unbearably cramped. We put on miner's hats and lamps and in we went.

The entrance is too narrow to walk two abreast. After we have gone maybe twenty meters, Jean Clottes shines his lamp on the wall to our right.

"Those black marks were made by a burning branch that was used as a torch. The visitor stopped momentarily and rubbed his torch against the wall to control the flame. We've dated the carbon in the marks; they were made about twenty-eight thousand years ago."

Someone came into the cave to see works of art that were already older than the Great Pyramids are today.

A strip of plastic sheeting has been laid down on the floor of the cave.

Again, in the interests of minimizing damage and contamination, no one is allowed to walk haphazardly across the precious floor. But I'm finding it hard to stay on track. To avoid hitting my head on the overhang, I have to duckwalk every now and then, and I'm distracted by the cave bear skulls and the stalactites. I mostly manage to stay on the plastic, but my clumsiness is difficult for Jean Clottes to endure.

Then we reach our first paintings. A bear drawn in red ochre, statically posed, about a meter across. The portrait is flat, crude.

This is not why we're here.

We continue along, and here is a hyena. The portrait is also about a meter across—and it is also flat, crude, and red.

It is OK for something so old, but not anything I would want to risk my marriage about.

We encounter another few crude paintings along the way, then pass a cave bear skull that seems to have been positioned on the chamber floor like an altar. A little farther on, and here is a set of fossilized cat prints encrusted with calcite crystals—Jean wants to know if they might be lion prints, but I say, "If they were lions, they must have been cubs. Otherwise, they're more likely to have been from a leopard."

Then we reach a broad panel in an alcove. The art suddenly leaps out from the shadows and shade like an aquatint by Goya with the lines of a Japanese print.

The images depict a group of horses, their bodies bursting with energy, their charcoal faces peering expressively, intensely, from the white limestone walls. Their heads are stylized—the muzzles smaller than in real life, the necks thicker—but they convey the essence of what makes a horse a horse. They are not frozen in time: they move.

The work of a true master, as fresh as this morning, but the panel is thirty-six thousand years old. Jean Clottes has confirmed the age with carbon-14.

Then immediately to the right of the horses, the reason why I'm here.

Two lions.

Cave lions, *Panthera spelaea*, extinct for twelve thousand years. A different species from *Panthera leo*, the African lion. But alike enough that it is possible to recognize that these two animals are not posed haphazardly. They are engaged in a complex interaction with each other.

The lion on the left is standing tall with its ears pressed back against its head, behaving dominantly toward the second lion, who is crouching, sitting on its rump, snarling defensively.

Whoever painted these animals had watched lions at least as much as I had.

The artist captured a slice of life with his charcoal and his ochre—and he must have been a very brave man.

I've only ever watched lions from a Land Rover. Lions don't pay any attention to me as long as I'm inside my vehicle. But this man didn't ride in a lion-proof box. He didn't even have the *wheel*.

If I ever do watch lions while on foot, I always rely on the powers of my binoculars. Our stone-aged artiste had no glass.

The most stunning feature of this painting is how the two lions are paying absolutely no attention to the artist. Their portraits are perfectly naturalistic. They are completely absorbed in each other. The artist had somehow spent enough time in the French equivalent of the Serengeti to understand fundamentals of lion behavior—and to recognize the significance of their postures, the positions of their ears.

We leave the lions to enter a very narrow passage—above the entrance is the stylized representation of a woman's vagina drawn with charcoal.

A few meters inside, we reach the next great panel. Two lions are walking side by side. They have been drawn concentrically. The inner figure portrays a female on the near side of the male. She is drawn entirely within the limits of his silhouette. I can tell the larger animal is a male because he has a scrotum. He is walking along with his head bowed slightly. The female is walking in a more normal posture.

The male is consorting her: he is eager to mate, and he is herding her slightly, keeping her away from potential rivals.

Jean's team hadn't realized that these are two different animals. They thought the larger outline was a copy of the inner lion.

"These are definitely two different individuals—male and female—they're a mating pair. Modern lions show this same behavior in the Serengeti."

And the most important lesson from this illustration? The cave lion had no mane.

Rock art portraits in Africa always emphasize or exaggerate the size of the lion's mane. Here is a species that no zoologist has ever seen in the flesh, and suddenly we have photographic evidence of their behavior, of their physical features that were never preserved as fossils. Interesting, too, in light of Peyton's study that when lions lived in ice age Europe, the ability to "take the heat" was not nearly as important as in tropical Africa. Female cave lions had no incentive to select their mates in a way that ultimately imposed long manes on their boy toys.

The most visceral aspect of this portrait, though, is that the lions have been drawn life-size in this womblike passageway. They are nearly three meters across, and I'm admiring them from only a few feet away.

They can only be viewed from a few feet away.

The artist clearly chose to display the male's gentle possessiveness of the female rather than the act of sex: there is nothing sensationalist in the execution of this drawing. The lions' expressions are neutral. They exist in a world of their own.

We are descending deeper now, down to the grand panel in the lowest part of the cave. The passageway opens out into the largest chamber in all of Grotte Chauvet. We have now entered the Sistine Chapel of Paleolithic art.

Incroyable.

Up along a high wall are dozens of wooly mammoths, wooly rhinos, reindeer, bison, and lions, lions, and more lions.

In the middle of the main panel is a set of four lions all staring in the same direction—drawn with such detail that you can see their whisker spots. In the Serengeti, we have always drawn the lions' whisker spots to identify individual animals, as their arrangement is as unique as a fingerprint. I very much doubt that the ancient artist used these patterns in a similar manner, but the subtle details—the very lives—of these animals are up on that wall for all eternity. And the artist has also given us the big picture about this extinct species.

Here we can see that cave lions were social, just like their African cousins. Those four cats were looking at the same thing in the same place at the same time. They were members of the same pride.

Above the attentive quartet, a single lion appears to be running after something. Its legs are extended forward, its expression focused on something somewhere up ahead. Even farther above the quartet are a series of crude caricatures—awkward sketches drawn by someone who had never watched lions in real life, someone who had only ever seen them in paintings. Some of these figures look a bit like snaggle-toothed hippos; one even looks a bit like Richard Nixon. They must be the work of an apprentice.

Then, over to the right, is a realistic lion being watched by a wary bison—in the same pose that wildebeest watch Serengeti lions. The lion is looking off at something else. The bison is looking straight toward the lion, its front left leg swiveled outward at an odd angle, ready to pivot sideways and turn tail at any moment.

We shine our headlamps farther off to the right, toward a second series of crude cartoons. The bottom figure is easily recognizable as a lion, but the two iterations above become increasingly distorted. The topmost image might even be half lion/half bison.

A lone stalactite descends in front of this particular panel. The tapering column is painted with a bison's head at the top and a woman's vagina immediately below. This part of the cave provides Jean Clottes with strong evidence for the importance of symbolic art—something that might convey a belief in magic. He hasn't been too thrilled with my constant comparisons between the high art of Chauvet and the paintings of James Audubon, but these final images are fundamentally different in tone and execution.

Audubon tried to capture the real behavior of real animals. The same spirit was obvious in the two lions in the dominant/subordinate interaction, the consort pair, the whisker-spotted quartet, and the lone lion being watched by the wary bison.

The hippos and Richard Nixon were just goofy copies by a bunch of wannabes with too much time (and charcoal) on their hands.

But these peculiar three "lions" are more chimeric, abstract. And chimeras are a common theme in rock art all around the world. Half man/half animal apparently signifies shape shifting and the supernatural powers of a shaman.

Certainly this final low chamber would have been a good place for an out-of-body experience. Stalactites and stalagmites are formed from dissolved calcium carbonate, which emits quite a lot of carbon dioxide, and the artists' work fires would have emitted a whole lot more. Carbon dioxide is heavier than air, so it sinks to the lowest depths of a cave.

We're not allowed to remain down here for more than fifteen to twenty minutes. It is easy to get dizzy in this gassy atmosphere, and there hasn't even been a fire down here in a few thousand years. (The original bits of charcoal are still here—they've run the tests.)

So imagine if you were to have visited this room back in the day, spent a few hours with a steady fire and maybe exerted yourself with a bit of chanting and dancing. The experience must have been overwhelming; the artwork must have been overpowering. Shapes must have shifted; visions must been vivid.

This place must have been holy.

We emerged from the cave into the light of a French afternoon, and my mind was on fire. *That* was an *experience*. That was *profound*.

And I'm utterly convinced that I know how our ancient master gained his insights into lion behavior.

We rejoin Susan back in an olive field by the edge of the Ardèche River. I've already lost any residual guilt for having abandoned her this morning.

I take her hand, tell her what it felt like to travel so far back in time, and then start outlining the origins of prehistoric lion science. The secret is right behind her in the oxbow bend of the Ardèche River, in the alignment of the natural bridge, which points straight at the original entrance to the Grotte Chauvet.

There would have always been a footpath up to the cave entrance. There would have always been a safe ledge from which to look down on the animals that came to drink at the river's edge. The oxbow would have been a lion hot spot. Upstream, the Ardèche River runs through a series of deep gorges, but in this part of the valley an entire pride could have lounged at the water's edge day after day, waiting for the thirsty mammoths, elk, horses, deer, and bison to drink from one of the rare stretches of river that wasn't hemmed in by narrow limestone walls.

And whenever the Pont d'Arc lions caught anything big, the humans would have been ready to chase them off with their spears and scavenge large chunks of meat. Humans have scavenged from big cats since the year dot. Fossil bones frequently show evidence of having been chewed by lions before being scraped by stone tools.

Of course, the Chauvet artist was interested in watching lions. He was even more fascinated than I am. His diet—his life—depended on understanding lion behavior.

And what insights he had! He may have started off just watching lions capture their prey so that he could eventually steal their meat, but his interests broadened. He enjoyed watching lions mate. He was interested when they fought. He was intrigued by their sociality.

A few months later, I took Susan on her first trip to Africa, and we endured the armed robbery at Barbie's. She heard enough horror stories about carjackings, murders, and mayhem over the next two weeks traveling through Kenya and Tanzania that she was convinced we were all crazy to live with such danger. But she also accepted that I would always have to work in Africa and that she couldn't always come with me—until we started Savannas Forever.

After the disaster with TAHOA, we decided to switch directions by, say, thirty degrees. Instead of working with hunters on their fake community development projects, we would turn to the big development agencies. The U.S. Agency for International Development, the United Nations Development Programme, the United Nations Educational, Scientific and Cultural Organization, the relief agency CARE—these organizations really did want to lift rural Tanzanians out of grinding poverty, really did want to improve their food security, really did want to improve their health. And they had huge budgets to do so—nearly a billion dollars of aid came to Tanzania each year.

And wouldn't it be marvelous if the big aid agencies really did improve the quality of life for rural Tanzanians? The one best hope for long-term conservation would be if Tanzania went through the demographic transition—once women attain a certain level of health and security, birth rates quickly fall from six or eight kids to the archetypal two-child family of the West. And if population growth finally abates—and, even better, if rural Tanzanians can become competitive for jobs in the cities—ecological pressures on places like the Serengeti might not only abate but possibly even recede.

Everyone has his or her own favorite story of wasted aid, of the cynicism of much of the aid industry. What if someone were to measure and then report the effectiveness of these aid projects so that everyone could learn from each others' mistakes and actually reach the point where they really did reduce poverty, improve food security, and so on and so forth?

We started hanging out in Dar es Salaam at the offices of USAID, the European Union, and the Norwegian Agency for Development Cooperation.

And what a team we made: Susan could conceptualize a strategic approach, and I could crunch a few numbers and find colleagues who knew how to study these problems properly. And we were ready to provide this amazing new service, which now went by two names: Savannas Forever in Tanzania (where we had become an officially registered nonprofit during the hunting-certification days) and the Whole Village Project in Minnesota (which more accurately reflected our focus on all aspects of village development).

My home institution helped us to an extraordinary degree: the university's Institute on the Environment awarded the Whole Village Project a four-year grant totaling nearly half a million dollars, and the president of the university provided additional funds to hire a full-time director for the Whole Village Project.

After Severre's departure from the WD, we finally received all the research permits necessary to go anywhere in the country and measure anything about village life that might reliably indicate health, wealth, and vigor of individual households. We just needed buy-in from the aid agencies and, of course, an additional million dollars or so each year to send dozens of well-trained researchers to remote parts of Tanzania and to analyze all the resultant data properly.

We needed the sort of funding one might receive, for example, from a big USAID contract to measure the impacts of their Feed the Future program.

So last year when I first came back to Tanzania just a few days after meeting with the State Department and USAID, I called the U.S. embassy in Dar and found that the head of Feed the Future had been unexpectedly transferred to the Joint Chiefs of Staff—yes, he was in Dar es Salaam at the moment, but he was completely tied up for the next few days. He wouldn't be able to meet with us.

And, yes, there would be a call for proposals for the monitoring and evaluation contract for Feed the Future, but the Whole Village Project would have to apply through an open competition along with everyone else.

The call for proposals came out in mid-April, and we were pleased to discover that USAID aimed to give preference to applicants from American land-grant universities with local ties in Tanzania that worked in close collaboration with Sokoine University of Agriculture in Morogoro.

Susan and I moved down to the Hilux Hotel in Morogoro and spent weeks drafting a proposal with three Sokoine faculty members who worked on childhood nutrition, agricultural economics, and the science of development. We knew that various other American organizations were planning to apply for the Feed the Future contract, but no one else had shown up at Morogoro.

We had the field to ourselves.

The U.S. Agency for International Development also wanted to build the capacity of the National Bureau of Statistics. Senior officers at the bureau told us how rural officials fabricated agricultural production data in their districts. We talked about how to build a countrywide system for measuring aid impacts beyond the end of the upcoming program. We were a bit puzzled, though, by what the statisticians told us about their ongoing discussions with USAID. Something seemed to be happening behind the scenes.

The U.S. Agency for International Development insisted that all monitoring and evaluation data be collated through a centralized computer portal. Our computer colleagues in Minnesota could do that sort of thing in their sleep. Everything USAID wanted, we demonstrated how to do better and more comprehensively than anyone imagined.

We e-mailed USAID our application from Morogoro and went back home to Arusha. It was the dry season by then, but we decided the best way to wait for USAID's decision was to start gardening. The lawn of our new house was still bare, the flowerbeds unplanted.

Arusha's main streets are lined with impromptu nurseries selling lilies, irises, hibiscus, hydrangeas, cycads, bamboo, daisies, ferns, cosmos, zinnias, geraniums, nasturtiums, and pentas. Our flower garden would run riot if we could only raise enough business to keep Savannas Forever alive for another few years.

Judgment day came, and we received our rankings.

The U.S. Agency for International Development graded our proposal on five different dimensions. The monitoring and evaluation activities outlined in our proposal won the highest score of any applicant. We received acceptable scores on three other dimensions, but USAID marked us down just enough on "administrative capacity" that we lost the contract by half a point.

The whole purpose of this contract was to provide USAID with independent measurements of project performance rather than to just write bland, self-serving success stories.

The winning bid was submitted by a Beltway bandit with an office ten blocks away from USAID headquarters in DC.

Everyone who has ever heard this story says the same thing. The U.S. Agency for International Development had already decided which company would get the Feed the Future contract before even advertising for it. The big aid companies have sizable lobbying budgets; they don't just yoyo into DC from Minneapolis and meet with random congressmen whenever they get a bright idea.

They actually know which strings to pull.

Although we had wanted to believe that we might be the lucky winners of the Feed the Future contract, we always knew our chances would be slim. After a few days of soul searching and heartache, I attacked our Whole Village data with a vengeance. This was not statistics for the faint of heart.

We had conducted a fifty-six-village survey for the President's Emergency Plan for AIDS Relief in 2009 and 2010. We had collected household data on socioeconomic status, food security, health, governance, and natural resource status. The idea of the Whole Village Project—as indeed it had been for our biocomplexity project several years earlier—was to start with a select set of ecological drivers and identify the intervening linkages between different aspects of village life.

I asked the data crunchers back at the University of Minnesota to e-mail the village averages for hundreds of different variables, and I sat down at my computer in Arusha, working day and night through two months of power failures and intermittent Internet.

I must have run a thousand statistical tests. But I wanted to show people what a holistic approach really looked like—and I needed to know what one looked like myself.

The key ecological drivers in northern Tanzania are rainfall and standing water. If your village has enough of both, you can be a farmer—and with sufficient rain, you can even grow coffee. Rainfall is a material blessing: with enough rain you can afford to buy cell phones, bicycles, radios, and tin roofs. If your family has wealth, you will have been sent to school as a child, as an adult you can provide your children with a better diet. If you are better educated, you know a lot more about HIV/AIDS, you make sure to send your kids to school, too, and you not only have a better diet but you are more resilient in the face of temporary food shortages.

But if you live in a dry environment, you live the life of a pastoralist. Your life revolves around your goats and sheep and cows. If you are a pastoralist, you can't let your kids go to school because you need them to tend your herds. But your herds are living in such dry areas, that they don't produce enough milk to properly feed your children. Over the millennia your ancestors have developed certain cultural practices: you rarely eat your livestock. Cows are your wealth, no matter how scrawny and disease ridden, it is quantity that counts, not quality. And you have a taboo against birds. If you were to hold a chicken, for example, you might not be allowed to touch your cows. But if you don't keep chickens, your kids don't get any eggs, and none of your family eats much chicken—the lack of poultry is one of the biggest problems affecting food security in Maasai villages.

So all these aid projects over the past fifty years have tried to improve livestock husbandry in Maasailand, and they've restricted themselves to cattle. But more cows mean more wealth to a Maasai not more food on the hoof. In fact, cows were already overgrazing and causing soil erosion across much of our survey area. Our statistics suggested another approach that would improve nutrition and leave far less of a footprint on the landscape: chicken farming.

Village life contained a number of other twists and turns that no one would have guessed without a systems ecology approach, and we shopped our results around to the donor community in Dar. By the end of the year Susan and I had managed to land a couple of small contracts with the United Nations and a British aid agency in Dar es Salaam. None of these new surveys would conform to the exalted vision of the Whole Village Project or lead to the la-la land laboratory of a 240-village grid, but they would keep our survey teams employed and finally offer Susan a modest wage.

We returned to Tanzania at the beginning of February. We were barely settled in Arusha when the horrible screech rose from our bedroom. Her shoulder hadn't been dislocated: her upper arm was shattered near the head of the humerus. The next day, we flew home to an orthopedist in Minnesota.

She hadn't needed a yank of the arm. She needed nine metal pins and a titanium plate.

SELOUS, 19 MARCH 2012

"They keep coming back again!" I'm sitting in the back seat of Henry Brink and Kirsten Skinner's ancient white Land Rover. We are driving on a dirt road at maybe twenty miles per hour, the sunshine is intense, the grass is green, green, green, and the sky is as blue as it gets.

Outside my window a dozen carmine bee-eaters are swooping down toward the front wheels of the car, catching insects disturbed by our steady forward motion. They are an impossible combination of scarlet red and metallic green, the edges of their wings are black: they look like tiny Japanese kites. They are the most beautifully colored birds I've ever seen, and they are flying along just a few feet from the car, swooping past again and again.

Henry says, "Why don't you sit on the bonnet?"

The car stops, and I hop onto the hood. Henry starts off again, and the carmine bee-eaters resume their routine. Small packets of absurdly bright color swoop down in front of me from over my shoulder. Formations of miniature space ships swoop in from the left. They dart after unseen insects close to the ground, pull up into the air again, and fall back behind the car for a few seconds before swooping past another time.

The experience is utterly exhilarating. I can't stop giggling. I am drugged, giddy. I've never felt better in my life.

The Selous Game Reserve is fifty-five thousand square kilometers—an area the size of Switzerland. We are in the Matambwe photo-tourism sector of the Selous. We have already seen a large pride of lions lounging in the shade of a flat-branched terminalia tree, encountered two packs of African wild dogs, and found three German rifle cartridges from a World War I ammunition dump near the gravesite of Frederick Selous. The weather is stiflingly hot, but the air is clear and the colors impossibly bright: every bird, mammal and tree is indelible.

Matambwe is so breathtakingly beautiful that even the WD was forced to acknowledge its value from photo tourism. The Matambwe hunting blocks were reassigned to tourist lodges in the 1990s, and the lions and wild dogs are now as tame as in the Serengeti—as, too, are the impala and hartebeest, though the buffalo (the meatloaf of the hunting industry) get nervous if we stop to look at them.

Yesterday we spent the day on a motorboat on the Rufiji River. The river was in flood, a mile wide, its surface dotted with water cabbages and driftwood. The rains have only been moderate this year. Given that, the river is surprisingly high. So many upstream watersheds have been deforested that all the rainwater that once soaked into the highland soils is now racing toward the Indian Ocean, eating away the high sandy banks on either side of the river—the roots of tall trees atop the crumbly cliffs dangle out uselessly above the rushing waters.

We landed the boat on the south side of the Rufiji and went for a walk along the dry channel of a sand river. We found a poachers' camp a few hundred meters inland. Tattered nylon nets were draped along the ground. A large green tree had been hacked into firewood for smoking fish. Bleached white hippo skeletons dotted the channel floor. All the tusks had been removed.

This side of the Selous is the iconic stronghold of the Tanzanian hunting industry. The oldest protected area in Africa, it was first set aside by the German colonial administration in 1896 and became a hunting reserve in 1905. Today, Gerard Pasanisi controls about a third of the Selous hunting blocks, including this shallow valley, but there is no sign of law enforcement, nothing but unguarded wilderness.

We pushed off from shore again and boated into an oxbow lake along the Matambwe side of the Rufiji. The mirror lake reflected every detail of every cloud in the sky. Palm trees towered above the brushy woodlands on either side. The watery landscape reminded me of the Everglades but much bigger and much brighter. Herds of impala dotted the green grassy shorelines. Giraffe browsed the lush green trees.

Two days ago, Henry, Kirsten, and I accompanied Harunnah Lyimo to the village of Mloka, just outside the eastern boundary of the Selous Game Reserve. We brought along Swahili posters illustrated with pen-and-ink cartoons showing how to protect yourself from lions: avoid walking alone at night, especially when the moon is below the horizon; encircle your homestead with bamboo walls to hide your domestic activities after dark; build barriers to protect your crops against bushpigs so you won't have to sleep in your fields.

Harunnah spent two years working with Hadas Kushnir and me on a bushpig-exclusion project in Rufiji. We designed four experimental treatments: low wooden fences strong enough to exclude hungry pigs, meter-deep trenches too wide for the pigs to jump, a system of metal drums that clang loudly when pigs hit the ropes, and a set of flashing red lights like those that repel white-tailed deer from suburban gardens in the United States.

Harunnah first came to Rufiji in 2009 to work in the villages of Kipo and Ndundu, located about thirty kilometers east of the Selous. He organized construction of about thirty treatment plots and demarcated about a dozen control plots. The results were encouraging: crops were essentially intact in the plots with fences or trenches, whereas a third of the control plots suffered severe bushpig damage.

Harunnah returned in early 2010, repaired nearly two dozen treatment plots and added another forty. Hadas and I had visited him in Rufiji that May, and while the fences and trenches had clearly worked, they also required a considerable amount of labor, and the farmers in Kipo wanted someone to build the barriers for them. In contrast, though, villagers in Ndundu had been so impressed with the fences that they built nearly two kilometers of fencing by themselves—the project in Ndundu was starting to go viral.

It was hardly earth-shattering that a fence could exclude a pig-sized animal. The question was how to scale up the program. Hundreds of farming communities live in the man-eating areas of Rufiji and Lindi. How could we push more villages to show the same initiative as Ndundu?

It rained off and on the whole day that Hadas and I visited Kipo. The villagers loved their fences but complained about the difficulties of maintaining the trenches in the sandy soil.

I asked everyone: "What would it take for you to build your own fence?"

The replies varied from, "In the rainy season, we are too busy farming; in the dry season, we are too busy fishing" to "We can't afford the building materials."

"What about trenches? They cost nothing."

No, no; it takes a team of people a week to dig a trench. Then the rains come, and the walls collapse. We would have to re-dig them every year.

"Can't you collect your own wood to build the fence?"

We'd still have to buy the nails.

"Nails can't be that expensive."

Harunnah tells me nails would cost about twenty dollars to enclose an entire field.

"Surely, that's not too much money."

"That's the problem," he says, "they don't have *any* money. These people are subsistence farmers; they grow just enough food to survive until the next crop. There is nothing left over to sell."

We greeted rice farmers as they reaped their fields with twelve-inch knives. We saw women winnowing rice on woven mats. We watched a man stack piles of rice beneath the roof of his *dungu*—an A-framed shack on flimsy wooden stilts where people sleep during harvest time. The farmer climbs down from his *dungu* whenever bushpigs come rooting around his field, but a lion can easily climb up and grab a sleeping farmer or member of his family.

People must have been farming along the Rufiji since the invention of agriculture, yet they somehow missed the invention of the fence. And it didn't seem to occur to anyone in Kipo that things could be any other way.

The next day we went to Ndundu, and, sure enough, for every fence that Harunnah had built, the villagers had added one or two of their own. They didn't worry about the quality of their poles: they went into the forest and brought back any old stick that was at least two meters long. They didn't worry about nails: they lashed the poles together with homemade twine.

Ndundu had twice as many bushpigs as Kipo and about twice as many lions snooping around their fields—no wonder the farmers in Ndundu took the problem so seriously. But why had they built so many fences? Instead of erecting a community fence around the perimeter of adjacent fields, each farmer carefully separated his own field from his neighbor's.

Harunnah wanted Hadas to meet one particular old man who had built four hundred meters of fencing. The *mzee* (old man) beamed with pride as he showed her his work. In some places, the sticks were woven together like basketwork. The wall was barely two feet tall, but it was enough to keep out the pigs. Looking down from the far corner of his fields it seemed odd that he had zigzagged through the patchwork of plots, so I asked him, "Why don't you work with your neighbors to build a perimeter fence that would protect the entire community? A village fence would require a lot less work than enclosing each separate field."

"That man over there," he said, pointing west, "is a good man. I would be happy to share a fence with him. But that family over there," he pointed east, "is all lazy. They would never look after their fence."

Mloka is the last of eleven villages that Harunnah, Henry, and Kirsten have visited this year, spreading the word about man-eaters, fences, and bushpigs. They were worried that today's meeting might go badly—two men from Mloka were shot by game rangers while poaching in Matambwe the week before. But the village leaders have been courteous and welcoming, Henry has fired up his portable generator underneath a nearby mango tree, and the dilapidated hall is packed. Sixty men and women are sitting on rickety wooden benches and chairs in the stifling heat. About half the men are wearing embroidered Muslim caps. Most of the women are dressed in headscarves and brightly colored *khangas*—large printed squares of cotton cloth. I'm standing outside the door behind a crowd of school kids in blue skirts and blue shorts.

Harunnah breezes through his PowerPoint presentation, listing the dangers of the darkness and the best ways to protect farm fields against marauding bushpigs, but his manner is a bit too mechanical. Half of his audience listens intently; the other half seems to have fallen asleep. My

Swahili isn't good enough to follow his words, but I recognize his graphs and photos.

He finishes after about forty minutes and asks his audience for questions.

Silence.

He looks over at me, and I walk in to the front of the room.

I address the crowd in Swahili: "Hello, everyone."

"Hello," they reply.

"How are you this morning?"

"Fine!"

"I'm sorry; I only speak a little Swahili. I want to say thank you very much to everyone for coming here. May I ask you about lions and bushpigs?"

My Swahili is too rudimentary, and Harunnah has to translate. Everyone agrees to be asked.

"Thank you! Now, how many people here are farmers?"

Sixty people raise their hands.

"OK, thanks. How many of you have problems with bushpigs eating your crops?"

All hands rise again.

"Do you think a fence would help?"

Only about a quarter of the hands go back up.

"Why not?"

"Someone would need to build the fence for us."

"Can't you do it yourselves?"

"We don't know how to build a fence."

"Did our pictures from Ndundu help?"

"No, we need to see a real fence in front of us."

"Our fences are still up in Kipo and Ndundu; you could go there to see them."

"We can't get there."

Kipo is thirty kilometers away; busses run from Mloka daily. "Can you take the bus?"

"We can't afford the bus."

"I guess we need to build a demonstration plot here in Mloka," I say, wondering how to pay for it.

"Ok, next question. How many of you have ever seen a lion?"

No one raises a hand.

"Have you ever heard lions roaring at night?"

Everyone raises a hand.

"Have you ever seen any lion prints around your fields?"

Almost everyone.

"Would you like to remove all the lions?"

About half say yes.

"Would you like to feel safe from lions?"

Of course.

"Fences won't stop lions altogether, but if enough of you had them, they might help."

But how to achieve that? Our funding for the Rufiji project came from Panthera, an American foundation that wants to conserve as many lions as possible, and we just spent our final few dollars to pay for these village meetings.

Henry and Kirsten have fitted several Matambwe prides with GPS collars. Their study animals rarely leave the reserve. The roaring lions around Mloka appear to be excess animals—subadults, nomads—that have fled from the powerful prides in the Selous and found refuge in the empty spaces along the Rufiji River valley. If people were to eradicate every potential man-eater from their village lands, the Selous population would be essentially untouched. The Rufiji lions have no real conservation value.

Except.

Except that people hate lions: at least half the people here would like to exterminate them. Villagers vote. Every time a lion kills a human being, someone asks why the Tanzanian government protects its lions against its citizens. Tanzanian MPs ask conservationists why they love wildlife more than people.

I frankly don't care if the villagers in Ndundu, Kipo, and Mloka eradicate all the lions in their midst—if a lion ate my family, I'd want to kill it, too.

But I do care about public perceptions, so we came to Mloka today as lion conservationists whose primary interest was in the safety and well-being of the villagers themselves. Lion populations in places like the Selous Game Reserve and Serengeti National Park will only survive the oncoming tidal wave of starving farmers if the Tanzanian public continues to tolerate wildlife conservation.

A prominent European conservationist once asked me not to publicize how many people were eaten by lions each year as it would be much harder to raise funds for lion conservation if donors realized how horrible lions could be.

But it isn't up to Western donors to decide the future of these lions.

It is up to the people in this crowded room.

I doubt whether Henry, Kirsten, Harunnah, or I will ever be able return to Mloka to build a demonstration plot. I can't expect Panthera to divert any further funding from protecting lions inside the parks and reserves.

I've tried talking to the international aid agencies about the serious threat that bushpigs pose to food security in these communities. But agricultural aid programs focus on pesticides and fertilizers, not problem animals.

I wish I knew what to do.

DAR ES SALAAM, 27 APRIL 2012

I watched too many jungle movies on TV when I was a kid. Some white guy with a pith helmet would inevitably step on what looked like solid ground and then start sinking—and the more he struggled, the quicker he sank. If he was a bad guy, down he went until only his pith helmet was left floating on the surface. If he were a good guy, somebody would eventually toss him a vine and pull him out to safety. Either way, peace would be restored, and every monkey in the jungle would be chattering happily by the final reel.

Many years later, I read somewhere that the risk of dying in quicksand was very low—as long as you relax and lay on your back, you can float.

Yesterday I decided to walk along the white sandy beach of Dar es Salaam, aiming for the Msasani Peninsula about a mile or so up the coast from Jane Goodall's house. I kept in mind the fetid stream that enters the ocean right between her house and President Nyerere's compound. Whenever I stayed with Jane in the 1970s–90s, I avoided swimming in the sea because of the odors emanating from the incoming sewage.

But when starting my seaside trek from the Italian hostel near the Mbalamwezi Beach Club, the journey was too tempting, and there were endless cowry shells and seashells that beckoned me forward along the gleaming white shore. So when I eventually reached Jane's fetid stream, I thought, "There's got to be a way around this thing."

I walked fifty meters inland atop a ridge of soft white sand, trudging over tattered plastic bags, empty bottles, old boots, used condoms, piles of seaweed, sticks, and junk, until I reached the fence of one of Jane's neighbors, and, to keep going upstream, passed through a hobo camp, stepping around palm-log benches and charcoal-strewn fire pits.

I had just finished reading the *Hunger Games*, and I've been assaulted enough times in Dar to be thinking, "This is pretty stupid." And even though the hobo camp was empty, I'm thinking, "Who knows what might be lurking behind the next mangrove tree?"

Oh, yes, the mangroves. They did provide a fair amount of shade, but they also spawned a tiny forest of air roots that grew straight up from the wet ground—hundreds of three-inch spikes every few inches for about ten yards, then a gap, then another micronursery of mangrove spikes. The stream narrowed, and I thought that maybe I could cross just over there.

The water was milky, the ground was bare, and there weren't any stepping-stones, but I just might be able to leap across, maybe, right about here. . . .

Then I start sinking into the ground. I am so determined to cross the stream that I keep going closer and closer to the water's edge, and I'm striding along maybe only sinking an inch or two—then *glooop*: one foot sinks down maybe six inches. My momentum propels my other foot even closer to the stream, and it sinks down maybe eighteen inches.

Now this is getting weird. I'm wearing my Chaco sandals, and, as I sink, they anchor my feet beneath the surface. The suction from trying to pull up one foot pushes the other foot deeper into the muck.

I finally extract the first foot, and I think, well, the sand in the bottom of the streambed must be packed down, right? So as I get closer to the water, the ground will be firmer, right?

Uh, no, quite the opposite. So the next step and I'm in almost up to my knee, but I'm so convinced now that the sand in the streambed will be packed solid, I finally extract the other foot and confidently step out into the stinking stream.

Big mistake.

Now I'm in up over my knees—both knees—and the sheer effort of trying to lift up one foot and sandal pushes the other leg deeper into the muck. And neither foot seems to have hit bottom. There is no solid foundation down there, and the muck is clinging tightly to both feet. It's a scary feeling and very primordial, something encountered in a past life.

I hadn't wanted to wade across back at the stream mouth because of all the sewage flowing into the sea. While this stuff probably counts as quicksand, there doesn't seem to be that much sand in the muck at all. It is pitch black, tarry, sticky, gooey, and, oh lord, does it stink.

Now, if the only way to survive is to lie on my back and relax, I'm sorry—that is not going to happen. I've already sunk down to my shorts; I can't stand the thought of exposing another inch of my body to this god-awful stench.

I collect myself, relax, breathe deeply and stand up very straight. Very calm and perfectly still, I start wiggling my right ankle, and after a few moments my foot comes loose from its sandal. I manage to pull my right leg up from the glistening black crud and try to step a little farther away from the stream.

I don't want to abandon my shoes—if I manage to escape, I'll still have to walk home. So I reach down to my elbow with my right hand and fish out the sandal. I start to repeat the whole procedure with my left ankle, but the left sandal is down deep, really deep, and when I first reach all the way down with my left hand, I mistakenly pull up something hideously slimy from the stinking depths. Ugh, a large plastic bag. Reaching down even deeper, I finally locate the sandal, pull it out, step away from the stream, and my bare foot only sinks in about six inches. After a few paces, I make it back up to the solid ground by the mangrove roots and eventually return to the white sands of the beach.

I looked like the Tar Baby from the waist down. And I smelled unbelievably bad. I cleaned off my hands in seawater to immortalize my stupidity with a few iPhone photos, by which time the muck had mostly washed off my feet. In the pictures, I look like I'm wearing a wet suit. But that isn't neoprene rubber, kids, it is excrement.

I never did make it to the peninsula. I returned to the hostel and showered, showered, showered, and showered some more.

In my youth, I visited the La Brea Tar Pits and puzzled over the remains of the saber-toothed tigers. They say that saber-tooths were social, like lions—some of the evidence derives from the sheer number of fossilized saber-tooth bones compared to other large carnivores from the same time period. The idea is that social carnivores will readily respond to distress calls of large prey whereas solitary predators prefer to dine alone. So when one member of a saber-tooth pride heard a struggling mammoth, they all ran in and got stuck, too.

I don't know whether my primordial fear in the muck was due to an atavistic response to quicksand or to tar pits. The color was right for tar, but if it had really been tar, I'd still be there.

Saber-toothed tigers survived in Los Angeles until the arrival of the first Native Americans. Ancient humans were never swallowed by the tar pits. Maybe the men always hunted in packs. If someone got stuck while scavenging entangled prey, surely his comrades would have thought to throw him a vine.

Or maybe the men always traveled with their families.

Maybe the next time I try to walk to the peninsula from the Mbalamwezi Beach Club, I should bring someone with me. If I'd gone with my wife or my daughter, there is no way they would have let me do something so stupid.

I couldn't help wondering if the consequences of male recklessness might emerge from a statistical analysis of our man-eating surveys. Perhaps unconnected males are more likely to get munched than anyone else: teenaged boys and old men behaving thoughtlessly with no one to rein them in.

This morning, I fired up my computer and started looking at Dennis and Hadas's man-eating data from Rufiji and Lindi. They recorded the age and sex for hundreds of victims. Looking at sex ratios across different ages, it turns out that lions attacked equal numbers of boys and girls up to the age of eight. At all older ages, boys and men were attacked more frequently than same-aged girls and women—not very surprising since males generally engage in more outdoor activities than females do.

But the sex bias in lion attacks clearly peaks in two specific age groups: seventeen to twenty-four years—when young men are notoriously irresponsible—and men older than fifty-six years of age. Lions attacked men in both age groups about three times as often as women of comparable ages.

Alas, I belong to the older age group, and I imagine that many of my Tanzanian counterparts are similarly inclined to wander off from the protection of hearth and home. Think of the widowers, divorcees, or cantankerous old coots who refuse to be tied down by the cautious fussing of their families—ornery old geezers who refuse to behave themselves after dark, even during outbreaks of man-eating lions.

For it is darkness, you see, that gives the lions their courage around people. It is darkness that allows their superior senses to overcome whatever technological advantages we might possess. I've watched plenty of lions at night in the Serengeti, and they are like a different species when they encounter humans in the dark. During the day, they remain hidden in the bushes when pedestrians pass by, and lions will run like crazy if there is no cover, no sanctuary from anyone carrying a weapon.

But there is nothing secretive about a lion's behavior on a moonless night. There is no skulking, no need to hide. The lions own the darkness. Darkness gives them strength.

The Kalahari Khoisan still live the traditional hunter-gatherer lifestyle of our ancestors, and they frequently hunt at night—but only when the moon is above the horizon and bright enough for human eyes to detect shapes and movements. Without the light of the moon, say the Khoisan, the night belongs to the lions. So they divide the night with the lions according to the phase of the moon.

And if there is a lunar eclipse? That is just a hungry lion, placing her paw in front of the full moon, stealing a little extra darkness.

Like in many other parts of Africa, the lions in the Serengeti and the Crater don't feed very successfully on bright moonlit nights. One of my PhD students, Ali Swanson, found that lions had fuller bellies around the new moon and were thinnest at the full moon. To compensate for the hardships of moonlit nights, the lions were more likely to scavenge or hunt during the day. Serengeti herbivores are available at all hours of the night, so the lions were just as lean in the last few days before the full moon as in the first few days thereafter.

Maybe you've never spent many nights contemplating the full moon. During the most intensive field study of my career, I took turns with David Scheel, watching lions twenty-four hours a day for four days in a row just before the full moon then a day off for the full moon then a second four-day follow the next night.

I did the seven-to-one shift—7:00 in the evening to 1:00 in the morning and 7:00 in the morning to 1:00 in the afternoon. I loved the first four-day follow each month: the moon was always up at sunset and grew brighter each night. With my primate eyes, I could look around, admire the scenery and more or less watch the lions without having to use our army surplus night-vision device.

On the nights before the full moon, the moon sets a few hours before dawn, leaving several hours of darkness before sunrise. But I was almost always asleep by then—as are most of us, most of the time. Humans universally stay up until around 10:00 P.M. or so and then we sleep until sunrise. If we're going to be out and about at night, we'll be out in the evening, not in the hours before dawn.

The full moon was our night off, and I could spend the evening at home with my kids. You can practically read a book in the light of a full moon in the Serengeti. It was a nice break, but there was always something cold about the full moon, something portentous.

The second four-day follow started the very next night after the full moon, and I always dreaded the evening shifts during round 2 each month. At first, the moon wouldn't rise for nearly an hour after sunset; by the fourth night, the darkness persisted for nearly three and half hours before the moon finally rose.

We were trying to observe every tiny thing the lions ate, so I had this chronic gnawing anxiety that I would miss a kill during that lengthening window of darkness each night. These were the first hours of real darkness in several days, so the lions were hungry, and they got moving shortly after sunset.

After Ali found the statistical trend between moon phase and food habits of the Serengeti lions, I analyzed Dennis and Hadas's data on the precise timing of lion attacks. People remember these events so vividly that over 90 percent of informants could recall the precise time and date of the attack, producing a sample of 450 cases.

The results were stunning. The vast majority of lion attacks occurred between sunset and 10:00 P.M., and while the last few nights before the full moon were the safest, the first few nights *after* the full moon were three-and-a-half times as dangerous. After enduring the bright evenings prior to the full moon, the lions were hungry, and they mostly attacked people on the return of evening darkness.

So what if the full moon doesn't actually make you crazy but just makes you a little nervous? Puts you off going outside for a couple of nights? What if your reaction to the full moon kept you out of harm's way?

The full moon isn't dangerous in itself, but after a few million years of dividing our nights with lions, it would be surprising if we *didn't* somehow sense the monthly dividing line between our time and lion time.

DAR ES SALAAM, 3 MAY 2012

This afternoon, I caught a taxi in downtown Dar es Salaam at around 5:15 P.M.—the worst possible time for traffic—and I asked the driver to follow the slow trickle of cars north for maybe eight miles to the hostel where I've been lodging for most of the past month.

My cell phone rang. I took one call and then another, and, next I knew, a couple of hours had passed. The sun had set, the evening was dark, and we had only traveled a few miles. We were barely at a crawl, when the driver turned around and said something incomprehensible. It sounded like he was asking about "a line"—and I assumed he meant something about the phone line. But he shook his head and said, "Do you know in English, 'uline?'"

After a second of puzzlement, I realized he wanted to say "urine." We had been in the car for nearly two hours. He was a quiet pleasant man in his forties, and it must have been extremely uncomfortable for him to raise such a humiliating subject with an *mzungu* passenger.

So I said, "You need to urinate?"

He nodded vigorously.

"Oh, OK, would you like me to drive?"

We were hardly moving, so I hopped out of the car and opened the door to take his place behind the wheel. He walked briskly along the side of the road, looking for a bit of cover. He was wearing pale trousers and a pale shirt that stood out in the headlights of the cars behind us. He continued on for about thirty seconds then disappeared behind a clump of bushes.

Meanwhile I drove the car forty, fifty, sixty meters in the slow flow of traffic.

He reappeared from the darkness and hopped in to the passenger's seat. We both laughed with embarrassment as I drove another hundred meters or so. He was so surprised and relieved that I think he would have let me drive all the way to the hostel, but I didn't have his patience for endless traffic.

ZANZIBAR, 13 MAY 2012

The two-story houses of Stone Town are stitched seamlessly together along either side of a maze of cobblestone lanes so narrow that the walls block the rays of the midday sun. Once upon a time, the stucco walls were snowy white, but they have been streaked gray by the relentlessly stifling humidity and two centuries of tropical rainstorms. Heavy wooden beams frame the iconic Zanzibar doorways, some of which are carved with stylized flowers and vines, while others are more geometrical and abstract. A few have passages from the Koran in flowing Arabic script. The doors themselves comprise a grid of heavy teak panels that are often punctuated by pointed brass knobs meant to deter foreign invaders from using elephants to push their way inside.

As far as I know, no elephant has ever set foot on the island of Zanzibar, but it's a nice touch.

Local people crowd the narrow streets: men in long white cotton gowns and white skullcaps embroidered with gray or brown geometric designs. About half the women wear long black robes while the other half wrap themselves in colorful *khangas*. Every Zanzibari woman covers her head, and many of the black-robed women brighten their outfits with headscarves of orange, green, purple, yellow, or blue. But a few women cover themselves completely in black, including a *niqab* (face veil).

Though Islam is the dominant religion here, it has an unfortunate history in Africa. The slave trade from western Tanganyika ran through Zanzibar to Arabia for a thousand years. Arab merchants built Stone Town. The British came in the nineteenth century to quash the institution of slavery, but they accepted the sociocultural status quo, leaving the Arabs (and, increasingly, the Indians) to run the place. When the British departed in December 1963, the newly independent country of Zanzibar held its first elections. The political party of the black Zanzibaris won the popular vote, but electoral districting (and political chicanery) left the Arabs and Indians with the majority of representatives and a continued grip on the island's economy.

Tanganyika had gained its own independence in 1961, but British military officers remained behind to serve in the Tanganyikan army until an African elite could be trained to replace them. In January 1964, African soldiers mutinied against their British officers in the barracks of Dar es Salaam, mostly over pay disputes with the Tanganyikan government.

Inspired by the mutiny on the mainland, the disenfranchised black Zanzibaris went on the rampage and slaughtered thousands of Arab and Indian men, women, and children.

The army mutineers on the mainland then killed a hundred Arabs and Indians in Dar es Salaam. At the request of President Nyerere, a British naval force quelled the army mutiny the following week. Tanganyika merged with Zanzibar a few months later and became the United Republic of Tanzania.

A Tory MP in the British Parliament claimed that both the army mutiny and the Zanzibar revolution were part of a larger Communist plot. Though the British press quickly dismissed his views, the damage was already done. Such suspicions in London helped drive Tanzania down the path of nonalignment and flirtation with the Eastern Bloc. Zanzibar was the first country in the world to grant diplomatic recognition to East Germany, and, to this day, Cuban doctors still help train health-care workers in Zanzibar.

Few tourists travel here in April and May because of the rains. Most wet season visitors are vacationing aid workers. You can go to the House of Wonders, which accommodated the administrative offices of the Zanzibar sultanate. You can pay three dollars to visit the Anglican Church, which was constructed by the British at the site of the slave auctions. Inside the grounds, there is a sunken monument with statues of half a dozen Africans in chains.

There is even a room below the priory that claims to have been the cramped chamber where fifty to a hundred slaves at a time were kept until they could be sold and shipped to Arabia. But I've also heard that this particular room was actually a wine cellar built by the Anglican clergy after slavery had been abolished.

I enjoy wandering the narrow maze of streets of Stone Town adorned by the Zanzibari women in their headscarves and robes. Most of the shops in the main tourist drag sell cheap tacky souvenirs—boring woodcarvings and gaudily wrapped packets of spice—though a fair number sell fine Zanzibar chests that echo the carved woodwork and brass knobs of the Zanzibar doors. A small selection of clothing stores offers fine local fashions, and a women's cooperative sells patchwork pillow covers with intricate Islamic designs.

And if you walk all the way up to the end of this narrow lane, you'll see a sign that reads:

MALCOLM

Bro. **X** St.

Go right for about ten paces, and you'll find an open door draped with t-shirts ("Tin Tin in Zanzibar" and "Tin Tin in Serengeti") and a collection of painted metal signs: the Fantastic Four (Mr. Fantastic, Johnny and Sue Storm, and the Thing); an American girl from the 1950s drinking a Coca Cola; three German nuns drinking Afri-Cola in 1968.

Step inside and the entryway is littered with shiny things: tinsel-covered children's toys, papier-mâché bowls, brightly colored *stuff*. But then go past the cash register, and the shelves are lined with—what?—antiques? junk? treasures?

On the wall, a collection of old padlocks—really old padlocks from Asia and Arabia. Padlocks with weird keys, weird designs—brass padlocks, iron padlocks, rusted, corroded. Some still work. Some cost $500 and might even be a bargain, others might be cheap copies of a traditional design. Dozens of old license plates line the walls, more metal signs, wooden carvings of Tin Tin and Captain Haddock. A brightly lit glass case filled with Damascus steel daggers with jade handles and gold inlays, silver necklaces, pewter candlesticks, beaded jewelry, and more and more and more.

The dimly lit side rooms are loaded with dozens of worn-out wristwatches, their crystals shattered or clouded with age. Here's a weird brass dinosaur on wheels; one of the wheels is missing. There's a cast-iron clown's head, garishly painted and a hundred years old. Old Coke bottles, snuff tins, samovars, film cameras from the 1950s, wooden carvings from Africa, India, and the South Pacific, Chinese plates, tarnished silver spoons, antique glass bottles, black-and-white photographs in Zanzibar frames, brass diving helmets, brass doorknobs, teacups, medallions, brass things, bronze things, thousands and thousands of things piled high on every shelf, like a mad hybrid between a flea market and Aladdin's cave.

The overwhelming clutter and the dense musty atmosphere may make you dizzy, but if you find something in here that really interests you and you show enough curiosity, you might end up talking with Salim, one of the two brothers who run the place. Salim might even take you to see the good stuff in his special storerooms upstairs or across the street—the museum quality carvings and bronzes from Benin, Nigeria, Ghana, Cameroon, Chad, Burkina Faso, Congo, Sudan, Angola, and the mainland of Tanzania—where every piece has a story.

This bronze figure conveyed the power of a chief, that terracotta figure cast a magic spell; this wooden carving not only makes music but also has a hidden chamber for preparing deadly poisons. Some of the carvings are two meters high; one of the wooden masks looks like it emerged from the center of all bliss; two of the bronze leopards are almost life sized. The shelves here are orderly, organized by material and country of origin. Everything is special, no missing wheels and no junk.

It is like going into the Smithsonian and being allowed to handle everything, feel the heft of the metal, stroke the texture of the wood.

As you tiptoe between the large objects on the floor, Salim provides a running narration, happily explaining the origins of every single item, its cultural significance—who made it and why—and the technique by which it was made. He is a walking encyclopedia of anthropology, metallurgy, and colonial history. Salim is a sixth-generation Zanzibari, black-haired and olive-skinned. His ancestors have been shopkeepers here for hundreds of years. He is thirty-six years old, handsome, proud, gently smiling, and serenely sincere in his enthusiasms.

I say to him: "I have some friends in America who make TV documentaries; they would find your shop utterly fascinating, but they would be especially interested in you—in all the things you know about all these pieces of art. You would make a great story. Would you mind if I told them about you?"

He is horrified. "No, no, no. I would never allow that."

"Why not?"

"You see all these things in here?" he asks with a lilting Indian accent. "They are my life. It took me years and years to find all these things—these are real—the real stuff. No one makes them anymore; I travel across Africa every year, and every year I find less and less. If someone came to Zanzibar and made me famous, there would be a big rush and everything would be gone.

"And then what would I have to do?

"I'm not here for the money. I enjoy what I do," he beams. "I am much happier selling one thing at a time. If people just want something without appreciating it, I don't even want to sell it to them. But if they do appreciate it, if they really want to understand what it is all about, then I will be happy."

A half mile long the main from the House of Wonders is a narrow building called Mercury House, where the sign claims that Farrokh Bulsara lived until 1954 when he went to boarding school in India. Bulsara returned home in 1962 but fled to England with his parents and sister in the wake of the Zanzibar Revolution in 1964. He later became known to the world as Freddie Mercury, lead singer of Queen.

In 2004, on the other side of the House of Wonders, someone tossed a grenade into the Mercury Restaurant, another spot that claims to have been the home of Farrokh Bulsara. In 2006, a local Islamic organization demanded the cancellation of a planned celebration of what would have been Mercury's sixtieth birthday, complaining that the notorious bisexual had degraded Zanzibar "as a place of Islam."

Carry on farther up the road past a number of high-end tourist shops, and here is Memories of Zanzibar, the largest shop in town. The showroom is open, airy, and filled with racks of beachwear and clothing from India and shelves and bins stocked with run-of-the-mill souvenirs and mass-produced trinkets. Scattered about are a few antique wooden carvings.

The carvings are midlevel: not the usual junk in most of the mainstream stores, but not up to Salim's private stock, either.

I find a vaguely interesting antique that doesn't have a price sticker. The Indian manager is wearing a blue-green sari and black-rimmed glasses. She quotes me a price, but I'd rather ask her about Salim.

She knows him, of course. Zanzibar is a small town—especially for Indians. She tells me that Salim and his brother are both unwell. They have diabetes. She buys her antiques from Salim. He also has the heavy equipment to manufacture plastic bric-a-brac, so she gets most of her shiny things from him, too.

I mention Salim's philosophy of slow steady sales instead of get-rich quick.

She agrees. "My brother and I run this place. We used to have multiple shops around town. But then we just had one other shop over by the House of Wonders. It was too much, so we closed it down. Who wants to work all the time? To get rich? Why? I wouldn't have any time left for my family."

I go for lunch in a vegetarian restaurant on the edge of Stone Town. A South Asian couple comes in and sits at the next table. They are in their early thirties. They are wearing Western clothes and have no trace of an Indian accent. They struggle to choose between the selection of dishes, so I take the liberty of recommending a few favorites.

They invite me to join them at their table. Amin and Mia have flown down from the United Arab Emirates—their families originated in India, moved to East Africa a century ago, but then emigrated to Canada. They grew up in Toronto.

Tanzania's race relations were only superficially calmed by the aftermath to the mutiny and the revolution. Frustrated by the continued disparities in wealth between Asians and black Africans, the central government selectively harassed small Asian businesses and placed strict quotas for Asian students in institutions of higher learning. A large proportion of the Asian community left for the United Kingdom and Toronto during the 1970s.

Those who stayed tend to keep one foot out the door in case the Africans decide to make life too miserable for them. Tanzania provides no incentive for the Asians to reinvest any of their wealth into the well-being of the country, producing a pernicious diaspora. Every Tanzanian Asian has family in Canada, Britain, Australia, Pakistan, or India, and every family has a bank account somewhere safely outside Tanzania.

Amin's parents got out forty years ago, and he has only visited Tanzania a few times in his life. He wants to develop online science curricula for school kids in the U.A.E. that could be directly related to the ecology of Tanzania's national parks.

I tell him about our new camera-trap surveys in the Serengeti. The cameras generate a million pictures a year—we have posted our images online so that people can help identify all the species in each shot, and we want to organize the IDs as part of an ecology curriculum for high school kids. We're so far only planning to work in America, but it would be great to establish parallel efforts abroad. . . .

Amin says he would like to bring students and teachers to the Serengeti to work hands-on with the camera traps. The U.A.E. is rich—they could easily support trips to Tanzania. It's not far from here, after all. . . .

Yes, it's a great idea, and I'm excited. All well and good.

Today I'm far more interested in talking with Amin's wife. She enjoyed her lunch; she is pleased that I've bonded with her husband.

And she has useful contacts in the highest levels of the U.A.E. government.

In 1990, I asked for a temporary ban on lion hunting in Loliondo Game Controlled Area; the now-defunct hunting company had exceeded its annual lion quota by the second week of the season. The Tanzanian Parliament intervened. Markus Borner organized a helicopter survey that found all of our Serengeti study lions but not a single lion in Loliondo. The Loliondo operator, a rogue white hunter from South Africa, lost his block to Brigadier M. A. al-Ali, the vice minister of defense in the U.A.E., who had close ties with Tanzania's second president, Hassan Mwinyi.

The brigadier's hunting company proceeded to antagonize the Maasai living in and around the hunting block. Unlike a game reserve, where people are excluded in the same way as a national park, local people are allowed to live inside a game controlled area like Loliondo as long as they don't convert too much of the land to agriculture.

But the Arabs quickly transformed a large part of the game controlled area into a no-go zone. Maasai told local news reporters about being forcibly relocated out of the Brigadier's core hunting area and about having their houses burned down by the Arabs' guards.

The Arabs were obsessed with privacy. None of the other hunting companies knew anything about them. When Karyl Whitman tried to survey an area close to Loliondo in 1999, the brigadier's game scouts fired shots over her head. A few years later, I went by Loliondo with Sarah Cleaveland while she was setting up the rabies vaccination program around the Serengeti. She wouldn't even consider taking a shortcut through the Arab concession back into the park.

The secrecy fueled the rumor mill. The Arabs came to Tanzania to escape the harsh strictures of Islamic morality and brought hookers and vast quantities of alcohol. They shot animals with machine guns while driving drunk in their SUVs. They shot animals inside Serengeti National Park. They crossed the border to Kenya and shot animals inside the Maasai Mara National Reserve. They had a C-5A transport plane that brought all their supplies directly into Loliondo without going through customs. The U.A.E. was one of the few countries in the world without prohibitions against importing unprocessed meat from Africa—their transport plane flew home each year stuffed full of antelope meat.

Markus Borner actually got to know the aide de camp of the brigadier, who assured him that the royal family only came to Tanzania with their falcons. Falconers only ever catch rabbits or other small mammals and made no significant impact on the ecological balance. In fact, by removing the Maasai and their livestock, the Arabs had effectively expanded the limits of the Serengeti National Park.

I met the brigadier's Tanzanian business manager a few years ago in Arusha. He was a tall gracious middle-aged African who complained bitterly about his company's poor reputation within the country. The Arabs had invested millions in the local communities. They had built schools and clinics, and all they got back from the villagers were complaints and court cases.

Our first round of village surveys for Savannas Forever in 2007 included two communities that were affected by the brigadier's rural development projects. We were surprised to see that the Loliondo villages had the most positive attitudes toward hunting companies—even higher than villages associated with TGT or Paul Tudor Jones. But then TGT and Paul Tudor Jones's management staff arrested hundreds of poachers each year, whereas the Maasai don't eat bushmeat so there may have been less reason to resent the Arabs' antipoaching units.

When we obtained the block-by-block hunting offtakes from Tarimo, we found that the brigadier's lion and leopard offtakes were well within the recommended limits and their harvest trends appeared to be stable. Assuming that they weren't smuggling out anything illegally in their C-5A, we saw no evidence of overhunting. At the same time, their lion and leopard offtakes were indisputable evidence that they didn't come to Tanzania merely to exercise their falcons.

There is no doubt that their acquisition of the Loliondo block had been outside the normal rules. Severre never seemed to have any direct links with the Arabs. They seemed only to be beholden to each successive Tanzanian president and his inner circle.

In 2005, Hassan bin Talal, prince of Jordan, brokered a deal on behalf of Hamed bin Zayed, son of the sultan of the U.A.E., and Mohammed bin Zayed, head of the U.A.E. airport. The bin Zayed family was allotted a hunting block in Yaeda Chini (located just south of the Serengeti) in return for a large campaign contribution to CCM. The bin Zayed hunting company was simply called UAE.

Extremely dry and inaccessible, Yaeda Chini holds a few hundred hartebeest and gazelle, but it is the last home of the Hadza, Tanzania's only remaining tribe of hunter-gatherers. I once met an anthropologist who lived, hunted, and ate with the Hadza. He told me how the entire clan shared the same space at night, talking in the darkness until everyone dozed off. Every now and then someone would wake up, and conversations would resume until everyone went back to sleep, over and over again until dawn. I wish I had known enough about man-eaters in those days to ask him if the Hadza, too, divided the night with lions.

The Hadza once roamed far and wide, but they had no sense of land ownership and lost their best lands to growing populations of farmers and pastoralists. When they came into close contact with agriculturalists, they showed the same dispiriting tendencies toward alcoholism as the Aborigines, Khoisan, and Native Americans.

Assuming that the Arab hunters would export antelope meat to the U.A.E.—as they supposedly did in Loliondo—Yaeda Chini would be stripped of its few remaining herds of wildlife, and the last stronghold of traditional Hadza lifestyle was at risk of extinction. Human rights organizations were prepared to take the case to court.

The bin Zayed family was extremely wealthy. Their agents promised to protect the remaining herds and to assist the Hadza with economic development. But whether UAE would have followed through on these promises will never be known.

The Yaeda Chini scandal generated so much bad publicity that UAE backed out, leaving the traditional Hadza hunting grounds vulnerable to the growing demands for agriculture. Over the following six years, farmers from neighboring communities converted much of Yaeda Chini's wildlife areas to onion fields.

The UAE hunting company had more than enough money to have restored the wildlife of any block they were allotted. With the proper oversight they certainly could have afforded to maintain the highest conservation standards.

"Look," I said to Amin's wife, "I have no idea how corrupt the Brigadier and UAE might have been in obtaining these blocks, but there has been so much bad press—and such a long history of suspicion toward Arabs in Tanzania—that it would be a good idea for the big shots in U.A.E. to try to open things up a little. Transparency and open dialog would surely help win over a few hearts and minds out here."

She assures me she knows high-level staff in the prime minister's office back home who are strong advocates for conservation, and she knows another man who is called the green sheikh. She'll follow up when she gets back home.

I thank her very much, but if I actually believed that she could get me inside the royal house of the U.A.E., I could also make myself believe that Farrokh Bulsara had been born twice on the same day, a mile apart, at Mercury House and the Mercury Restaurant.

6 Fences

ARUSHA, 14 JUNE 2012

In the spring of 1995, I was invited to Las Vegas by MGM Grand Hotel to develop a corporate connection with lion conservation. Leo, the MGM lion, was one of the most famous logos in the world, and the Grand kept a family of African lions in a glass-walled display area surrounded by banks of slot machines and a stage devoted to amateur Elvis impersonators.

This was the year following the distemper die-off in the Serengeti, and MGM Grand wanted to help protect Africa's lions. Their fund-raising strategy consisted of charging visitors $10 apiece for being photographed in the casino with a lion cub.

They needed a worthy recipient for their philanthropy, and they also needed an outlet for all those photogenic cubs—did I know anywhere in Africa that could use a steady supply of subadult lions?

I met with the CEO in his office at the MGM Grand and gave him a slide show about our lion research projects, our plans for a dog vaccination program outside the Serengeti, and a brief outline of a few other topics I wanted to explore.

He looked me in the eye, pointed his finger at my chest, and said, "I want you." The Serengeti was the most famous lion park in the world, and I was the best-known lion scientist. He would donate funds in the next few days.

Then we got down to the real reason for my visit.

MGM Grand has the lion logo, and they have the lion display—they want to be the premier destination for big cats in Las Vegas.

But they only rank a distant second on the Vegas Strip.

The Mirage has signed Siegfried and Roy (Masters of the Impossible!) to an extended contract for over $50 million per year. Siegfried and Roy's act features disappearing tigers and materializing lions: they are the top animal attraction in town.

MGM Resorts have an exclusive contract with Cirque du Soleil at Bellagio, their newest hotel and casino in Vegas. A creative genius at Cirque du Soleil wants to combine animals and acrobats for an innovative new show at the MGM Grand—a totally new concept that could potentially provide an inexhaustible source of funding for my lion research. . . .

That night, I walked over to the Mirage, past the sky-high water fountains and the multistory neon signs, escorted by a handler from the MGM Grand who had once been an agent for Siegfried and Roy and didn't want to be seen working for their cross-town rival. So Mr. Serengeti Lion, here, is now traveling incognito, working as an industrial spy at a dinner table in a Vegas showroom as the floor lights go down, and there on the spotlit stage is a long line of seminude young women prancing around with rhinestones and pearls when—poof—Siegfried and Roy appear from nowhere—and these two Germans strut across the stage in their sequined outfits, exaggerated codpieces and pageboy haircuts.

They take turns talking about how they are such great conservationists since they have done so much to preserve the world's few remaining white tigers and white lions (*"Extinct in zee vild!"*). And among all the chintzy Vegas glitz and hype, they would, at regular intervals, do something absolutely incredible with a white tiger or make a white lion (snow white mane, snow white body) suddenly appear in midair against the coal-black backdrop of the stage curtains.

After the show was over, my fellow spy and I filed out of the Mirage with the rest of the crowd, and we passed the holding area for the white tigers: a mock-up of the Taj Mahal decorated with paintings of voluptuous women in veils and saris that somehow implied the Royal Palace was the natural habitat for these rare animals

I turned to my handler and asked, "What am I doing here?"

He looked puzzled, so I said, "That was gloriously entertaining, the audience loved every minute of it, but I am a scientist." *Am I supposed to come back to Vegas wearing a codpiece?*

"Vegas is changing fast," my handler explained. "Next year we expect to earn more money from entertainment than from gambling. We're trying to turn the city into a permanent world's fair, a place where people can come again and again, year after year. And if they leave here feeling like they've learned something—like they've gained a bit of education—they'll feel even better about coming back."

When I went back to Vegas this January, I didn't see much sign of educational uplift for the world at large, although MGM did donate a hundred thousand dollars for my research before the CEO was replaced amid an economic downturn, and the new boss didn't see any virtue in lion conservation.

The animal-acrobat show from Cirque du Soleil never got off the ground and the Elvis impersonators seem to have fallen out of vogue, but the glass-walled lion area in the MGM Grand Hotel is still there. I spent half an hour watching a female lion entertain herself by stalking children through the glass wall of the gift shop. Siegfried and Roy are gone now, of course. Roy was severely injured onstage by one of his white tigers in 2003.

White tigers and white lions are not legitimate subspecies—they are genetic mutations that have been perpetuated in captivity because their coloration is so popular with zoo-goers. They aren't albinos—their eyes are blue instead of pink—but a white lion will only breed true if it mates with another white lion, so they are becoming increasingly inbred.

I don't know what MGM Grand ever decided to do with their excess cubs. There is no shortage of captive lion cubs in the world. The problem for lions is not reproduction—it is finding a place where they can live in which they won't get poisoned, speared, or shot. Should circumstances change so that lions could recolonize their former hunting grounds, the best colonists would be wild-caught lions from an African park or reserve—a lion that already knows how to fend for itself in the real world, not a lion whose idea of hunting consists of stalking children through glass walls in the gift shop of a Las Vegas casino.

The idea of releasing captive-born lions into the wild has resonated with the public ever since the 1950s, when George and Joy Adamson first released hand-reared Elsa back into the wild.

Elsa quickly got sick and died.

The Adamsons then released Elsa's orphaned cubs, Jespah, Gopa, and little Elsa, into the Serengeti.

The orphans faltered, sickened, and disappeared.

Joy hand-reared a cheetah and a leopard. George rehabilitated lions in the desert.

None of their cats survived as truly wild animals.

The *Born Free* strategy has been endlessly repeated in Zimbabwe, Zambia, and South Africa over the last fifty years but to no lasting effect. Individuals come and go; they blink out in a flash. It is the future of a *population* that really matters for the survival of a species.

But once we start talking about lion populations, everything gets too vague, too diffuse. Eyes start to glaze.

How I wish someone could pick up and cuddle the Selous or the Serengeti.

*

This morning I gave a presentation at the Kibo Palace Hotel in Arusha to about forty Asian, African, and European safari managers at the 2012 meeting of the Tanzanian Association of Tourist Operators. These are the businessmen who hire the safari guides who escort the clients who spend thousands of dollars for the trip of a lifetime to the Serengeti, Zanzibar, Kilimanjaro, and Selous.

The IT guy from the hotel had a bit of trouble getting my PowerPoint to work properly on his computer at first, but we're soon off.

First slide: cover photos of books by my predecessors on the lion project (*The Serengeti Lion* by George Schaller [1966–69]; *Pride of Lions* by Brian Bertram [1969–73]; *Lions Share* by Jeannette Hanby and David Bygott [1974–78]) and photos of my current field staff.

Second slide: the benign perception of lions in Western popular culture. Color pictures of the Cowardly Lion and Dorothy; cartoons of the *Lion King* with his bride-to-be and with his friend the warthog; photos of European children hugging fluffy lion toys; lions cute and cuddly.

Third slide: a collection of folk art from around Tanzania showing Maasai warriors spearing lions; fierce black men in loincloths bludgeoning lions with clubs, shooting lions with arrows, carting away decapitated lion carcasses.

Next slide: a photo taken in 2007 at a Maasai cattle boma near Tarangire where two cows were killed by lions—then click—photos of the three hundred spear-carrying morani who set out within three hours to kill the cattle killers.

I haven't got enough time today to mention the case from just last week where Maasai grazed their cattle inside a privately managed conservation area. Lions killed a couple of cows; the Maasai retaliated. The government investigated, and a morani confessed and was promptly arrested. The next morning, three thousand Maasai showed up at the jailhouse and threatened to burn the government buildings to the ground. The lion killer was promptly released.

Next slide: the results from Dennis Ikanda's research in the Ngorongoro Conservation Area. Some Maasai livestock herds are tended by spear-carrying warriors, others by children as young as seven years old. Lions preferentially attack the children's herds, as well as those tended by relatively few herders.

We have pointed out both these problems to Maasai elders, but this wasn't news to them. "Our morani have started leaving the NCA for employment in town: there is a shortage of adult herders. What else can we do?"

Next slide: Tarangire National Park only encompasses 20 percent of a vast migratory ecosystem. Bernard Kissui has found that the Tarangire lions only reside inside the park during the dry season, spending the remaining seven months in Maasailand, where they eat a lot of livestock. Leopards are too sneaky to get caught for livestock theft and hyenas run too far too fast, but lions stand their ground near the scene of the crime. For every cow killed by lions, a lion is killed in retaliation, and there are a lot more cows than lions in Maasailand. The Tarangire lions have declined by 25 percent in the past ten years.

Livestock attacks often occur at night: lions charge the cattle bomas, the livestock panic and break out into the dark. By selling a few cows, the Maasai can raise half the necessary costs to reinforce their bomas with chain-link fencing; Benny covers the other half with funds from the African Wildlife Foundation. After six years, Benny has reinforced about 130 bomas, but the Tarangire ecosystem is so large that complete coverage will probably require fencing another *eleven thousand* bomas.

Next slide: lion population trends in Ngorongoro Crater where tourists typically see their first lion. Back in the 1980s, we found clear evidence that close inbreeding in the Crater lions was having measureable effects on their reproductive health—as illustrated by the two-headed sperm shown on this same slide. However, inbreeding not only increases the rate of sperm abnormalities but also increases susceptibility to disease. If every animal in a small population shares the same genes through common descent, a pathogen can easily spread from one animal to the next, and the Crater lions have suffered a number of serious disease outbreaks in the past twenty years.

Whereas only eight thousand Maasai lived in the NCA in 1963, there are eighty thousand today, and they have become increasingly aggressive toward lions. In the past few years, Maasai speared at least five Serengeti lions that had followed the seasonal wildebeest migration into the NCA. Maasai have even started coming inside Serengeti National Park to hunt lions for ritual purposes. Lions once inhabited a large swath of the NCA, from the mountains of the Crater highlands to Olduvai Gorge, but now they must run the gauntlet. The Crater lions are isolated far more effectively by the ring of Maasai spears than by the two thousand foot walls of the caldera.

Next slide.

I didn't spend enough time preparing for this talk. I just slapped together two different PowerPoint presentations that had been written for two different audiences, and now I've reached the seam.

I pause to change the subject.

"OK, so what are the best strategies for successfully sustaining lion populations? We have two options, and I'm afraid that both are going to be difficult. First, we need to recognize the amount of money required to conserve healthy lion populations. Second, I must ask you to consider a tactic that has always been considered taboo in Tanzania: game fences.

"I have gathered lion population data from nearly a dozen different countries across Africa—from over forty different lion studies. My colleagues have developed an ecological model that can accurately estimate how many lions ought to be living in each park or reserve, based either on prey abundance or on local rainfall and soils."

Next slide: a set of nine panels showing *observed* annual population sizes in squiggly lines versus horizontal red lines that indicate the *expected* population sizes. The panels come from different game parks in Benin, Botswana, Cameroon, Ghana, and Kenya. The squiggly population totals are either dropping fast or have already fallen far below the respective red lines.

Next slide: nine more panels from Kenya, Mozambique, Namibia, and Tanzania. Some of these squiggly lines are also declining, others are also low and stable, but a few are either rising or staying close to the red line.

Next slide: nine panels from Namibia, South Africa, and Zimbabwe, where the squiggly lines either coincide with the red lines or are rapidly approaching the red lines.

"Most lion populations in Africa are in trouble. But these last nine reserves are all doing just fine. Why?

Next slide: the most recent conservation status of each study population plotted against the size of the respective management budgets. The data are divided into two categories. The black dots represent conservation success of fenced reserves; the red dots are unfenced. The two colors show that, for a given management budget, the fenced reserves keep their lions much closer to their potential population size than do the unfenced reserves.

"We have a strong aesthetic aversion to fences in Tanzania. But all of the parks in South Africa are fenced and so are a lot of reserves in Namibia and Zimbabwe—even Kenya has started building fences around some of its parks.

"But if Tanzania continues to leave its reserves unfenced, the slope of the red dots shows that, on average, an unfenced park can only maintain its lions at about 50 percent of its potential if the management budget is at least two thousand dollars per square kilometer per year.

"Only Serengeti and Kilimanjaro generate that much revenue, but TANAPA [Tanzania National Parks] diverts a lot of the cash from these two parks to subsidize all their other parks—and the Serengeti's own management budget is only enough to protect the central core of the park. TANAPA should spend about three times as much to protect all of the Serengeti.

"Think about the Selous: fifty-five thousand square kilometers—an area the size of Switzerland—the WD should be spending about a hundred and ten million dollars a year on management; their current budget is less than three million a year."

Next slide: the extrapolated proportion of fenced and unfenced reserves that would remain viable in future years if recent population trends were to remain constant.

"This graph shows that, whereas lions in all the fenced reserves across Africa will likely persist for the next one hundred years, only about half the unfenced populations will still be viable by 2030.

"The downward trends in the unfenced parks will hit hardest in the next twenty years—that's within the lifetime of your businesses. Your income depends on having lions to show to your clients. I wanted you to see the scale of the problem; TANAPA doesn't raise enough money to protect its parks—not even close."

The room is quiet as a tomb.

Next slide: the cover page of a paper titled, "The Significance of African Lions for the Financial Viability of Trophy Hunting and the Maintenance of Wild Land."

"Tanzania relies on two major economic activities to raise funding for lion conservation: tourism and trophy hunting. Hunters claim to give value to wildlife so that local people will tolerate lions, but this paper. . . ."

Next slide: details of the economics of trophy hunting revenues in Mozambique, Namibia, Tanzania, Zambia, and Zimbabwe.

"This is odd . . ."

I stare at table 5 from the hunting paper, and it doesn't make any sense. I've shown this table a half dozen times by now, but I can't recognize the numbers.

I wanted to use this table to emphasize that, even with lions on quota, concessions in Zimbabwe only generate $1,028 per square kilometer per year, whereas hunting income in the rest of Africa is only $130–420 per square kilometer per year—and it isn't likely that anyone spends all their earnings on management.

But the numbers onscreen today are impossible. Instead of showing gross earnings ranging from $130–1,028 per square kilometer, the totals are in the millions.

The room is spinning. I'm in a nightmare. I can't believe I'm standing before all these people trying to explain a string of digits that make no sense whatsoever.

And all these people are hanging on every word.

I can only stammer out, "I don't recognize these numbers . . ."

The offending numbers are in fine print on a wall too far away for most people to read, so I try to shift everyone's attention to the large-typed text box beneath the data table.

"Here is the take-home message from their analysis, 'returns from trophy hunting in most concessions are low, reducing available funds for anti-poaching . . . suggesting that hunting may generate [only] a fraction of the funding needed to protect lions effectively in the long term."

Still struggling to regain my balance, I move on to the next slide. "One of my former graduate students, Dennis Rentsch, has estimated that the six hundred thousand people living along the west side of the park eat a hundred thousand wildebeest a year. Poachers ride *pikipiki*'s [motorbikes] right inside the park to haul out the meat."

And now I'm in the middle of another seam. Why am I talking about bushmeat all of a sudden? The setup in the first half of the talk was about Maasai and retaliatory killings.

Oh, well, I can't recognize the numbers on my own slides, and I've abruptly changed subjects in midstream. Might as well pretend that nothing unusual is happening.

"I'm trying to provide a reality check. The costs of lion conservation are far higher than is generally recognized; TANAPA and the WD only generate about 6–20 percent of the necessary funding, and there's no easy way to make up the difference.

"Lions are essential for the well-being of your industry, and we are at risk of losing lions from most of our parks in the next twenty years. The government can't do this alone.

"But time is running out."

And with that, I thank everyone for their attention, ignore their applause, and step to one side, looking for a place to hide.

My Tanzanian Tourist Operators host asks the audience for questions.

An African tour operator raises his hand. "When you talk about putting fences around the parks, I think about Tarangire where the animals leave the park for so much of the year. Won't you be cutting off access to grazing lands when animals could no longer go outside in the rainy season?"

"Yes," I say, "if anyone ever tried to fence Tarangire, the ecosystem would collapse, likewise in any other migratory system like Manyara, for example. So, no, fencing there isn't an option. These areas will require budgets of two thousand dollars per square kilometer per year forever more.

"On the other hand, the area covered by the Serengeti migration has been almost completely enclosed inside the national park, the NCA, the Mara Reserve in Kenya, and the Maswa and Grumeti/Ikorongo Game reserves—these are all protected to some extent.

"No one is allowed to live in the game reserves to the west of the Serengeti National Park, but step just outside the Maswa, and it's all cotton fields. Step outside Grumeti/Ikorongo and it's all maize. The farmers have already destroyed the land, and none of the wildebeest go out there anymore.

"Fence the westernmost edges of these three game reserves, and you would effectively expand the size of the national park. Those hunting companies are owned by eco-philanthropists who spend a fortune on anti-poaching every year, but they are being overwhelmed."

Remembering something George Hartley told me last night, I add: "Oh, and Sukuma cattle herders treat the game reserves as open pastures. There are no grasslands left to graze outside the reserves. Just last month, Sukuma murdered a ranger from the WD that tried to chase them out of the Maswa.

"On the east side of the Serengeti, however, there is no farming, and it's all Maasai, so we need to get serious about scaling up projects like Bernard Kissui's. Ingela Jansson has started a conflict mitigation project in the NCA to try to reconnect the Ngorongoro Crater lions to the Serengeti—but it is going to be tough without a lot of money."

Time for two more questions.

Another African raises his hand. "I've visited South Africa and I've worked in Kenya, and I know that our visitors don't like to see fences. They don't want to feel like they're going into a zoo. I think we need to try harder to educate the local people about the value of wildlife, we need to get them to learn to live with these animals."

"I'm afraid that the time has come to start making some hard choices. I've worked here for forty years, and I never wanted to see fences either, but the sad fact is that fences work. Those lion data were from a dozen countries and dozens of parks all over Africa. Education and living with lions sound great in theory, but these strategies will take generations—and the data suggest we might lose half of our populations in the next twenty years. We can't wait that long.

"Even on top of the financial shortfall, there is a lack of political will in this country to arrest all the poachers—poachers vote, too, and so do their families. TANAPA doesn't dare arrest everyone; CCM would never let them. A fence would help the situation enormously—it might not stop all the poachers from entering the park, but it would be a lot more difficult. And most importantly, a fence can be presented as a way to protect local people against wildlife—every year the newspapers tell stories about how Jumbos are destroying people's crops, about lions eating people's livestock. Wildlife can devastate people's livelihoods, and these have always been awkward facts for conservationists—but a fence would show that the wildlife authorities care about people, too.

"Tourists may not like to see fences, but remember that the South Africans already have fences—that's just the way it is down there—and tourism is thriving in Kruger and the rest of South Africa's parks. Fences work: they protect neighboring communities from lions and elephants; they clearly mark out the land that belongs to the animals.

"If we want fences to have a low profile in Tanzania, we could build them so that tourists wouldn't even notice them—most tourist roads are already well away from the park boundaries, and the fences at the tourist gates could be discretely hidden behind trees or kopjes."

Time for one more question, but no one raises a hand.

Then came teatime, and I expected to be discredited by my brain freeze and assailed for wanting to fence the Serengeti, but instead I was surrounded by tour operators who told me how much they appreciated my talk—they had been complaining to the government and to TANAPA about the deteriorating conditions in the parks, about the lions being killed by Maasai. They were disgusted that most of their TANAPA fees went to Dar es Salaam rather than to the upkeep of the parks. They knew that fences were controversial, but the time had come to be honest about what was happening. They needed someone to stand up to Power.

After I got home from the Tanzanian Tourist Operators meeting, I reviewed the fatal data table on my MacBook, and it turned out that the "±" sign on the Mac version of PowerPoint showed up as a three-digit number on the Tanzanian Tourist Operators' PC at the Kibo Palace.

So 130±25 came out as 13011125—which looked a lot more like a phone number in Maryland than $130 per square kilometer. I felt a lot better about the state of my brain, but I still worried about how it looked to the people in the room. But when I asked a friend who had attended my talk, she assured me that it was OK. My confusion made me "look more human."

DAR ES SALAAM, 15 JUNE 2012

I've been back to Dar many times the past few years, and I've always stayed at the Heritage Hotel in the center of town or at an Italian hostel on the other side of the peninsula. But there must be a convention going on somewhere tonight, as all the midpriced hotels are fully booked. I've returned to the Econo Lodge for the first time in five years. The taxi drivers in the alley and the hotel manager still remember me, and they all ask how I've been, and how Susan is.

I climb up the same familiar stairs, turn the right-hand corner on the fifth floor hallway, and I'm back in room 513.

Back in the room where Luke Sidewalker nearly leapt out the window, back in the room where I fell to pieces during my son's last visit to Tanzania.

It has been five years. I think I can finally talk about it.

Two days after Luke was attacked on his way to the Econo Lodge from the Kilimanjaro Hotel in April 2007, Saleh Pamba, the timid little man who promised to take Luke's dossier to President Kikwete, was fired as principal secretary at the Ministry of Natural Resources and reassigned to a minor posting in Songea, the most isolated town in Tanzania.

Meanwhile, Severre carried on as usual in the WD.

A month later, Luke presented his dossier to the Donor Providers' Group in Dar es Salaam. The group included representatives from USAID, World Bank, The European Union, the Department for International Development (United Kingdom), Deutsche Gesellschaft für Technische Zusammenarbeit (Germany), the Danish Aid Agency, the Norwegian Agency for Development Cooperation, the Swedish International Development Cooperation Agency, and the Japan International Cooperation Agency. His PowerPoint highlighted Severre's murderous reign as director of wildlife, the hunting industry's complex web of linkages with corrupt ex-presidents, ex–prime ministers, and ex-ministers of natural resources, as well as most of the current incumbents. He outlined the code of silence imposed on TAHOA by Gerard Pasanisi, the greed of Sheni Lalji, and the scandals of the Arabs in Loliondo and Yaeda Chini.

The following day, a member of the Donor Providers' Group "anonymized" Luke's PowerPoint, converted it to a PDF file, and e-mailed it to every embassy in town. Within two days, the dossier had spread to universities, research organizations, and government agencies throughout the country.

Severre held an emergency meeting in the Ivory Room and instructed TAWIRI to investigate my files. He was convinced that I was the only person who could have written it.

Several people in Arusha told me that my life was in danger, that I should get out of the country as soon as possible. A friend in Dar told me not to worry: since everyone "knew" that I had written it, it would be obvious who had killed me . . .

At the height of all this, Luke Sidewalker stayed in room 513 of the Econo Lodge. His feet still stung from being scraped across the pavement during the attack outside the Kilimanjaro Hotel.

Defiant, he continued to use the Internet at the Kilimanjaro and walked back to the Econo Lodge around eleven or twelve P.M. each night, though he did alter his route so as not to pass through the unlit site of his attack. The shoulder strap of his computer case was now ruined, so he militantly gripped the case with one hand and conspicuously clenched the other fist as he strode boldly down the middle of the midnight streets of Dar es Salaam.

He climbed to the fifth floor of the Econo Lodge late one night, went into his room, and locked the door.

At about one A.M. he was awoken by a series of explosive knocks.

BANG BANG BANG.

All his lights were out. He could see shadows in the hallway beneath his door.

BANG BANG BANG.

He held his breath.

BANG BANG BANG.

If this were Severre's gang, he would be outnumbered, and they would be well armed—they might even try to shoot through the wooden door.

BANG BANG BANG.

No one was talking in the hallway. Maybe they didn't know he was here.

BANG BANG BANG.

Don't breathe, don't move.

The sound of footsteps faded down the hall.

Maybe they've given up. Maybe they've gone looking for a key. They can't be sure he is in here. Maybe they'll wait for him in the morning.

Minutes crawled by. Footsteps came back up the hall.

BANG BANG BANG.

He was now dressed and standing beside the window, but he was on the fifth floor and could easily be picked off by a marksman on a neighboring rooftop.

BANG BANG BANG BANG BANG BANG.

There had been no other banging on any other door: they wanted the man in room 513.

BANG BANG BANG BANG BANG BANG.

If only he had tied his bed sheets together, he could have made a rope and made his way down to the alley. . . .

Then nothing.

He waited by the window for five minutes, ten minutes, half an hour.

No more bangs for the rest of the night.

Luke stayed in his room until about eight A.M. He wanted to wait until he heard enough people walking around the halls to feel the safety of numbers.

He opened the room door and something fell on his feet.

His computer case.

He had left it out in the hall—the night watchman had been trying to return it.

Jonathan had come to Tanzania in June 2007 to spend the summer helping Dennis Ikanda with his redesigned man-eating project. I had convinced Dennis to give up on using loudspeakers to repel lions from rural villages. He now saw the virtue in pursuing the lion-bushpig link, and he wanted to collect more comprehensive data on pig damage in Lindi and Rufiji. Jonathan could survey the size and location of the farm fields with a handheld GPS.

But by then, Dennis was worried about being associated with me: the WD wanted to deport me. His job prospects could be damaged. Whereas Jonathan, Carrie, and Mimi had no problems working in the man-eating areas the year before, Severre had now blocked Jonathan from doing anything in any of the wildlife areas.

Jonathan loved the Serengeti when he was a kid. He loved driving Land Rovers and chasing baboons and hanging out with my field assistants. But he was twenty now, and he had given up the chance for a summer job in a stem-cell lab back in Minnesota. He was thinking about going to medical school, and the lab job would have looked good on his résumé.

Before he had flown out that summer, I had assured Jonathan and his mother that I could easily overcome any bureaucratic hang-ups in Tanzania—sure, Severre had blocked Savannas Forever's research clearance for the previous year, but this was just a kid looking at pigs in a cornfield.

I missed Jonathan's company way too much to listen to anyone who tried to talk sense into me, and he arrived to find nothing to do except sit in our house in Arusha or stay in a cheap hotel in Dar es Salaam. I felt unbelievably stupid and selfish.

After a couple of weeks e-mailing back and forth to the United States, he landed another lab job, and I arranged his return ticket to Minnesota. On his last night we stayed together in room 513 and watched *30 Rock* on his computer until I suddenly lost my composure.

Trying to set up hunting certification had been a stupid idea. Even if Pasanisi hadn't sabotaged us, the hunting industry could never have afforded to pay for "conservation certification"—and the bad guys were too well protected to fear exposure of their bad practices.

Bringing Jonathan to Tanzania during a blood feud with Emmanuel Severre was even more stupid—not to mention irresponsibly dangerous.

Emotionally, I had been carrying everyone along for the past two years with Savannas Forever, the lion project, and life in Arusha and Dar es Salaam. I was advising five graduate students with complex thesis projects who were all counting on my judgment. I had a wife who missed her daughters and who wondered if she was unsuited for life in Tanzania.

And I had just wanted to have a few weeks with my son while he was out of college for the summer, to have fun together like we had in the old days.

I just started crying and crying and apologizing to him and telling him all the things I regretted—including letting him see me cry, which must have been upsetting for him, too. . . .

The next day we took a taxi to the Dar airport, and I waved him through the security entrance and proceeded to walk the ten filthy miles back to room 513 in the Econo Lodge, where I took a long, long shower and swore that I would never let Severre get the best of me again.

Severre was a small man with an outsized personality. He always wore business suits and gold-rimmed glasses and dyed his hair jet black. He could recite long tracts of complex environmental legislation; he could humiliate larger men in a thousand ways; he could charm whomever he wished. He had his core group of lackeys who attended every important meeting, who did everything they were told, who were unwaveringly faithful.

In the middle of March 2007, I went the Ivory Room to discuss lion trophy trends with Tanzania's official CITES officer who was also a member of Severre's trusted inner circle. I then walked into an adjacent office building to do some fact checking with one of Luke Sidewalker's sources.

Over the past few weeks, the whistle-blower had grown increasingly terrified of exposure—worried of being murdered some dark night in Dar es Salaam—so I nearly jumped out of my skin when Severre spotted me emerging from our clandestine conversation.

I tried to act calm, greeted him warmly, and quickly outlined the same statistical analysis that I had shown his CITES officer: I now had clear evidence of overhunting in Botswana, Zimbabwe and Namibia in the 1980s and 1990s. I asked if we could discuss the implications in more detail, and he agreed to meet me in his office a few days later.

I saw Emmanuel Severre for the last time on the twenty-second of March 2007. The electronic device hidden in my bike bag recorded the entire meeting, but I never felt like listening to it until returning to room 513 tonight.

Hearing Severre's voice for the first time in five years, I am most struck by the very slow and deliberate pace of his speech. He stretched out his words to the point where it was often difficult to follow the thread of his arguments, and I wonder now if he was deliberately oblique. In transcribing our conversation, I couldn't fully comprehend his pronouncements until about the fifth listening. In real time, of course, I never dared ask him to repeat himself, so I always left our conversations feeling puzzled and uncertain.

The recording starts with me describing how the World Conservation Monitoring Centre had posted the CITES trade database from all around the world. The number of lion trophies exported from Tanzania, Botswana, Zimbabwe, and everywhere else was public information, so the antis could download the hunting offtakes of whatever species they wished.

I showed Severre and his crew the striking similarities in lion offtake trends across several African countries: trophy exports increased through the 1970s and 1980s, reached a narrow peak in the early or mid-1990s, and then suddenly dropped—never to recover again. Assuming that the hunting companies were shooting as many lions as possible each year, the patterns implied overconsumption and collapse.

The World Conservation Monitoring Centre data from all across Tanzania had just begun to show hints of a downward turn, but it wasn't yet clear if we were about to experience as severe of a drop as in Namibia, Botswana, and Zimbabwe. However, Rolf Baldus and his team had set up a comprehensive monitoring program for the Selous Game Reserve, and their data showed a much clearer pattern of rise and fall.

In contrast, leopard offtakes had been stable over the past thirty years across all of Africa, as well as in the Selous, suggesting that leopard hunting had been managed effectively.

The problem of overhunting seemed to be restricted to lions.

All these findings, of course, depended on the assumption that demand exceeded supply—that lion offtakes hadn't dropped because no one came to hunt them anymore. So could the WD please tell me their lion quotas and the number of clients who came to hunt lions each year?

During my presentation, a secretary came in and served everyone tea, but I kept talking above the clinking of cups until I had finished my spiel, and then I was done.

"*Karibu chai*," Severre said, indicating that I should drink my tea. "This is very interesting. There is a—I don't want to be defensive—but there is different ways of looking at this. One is that you can only hunt a lion if you . . . come in for a twenty-one-day safari. And a lot of . . . small companies . . . come in with clients who are not all that experienced and seasoned, and they don't stay on for twenty-one days. They're here, they come on for buffalo hunts: seven-day safari or fourteen-day safari [where] you cannot get a lion; it's just not . . . on your hunting permit . . . That in itself protects the lion here, and it can also contribute—can reflect—on how many lions can be pulled out from outside or inside the protected areas. So that dimension cannot be ignored. So one has to look at the data.

"The other one is we've been very strict ourselves, if you hunt a lion which [does not have a] big mane, right? You will have to double-pay that amount, we have to penalize you, and not just you who own the company . . . but also the professional hunter. That is not a very popular statement to make at the [TAHOA] meeting. . . .

". . . [S]ome guys want to come and hunt for twenty-one days. [But if he hasn't shot his lion in the first few days,] maybe he's got other meetings and stuff, so he says, OK, forget about the lion. . . .

"So what I'm trying to say here, comparing these countries with different—without common denominators—may not be—may not guide one very well. So . . . I'm not saying [that] should be a caveat, but I think this is something to take into consideration."

I agreed that "effort" must be measured, so I repeated my request: the number of twenty-one-day safaris would provide the best measure of hunting effort. If he could give me the numbers for every year since 1990, the analysis would be far more powerful, and could I also have the quotas each year?

"And you know, one other thing, Craig, is that I have experience. I was once [requested] to go and count lions in Loliondo area because someone wanted a very nice lion, blah, blah, blah [*he punches the palm of his hand*], and we said, 'No, no, no, no, some of these very nice lions could be coming out of the national park.' What? So I went and I came [to] the recommendation of no lion hunting.

"Mweka [wildlife college] had done that survey, and they said, 'No lion hunting.' I went also. I did it independently: 'No lion hunting.' Then Frankfurt Zoological Society was asked to go, and these guys took a helicopter, and they came back with the same answer."

Although I had never heard of his personal visit to Loliondo, I congratulated him: "You helped triangulate the results—that's the best way to do it."

He continued: "But what I'm saying here is that lions are very proud animals, and they don't like kills, they want peace. They . . . only come out into conflict when they must necessarily do so. . . . Most of the lions learned that there is hunting in this area, but there is peace at the other end. So most of them—beautiful—big lions, you find them in the parks.

"So for northern Tanzania, I will not be surprised, when in areas around Tarangire, Manyara, [etc.] . . . if it comes out that the client may not get the lion easily because these lions have retreated to the protected areas. . . .

"And that in itself is a parameter, and, . . . there is a lot of other parameters that you must factor in into these graphs so that you reflect exactly what could be more realistic. . . . So one has to do some more work rather than say, 'Oh, this is the bible of the day. Things are like this.'"

Although I'm thinking that it was unlikely that lions would have suddenly gotten too scared to stay in the hunting blocks after thirty years of big-game hunting, I merely agree that we should remain open to alternative interpretations and say that it is interesting that leopards didn't show the same pattern of rise and fall. Were leopards also restricted to twenty-one-day safaris?

"Yes, leopards—people had a rough time getting leopards on quota when we had two hundred fifty, when we climbed to a quota of five hundred, most people, you know, they sell safaris five, three, four years in advance. So most people still intrinsically have that feeling that selling a leopard may be risky. Then they still don't have the confidence. It takes time to get confident.

"And a leopard is a leopard, and that's it.

"And a lion is a lion, and that's it.

"Each of them presents a different taste to the different clients. So I wouldn't really compare these species, I wouldn't. Statistically I wouldn't, looking at it from a statistical point of view. . . . . You cannot compare populations and what is happening to leopards to what is happening to lions. If they make that proposal then we will reject it tomorrow morning." He slaps his knee.

"I think this very important. The taste for a leopard is very different than the taste for a lion."

I was thoroughly puzzled by now. "Amongst the clients, you mean?"

"Yes, the way I look at the leopard, given the opportunity, I'll go for a lion. Because [it's] a big thing and magnificent, big mane. The leopard is—apart from the size of the leopard and the sober attitude they present when they are seated and when they are relaxed—apart from that, there is no mane and the bite, the power to kill [*he hits his palm with his fist*], all these nice things about the lion, you don't see them in the leopard.

"And some guys go in with their wives, and the wives see the leopard, 'No, no, no, no, hit the lions first.' . . . It is very interesting, heh heh.

"Just to inform you like I told you the other day in the corridors, this behavior of graphs or not, is that we are soon going to produce other strategies with regard to lions, leopards . . . that is, how do you procure a leopard? How do you procure a lion? Some of these strategies may cause limitations, we may not be popular, but this is fine."

I say, "Well, if you're not popular you can always show your companies these graphs, and they'll say, 'Yes, that's a good thing!' Right?"

He laughs. "Within a year we should introduce some of those elements which will limit . . . the taking of these species."

I asked, "Do you mind telling me what you have in mind?"

His whole court burst out laughing.

"No, no, no, we have not discussed this extensively around, so we have been so far keeping the cats under the carpets. But within six months, I should say, we should be able to announce some of these strategies.

"The TAHOA people will not like it, I know, but you tell them we are talking of sustainability here. Of course these people say, 'There are plenty of lions, there is plenty of this.' They have to say that. [*a reflective pause*] Whether there is plenty, god knows."

I told him that he had just reminded me of the passenger pigeon, which was once the most abundant bird on the planet. Flocks of hundreds of millions of birds darkened the skies as they moved between the Great Lakes and the forests of the eastern United States each year. Each "roost" might have such tremendous numbers of birds so crowded and massed together that they frequently broke the limbs of the trees. People could walk outside, point their guns in the air, and kill a half-dozen birds with a single shot—you could buy a dozen for fifty cents. But then the forests were cleared: flocks were netted and shot and shot and shot. By the eve of World War I, only a single passenger pigeon was still alive, and she died in the Cincinnati Zoo in 1914.

He seemed genuinely shocked.

I suggested that maybe some of the hunting companies had a misplaced belief in the "inexhaustible supply" of nature. They needed to learn that natural populations were finite. I would love to help in whatever plans he might have for ensuring sustainable lion hunting. The WD already enforces size minimums on leopards and elephants. The best way to protect lion populations would be to establish an age minimum. But age estimates were more difficult than measuring the size of a leopard pelt or weighing the tusks of an elephant, so Karyl Whitman and I had recently published *A Hunter's Guide for Aging African Lions*. It was filled with photographs of lion noses, lion teeth, and lion manes—all from known-aged males. It had been distributed at the SCI convention in Reno in January.

Severre said, "I would love to get a copy, too."

All the operators, professional hunters, and hunting clients should receive copies. I had been asked to talk about estimating lion ages at the wildlife college at Mweka, whose graduates become rangers and wardens. Age assessments could become standard procedure in the WD.

"Your training at Mweka, I think, this must be welcome . . . I will be very interested to see what those Mweka guys . . . apply . . . in the field and . . . we will assist them at the points . . . where they . . . bring the [lions] to the

Bwana Mkubwa [big man], and . . . then they will be the inspected trophies, blah blah blah [*he punches his palm*], stamp them [*slap*], this is fine [*slap, slap*], but the guy will be there also, and I think that they will use these checks and balances. So, immediately, that output—we shall use it."

I told him that I would be speaking at the wildlife college in two weeks.

"Is that right?" To his staff: "Any of you guys want to . . . attend the lecture?"

"Yeah at least some of our guys should be there," somebody said.

Severre resumes, "It depends on how you look at it, really, because it's not just a lecture, 'Yes I am here. Hi, I attended the lecture.' You should understand . . . the applications . . . Heh heh heh. Oh, that Craig Packer is a good lecturer—it's not enough."

Not at all sure what he was talking about, but relieved that he was at least pretending to be interested, I started to wrap up our meeting and again offered to help however I could. I apologized again for not having a hard copy of the aging guide and offered to give him a PDF, even though the publishers had forbidden me to distribute soft copies. "Just don't tell anyone that you got it from me."

Severre laughs. "No, we have never seen you for the last one year."

I promised to send it to him in an e-mail.

"Thank you very much, Craig," he said graciously.

I promised to get him a hard copy as soon as possible.

Then I remind him of my central mission: could he permit his staff to give me the quotas and the number of twenty-one-day safaris?

He answers coldly, "Ah-hem, leave this with us, and we will discuss."

I started moving toward the door.

He turned on the warmth again. "Thank you a lot for coming, Craig, and for your time and for your information, eh?" he said cheerfully. "Take care."

I took care all right; I took care not to believe a single word he said.

The number of twenty-one-day safaris had increased by 60 percent between the peak in lion offtakes in 2000 and our meeting in the Ivory Room in 2007. There had never been any attempt to protect underaged lions. No one in the WD knew what was happening in the hunting blocks. Either they believed in the inexhaustible supply of nature or they only worried about this month's payoff from the hunting companies.

Shortly after the release of Luke Sidewalker's dossier in May 2007, all three of Luke's confidential informants were reassigned to "punishment posts" in other parts of the government, with concomitant losses in salary and pension benefits.

Severre spent most of his time locked away in his office at the Ivory

Room, revising the draft language of the new Wildlife Act. Several of the new statutes had been crafted by Parliament to curtail the power of the director of wildlife, but Severre revised the draft to strengthen his power.

Severre bought off the MPs in Dodoma, and he personally attended the next TAWIRI meeting so as to veto our request for research clearance for Savannas Forever, and no one from the WD showed up for my talk at Mweka.

Despite all the dire warnings about my safety, I flew home to Minnesota unharmed at the beginning of September and started teaching fall semester in Minneapolis.

In November 2007, Susan sent me a text message from Arusha: Severre had been sacked. He had been transferred to the College of African Wildlife Management, Mweka.

The chief mafioso was now a school principal. His former bulldog, Erasmus Tarimo, had been chosen to succeed him.

Severre turned out to be only the first domino to fall. A few months later, two of his primary allies were also gone: Prime Minister Edward Lowassa and the minister of finance, Mama Meghji, both now under investigation for grand corruption.

While the Tanzanian press had repeatedly called for Severre's ouster both before and after the release of Luke's dossier, the fall of Lowassa and Meghji came as a surprise. It was common knowledge that they had misappropriated vast sums of money (Lowassa had bilked the World Bank out of eighty million dollars; Meghji was part of a thirty-million-dollar scheme at the Bank of Tanzania), but they had always seemed untouchable.

The precipitating factor was the imminent arrival of U.S. president, George W. Bush, on Valentine's Day 2008. The Bush administration had developed an alternative to USAID, which they called the Millennium Challenge Corporation. Partner countries had to qualify for aid by taking concrete steps for controlling large-scale corruption, so Lowassa and Meghji's ousters were a Valentine's Day gift from the Tanzanian government to the American president in return for six hundred million dollars in American aid.

Severre moved to Moshi—the embarkation point for tourists seeking to climb Mount Kilimanjaro—but he refused to live in the principal's house on the campus at Mweka. He instead stayed in the most expensive room in the most expensive hotel in town. He stole as much as he could from college coffers and instituted a new rule that graduating students could not gain their diplomas until they enrolled in a few extra classes after the end of their formal education and paid the requisite fees, even though these "classes" involved no further classwork.

Following the removal of Prime Minister Lowassa, Tanzania's new prime minister was Mzingo Pinda, who also happened to be the chairman of the board of Savannas Forever.

I flew back to Tanzania in March of 2008 to attend a meeting at TANAPA headquarters in Arusha. Pinda instructed the heads of TAWIRI (Simon Mduma), TANAPA (Gerald Bigurube), and the WD (Erasmus Tarimo) to sit side by side on a two-person couch directly in front of a very large desk where the prime minister courteously but firmly instructed them to ensure that Savannas Forever was finally granted research clearance. Pinda was particularly clear to Tarimo that, if there were any further delays, Tarimo would have to answer to him personally.

In my experience, Tarimo seemed a totally different man as director of wildlife than he had been as Severre's bulldog. People said that he had become a born-again Christian. Or maybe he had just been biding his time. He told me that he was first and foremost a scientist: he didn't have the slightest qualms about giving us the lion and leopard quotas. Likewise, in the final months of her life, Miriam Zachariah also rose from being Severre's hatchet woman to becoming highly respected for her contributions to wildlife conservation.

As mentioned previously, Miriam died in a plane crash in 2008. Tarimo was sacked in 2011 for failing to allocate the blocks to undercapitalized Tanzanian hunting companies. Severre reached the mandatory retirement age in May 2012 and left Mweka without a word to anyone.

ARUSHA, 30 JUNE 2012

"Dennis! Welcome back! How was your trip to Germany?"

"Too long," he sighed, "Ugh, cold and rainy!"

The Convention for International Trade in Endangered Species organizes species-review meetings before each new conference, the next of which will be in Bangkok in 2013. The status of the lion will be high on the agenda, and Dennis is just back from four days in Germany where he represented Tanzania at the lion review.

The WD organized a meeting with TAHOA two weeks ago to inform the hunters about the new lion-hunting regulations. The WD initiated a pilot project in 2011 for assessing the ages of trophy males. The hunting companies had been requested to send mug shots, X-rays of teeth, and intact skulls; compliance by the operators had been nearly 100 percent.

I had visited the WD's age-assessment team in April with Henry Brink. The pilot system was comprehensive, and the hunters were cautious: back in the bad old days, lion offtakes peaked above three hundred in the year 2000. Overhunting had reduced this by half in recent years, but fear of the new regulations pushed the number down to eighty-six this past year.

However, the WD only allowed us to see their *system*, they hadn't shown us the trophies or their tentative age assessments. If the assessments were kept secret, the whole exercise could just become a cover for more lies and deception. I politely reminded everyone that I knew a few officials at USFWS, so if I could verify the WD's age assessments, I might be able to prevent up-listing of the lion on the Endangered Species Act and prevent any problems with the Lacey Act.

I had first met with the current principal secretary, Mama Tarishi, at the new headquarters of the Ministry of Natural Resources, a multistory building called Mpingo House beside the old Ivory Room. She arranged my initial meeting with the age-assessment team, but when I asked if I could see the trophies and assessments myself, she said I would need clearance from the new director of wildlife, Paul Serakike—whom I had last seen as Severre's slush-fund manager and delivery boy.

I expected to meet Serakike one-on-one, but a dozen people were seated around the conference table, including two senior staff that had been caught trying to smuggle live giraffe to the UAE in Qatari aircraft last year. Tarimo's successor was sacked right after the story leaked to the press—but these two guys looked just as comfortable as any scowling thug in Severre's era.

The age-assessment team was present, too, but the most surprising attendee was the former head of Selous Game Reserve, Benson Kibonde. Benson had been transferred to the WD's ranger-training school in Mwanza about the same day that Luke's dossier hit the Internet. After successfully resurrecting the school's infrastructure and staff, Benson was reassigned to the only game reserve in the country without large mammals—a biodiversity hotspot that had been set aside to protect plant habitat. Benson established a system for monitoring forest structure with standardized photos of individual trees. His punishment posting was now the best-kept reserve in the country.

After Serakike asked us all to introduce ourselves, I briefly summarized the implications of the new Tanzanian Wildlife Act in the context of American law and CITES and was about to discuss how to age a lion when Benson spoke up saying, "I may not be understanding things very well, given the limitations of my experience with these problems. But I don't think we can provide useful inputs without first hearing the background of the issues at hand."

I loaded a PowerPoint onto the WD's computer, explained the importance of restricting offtakes to older males, and outlined evidence for overhunting in Botswana, Namibia, and Zimbabwe, the Selous, and the rest of Tanzania. I further elucidated how we had used the Tanzanian data to estimate sustainable quotas in the absence of an age minimum, but how Tarimo had opted for the age minimum so as to avoid economic hardship for the smaller operators. I showed photos of teeth and noses from known-aged lions in the Serengeti, asked if I could help refine their age assessments at the WD, and invited everyone to visit the Serengeti to see known-aged lions in the flesh.

I had always viewed Serakike as a bit of a weasel. He had long sharp features, and he darted his eyes around the periphery whenever you looked straight at him. But now that he was the director of wildlife, he projected the cautious authority of an accountant who had just been promoted to the boardroom. He was clearly nervous that the U.S. government might prohibit the importation of Tanzanian trophies, so he didn't want to test the limits of international law.

In the end, everyone agreed that I could help improve their age assessments and take the good news back to USFWS. But proper procedures would have to be followed. The Tanzanian Wildlife Research Institute was the country's scientific authority on wildlife research. Estimating lion ages was a research topic. If TAWIRI were to ask for my help, the WD would not object.

Serakike seemed relieved, the smugglers relaxed their scowls, and as soon as I got up to leave the room, Benson looked straight at me with a grin as wide as Kansas and said, "The next time the distinguished professor comes to address the government about matters pertaining to governmental policies and regulations, I believe he has worked long enough in this country to address this group in Swahili."

Everyone burst out laughing. I smiled, too, and fired back in my fractured Swahili, "I am sorry. I have lived in Tanzania for forty years, but I am not able to speak good Swahili. I do not know if I will ever be able to speak good Swahili. But I know lions, so I hope to meet you all again in Serengeti to talk about lions."

Laughter echoed down the halls as I left Mpingo House.

Dennis Ikanda represented TAWIRI at the TAHOA meeting two weeks ago, and the WD told the hunting operators that the age assessments would become binding in 2013. Penalties would be modeled on the system in Niassa Reserve in Mozambique. The WD would confiscate any trophy male that was clearly too young, and the operator's quota would be reduced the following year, while quotas would be raised if a company's trophies were all at least six years old.

The foreign operators were resigned to the new age-assessment system, but many of the local operators were unhappy. "The Arabs and Asians were really pissed off," Dennis explained. "They were told they would have to follow your recommendations to shoot no more than one lion per thousand square kilometers in any of the new blocks. They got most of the blocks that came up in the latest allocations, so they got real aggressive and accused the government of treating them differently."

Aha: Was Sheni Lalji at the TAHOA meeting?

"Yes, and he was very rude! He got really angry. He kept complaining, 'My new block is only five hundred square kilometers; how am I supposed to shoot half a lion?'"

Before the most recent allocations, Tarimo was forced to free up land for more Tanzanian operators. But blocks could only be leased to companies with adequate capital, and too few Tanzanians met the requirements, so the rules were readjusted behind closed doors. Foreign operators were finally supposed to relinquish their excess hunting blocks—no one was supposed to hold more than ten—but Pasanisi still controlled twenty blocks in the Selous, and Sheni Lalji now held dozens of blocks in his mafia network across the country. Tanzania Game Trackers agreed to relinquish two of its blocks, but then the WD wrested away a third block to TAWIRI for "research" and carved off half of another and gave it to Sheni. Tanzania Game Trackers managed to reclaim the "research" block (which TAWIRI knew nothing about), but the WD told TGT to sort out their issues with Sheni in person. Sheni offered to sell back TGT's half block to them for six hundred thousand dollars—the lease had cost him about twelve thousand.

We had sat next to Sheni at the table of shame at Pasanisi's gala bash in June 2006, and I haven't seen him since—but it seems like he's everywhere. During Savannas Forever's village surveys around Tanzania, our teams interviewed thousands of local people—and the most vocal complaints were always about Sheni's hunting companies. Sheni never kept his promises, he never paid anyone, his hunters were bullies, and his clients were racists. Many of the nearby communities tried to extract their village land from the hunting blocks so they could switch to photo tourism—and never have to deal with Sheni again.

But Sheni no longer operated these outfits himself. He was now an absentee landlord who subleased all his blocks to South Africans and white Zimbabweans. Henry Brink's analysis of the Tanzanian hunting-block data clearly showed that subleasers were the ones who overhunted the lions most seriously since they only cared about their short-term profits.

So more than anyone else, Sheni has been feeding the beast. He is a Tanzanian citizen, but his ancestors came to Africa to supervise the slave trade. He was born in Kigoma, the trailhead for the slave caravans that trudged six hundred miles from the shores of Lake Tanganyika across the broad dry country to Zanzibar. Kigoma is close to where Stanley met Livingstone, and Sheni owns the best hotel in town—which often serves as a point of departure for tourists going to see the chimpanzees of Gombe. Sheni's hotel partner is Tanzania's third president; his hunting business partners include former directors of wildlife and ministers of natural resources. Mama Meghji appointed Sheni to the board of TANAPA, and he gave her several large houses in Dar es Salaam. Sheni was appointed to the Central Committee of CCM in 2009—becoming the highest-ranking Asian in the ruling party.

Since 1999, when a yacht and two of his houses were seized for tax evasion, Sheni has entangled numerous politicians in his web of business deals. He can't be taken out without dragging down half the ex-presidents, cabinet ministers, and Asian businessmen in the country. Meanwhile, his children all went to the best private school in Dar. His oldest daughter was admitted to the freshman class at Stanford University in 2010.

SERENGETI, 19 JULY 2012

When I first came to the Serengeti in 1978, I encountered a black rhino every few weeks while looking for lions. Rhinos always reminded me of medieval military machinery with two gunmetal spikes growing up from their snouts. They belonged to a Pleistocene landscape filled with mastodons and saber-toothed tigers. Our puny Serengeti lions stayed well away, so rhinos always seemed immune to all the normal hazards of daily life.

In those days, elephants knocked down so many big acacia trees along the Seronera River that the chief park warden assigned a special vehicle to chase them out of the tourist areas. The big bulls could only sneak into our front yard at night to strip off the bark from our shade trees. From Kenya to South Africa, wardens and wildlife ecologists had long claimed that elephants would eventually annihilate the woody vegetation inside their national parks. Kenya culled hundreds of elephants in Tsavo in the 1960s, and TANAPA contemplated the same strategy for the Serengeti.

But by 1984 organized gangs with automatic weapons had solved the elephant problem, and the black rhinoceros, *Diceros bicornis*, had essentially been exterminated from the Serengeti. The international outcry inspired Kenya to burn its ivory stockpile and CITES to ban the export trade in ivory in 1989. East African elephant populations soon rebounded, but the parks in southern Africa had always been better managed, their elephants were so numerous that they were hammering the vegetation, and the ivory ban was precluding a valuable source of revenue from all the South African elephants that had died from natural causes. I flew with research staff in a helicopter survey of South Africa's Tembe Reserve in 2003. We spotted skeletons of two elephants that had died of natural causes. The ground crew dutifully removed the tusks and put them in storage in case they could someday be sold.

In 2007, South Africa, Namibia, and Botswana received special exemptions from CITES to sell their ivory to the Far East. Tanzania wanted to piggyback on these efforts, but Minister Maghembe kept his promise to Sara Bambara and withdrew the WD's request to export seven tons of ivory. Tanzania tried again in 2010, but by then it was obvious that the WD had been unable to stem a renewed tide of ivory poaching. Middle-class Chinese may only want a piece of ivory the size of a lipstick case, but there aren't enough elephants in the world to meet the demand for a billion lipstick cases.

In the good old days, Selous held 110,000 elephants, the largest herd in Africa, but the population had dropped to 27,900 by the time of the CITES ban. The Selous population recovered to around eighty thousand by 2002, but the street price for ivory in Beijing is now three thousand dollars a pound, and illegal trade has reached the point where fewer than thirty thousand elephants remain today—and about thirty of them are being shot every day.

So while the CITES ban curtailed the *legal* trade in ivory, the *illegal* trade eventually swamped any efforts at law enforcement, and the legal ban on sale of rhino horn never had any effect at all. I helped immobilize a wounded female rhino in Manyara National Park in 1980. She had been speared in the side. If the poacher had killed her, he would have hacked off her horns and sold them to a middleman, who would have sent them to Yemen to be used as dagger handles or to Asia to be ground up as an aphrodisiac.

The thousands of rhinos that were killed in those years must have saturated the market for dagger handles, and Viagra has presumably wiped out the market potential for phony aphrodisiacs. Today, rhino horn mostly goes to China to "treat" fever and to Vietnam to "cure" cancer. When relying on magic, an apothecary only needs to add a homeopathic dosage of any particular ingredient, so rhino-horn analgesics are cheaper than Tylenol or Advil—even though rhino horn sells for over thirty thousand dollars a pound. At these prices, it is well worthwhile for poachers to hire helicopters to shoot rhinos inside the fenced reserves in South Africa and Zimbabwe.

Pound for pound, the price of rhino horn now rivals that of heroin, and ivory is as valuable as marijuana. Efforts to ban the international trade of ivory and rhino horn have been just as successful as the war on drugs.

In 1979, a Tanzanian scientist named Feroz Kurji came to see me in Seronera. He had just discovered that lion claws were being sold in Mwanza for fifty dollars apiece, making a dead lion worth eight hundred dollars, which at the time was about the same as a rhino horn. He thought that lions would soon join the same hit list as elephants and rhino. I've heard similar fears over the past five or six years, especially as Chinese road crews have started building so many new highways in Kenya, Tanzania, and Mozambique. The price of a lion claw, however, has remained fairly constant, so the demand for lion parts has never seemed much of a conservation threat. I have a lion claw back in Minneapolis somewhere. It is a handsome talisman. But no one is likely to break into my office to steal it.

The bones of big cats, however, have long been ingredients in traditional Chinese medicine. Tiger bone wine, for instance, supposedly cures rheumatism and various other maladies. Most of the illegal poaching of wild tigers is aimed at collecting their bones, and there are now fewer than two thousand tigers left in the wild. So the Chinese have established tiger bone farms where as many as forty animals are cramped together per cage.

Somewhat similarly, about a decade or so ago a peculiar subculture sprang up in South Africa. Landowners started breeding lions as trophies. Their clients wouldn't have to buy a twenty-one-day safari, they wouldn't have to suffer from tsetse flies or travel to remote areas. They could fly to Cape Town or Jo-burg and drive to a private facility with a fifty-square-kilometer fenced enclosure where their "trophies" sat waiting for them, having been released the preceding week. Peter Lindsey and his colleagues have looked at the economics of "canned hunting," and they found that a captive-bred lion only costs about thirty thousand dollars to shoot as opposed to seventy-six thousand dollars in Tanzania. Hunting success is nearly 100 percent in South Africa against only about 60 percent in Tanzania. Canned hunts only require about three days versus twelve days in Tanzania, and captive-bred lions grow much bigger than wild lions—so canned lions make very impressive trophies.

And, hey, in South Africa, you can even hunt a white lion (*"Extinct in zee vild!"*), if it suits your fancy.

Twice as many captive-born lions are shot in South Africa each year than all the wild lions from the rest of Africa combined. But once a canned lion has been shot, the client only wants the skin, the skull, and the paws.

Why not sell the bones to China? The sale of South African lion bones to Asia jumped dramatically in 2009, and Chinese tiger bone wine now clearly lists *Panthera leo* on its list of ingredients.

Animal rights activists are aghast, and political pressure is building to abolish canned lion hunting altogether. Although Internet access has been spotty in Seronera the past week, I've received half a dozen requests to sign an online petition, urging the South African government to ban the export of lion bones.

But I can only think of ivory and rhino horn and their convergence with illegal drugs.

Elephants and rhinos are slow breeders. Females take ten to fifteen years to reach reproductive maturity and only produce a single offspring every four or five years. Rhino "horn" is like a fibrous fingernail that grows continuously: a kilogram of horn can be shaved off each year without harm. Thus rhino could conceivably be farmed like ostriches or alpaca.

But elephants don't do well in captivity, and their tusks can't be shaved. About their only hope is if enough tuskless elephants survive the onslaught to sustain the species in the wild. Schoolchildren may someday see elephants, but they might only see descendants of mutants that are unable to grow ivory.

Captive lions breed like rabbits. There are already about thirty-six hundred lions in the canned hunting facilities in South Africa. In captivity, a female lion can start breeding before her third birthday. A farmhouse lion can produce a litter of three cubs every year, so the captive population could easily be built up to around fifty thousand in about only five to ten years.

Australia and India successfully conserved the saltwater crocodile by permitting the construction of crocodile farms to meet demand for crocodile skins. Similarly, Chinese and Laotian bear farms have sprung up to sell bear bile for medicinal purposes.

So what will happen to the lion? Will they end up like elephants? Or like crocodiles and bears? Will lion bones become like heroin—or like ostrich feathers? The choice may be dreadful, but whereas animal rights organizations give equal weight to the well-being of captive and wild animals, I would always value wild animals more than their captive counterparts. Wild lions contribute to the normal functioning of intact ecosystems, whereas captive lions are, well, not that different from bear-farm bears.

If flooding the Chinese market with canned lion bones were to keep the price low enough, there might not be any incentive for winemakers to come all the way to the Serengeti or Selous to harvest wild lions.

That's one way of looking at it. But closer comparisons with crocodiles and bears don't provide much comfort. Crocodile farming is successful because farmed crocodiles seldom suffer flesh wounds. Their unscarred leather is more valuable than the skins of wild crocs. There's no reason to expect the Chinese to *prefer* farmed lion bones. In fact, if a bone's magic depends on heroic behavior, a farm-raised lion would be unlikely to have earned much juju. Lion farmers would never be able to compete with the poachers.

And bear-bile farming hasn't done much to reduce poaching of wild bears. Bear farms produce so much bile that marketers have developed entirely new uses for the stuff—including shampoo, toothpaste, and cough drops—which, in turn, have stimulated entirely new demand. Bear bile is biologically active—the ingredients can even be synthesized—but bile from wild bears is more biologically active than captive or synthetic bile, and thus the advent of bear farms hasn't alleviated demand for poached bear bile.

Would farm-raised lion bone wine merely become Two Buck Chuck, while the wine from wild lions would still fetch the price of a 1978 Grand Cru?

The bioactivity of lion bones lies solely in the mind of the consumer. So even if shoppers preferred a 2014 Chateau-Serengeti, there would be no way to guarantee the legitimacy of the ingredients. Maybe the Chinese market could be flooded with counterfeit "wild" lion bones and remove any premium from the skeletons of wild lions. Then again, no one has ever managed to substitute fake rhino horn into Asian medicine, so counterfeiting might not be possible.

Of course, the best solution would be to eradicate demand altogether. Chinese celebrities have started coming to Africa to pose beside the carcasses of hacked up elephants and show Chinese children where ivory really comes from. But education will only slowly work its way up the age chain—and TV commercials about "bloody ivory" won't turn off the tough guys and old men in Beijing.

Rhino farming is tempting, but rhinos have become so scarce that any unintended increase in demand from legalized trade could easily spiral out of control.

But, come to think of it, rhino farmers wouldn't need international permission to sell horn for domestic consumption. So maybe the South Africans should try setting up rhino-horn spas where rich Vietnamese could come to take the cure in Johannesburg or Pretoria. Charge rich clients black-market prices per gram of horn and plough the revenues back into growing their rhino population.

And toast the visitors' health with a good domestic lion bone wine . . .

Strange fantasies, perhaps, but if these species are to survive, they will need something more imaginative than an online petition.

SERONERA, 27 JULY 2012

Dusk has fallen, and Markus Borner is sitting in an armchair in his front yard, watching a slide show of his youthful self studying Sumatran rhino in Indonesia, his early years in Tanzania, his first wife, his mop-top head of hair—first dark, then gray, then white—his children small, then fully grown, Markus with his past and current bosses at Frankfurt, past and present presidents of Tanzania, colleagues, pals, and girlfriends. He is seeing most of these photos for the first time. Some of them make him laugh.

Outside, farther beyond Markus, a ring of canvas chairs encircles a large campfire. Right next to the house, his son Felix and his successor, Rob Muir, are tending a barbecue loaded with chicken parts, pork chops, and sausages. A spotted hyena is lurking at the dark edge of the lawn, waiting for scraps.

Inside the house, the dinner table is stacked high with serving bowls of potato salad and pasta. Hundreds of chunks of mango, melon, and pineapple have been poured directly into a large blue-plastic picnic cooler together with several gallons of Pimm's punch. A wide-screen TV displays one image after another of Serengeti sunsets, giraffes, acacia trees, leopards, wildebeest, zebra, lions, elephants, and cheetah cubs. The music is loud but not too loud. Every few minutes, an occasional album cover interrupts the parade of wildlife photos: Loretta Lynn, the Traveling Wilburys, Roy Orbison, and U2. A thousand-gallon aquarium behind an adjacent wall throws a steady yellow-green glow into the room. Maybe a hundred Lake Tanganyika cichlids swim around the tank. Some are bluish, some are yellowish, most have black stripes either vertical or lateral. At least a dozen *wazungu* (white people) are standing around in the living room, dozens more people are crowding the veranda and spreading out toward the campfire. The Tanzanians mostly stick with each other but their interactions with the *wazungu* are friendly and relaxed.

By the 1950s Tanganyika's human population totaled less than ten million, but the western half of the Serengeti ecosystem was already threatened by rapid agricultural expansion, a vast checkerboard of manmade pits that trapped thousands of migrating wildebeest each year and long lines of wire snares that caught thousands more. In December 1957, the founder of the Frankfurt Zoological Society, Bernhard Grzimek, flew from West Germany to the Serengeti in a small zebra-striped airplane with his son Michael. They counted animals and mapped the extent of the wildebeest migration until early 1959, when Michael died, his aircraft having hit a griffon vulture while flying above the eastern plains. The vast popularity of Grzimek's subsequent book and film, *Serengeti Shall Not Die*, directly led to the current configuration of the Serengeti national park, and FZS has been the primary source of outside funding for the Serengeti ever since.

For over twenty years, Grzimek aired a weekly TV program where he showed wildlife films with the sound tracks turned off so that he could ramble excitedly about whatever topic was on his mind. He always ended his show with a stirring plea for his German audience to write out their checks to support the good works of Frankfurt Zoo in Serengeti. At fund-raising events, Grzimek would charm one little old lady after another into including FZS in their wills. When nothing could be imported during Tanzania's puritanical socialist phase, Grzimek filled his suitcases with spare car parts for TANAPA's antipoaching vehicles and carried them directly to Seronera. He was also famous for devoting nearly 100 percent of all those donations to actual conservation—almost nothing was diverted to administrative costs.

In the mid-1970s, the Tanzanian government sacked the colonial-era wardens in TANAPA and President Nyerere announced plans to construct a railway across the Serengeti National Park. Grzimek conducted shuttle diplomacy between Frankfurt and Dar es Salaam to dissuade Nyerere from building the railroad, but as conservation became ever more complicated and time consuming he realized he needed a full-time representative inside the country.

Dr. Markus Borner first came to Tanzania in 1978 to work in the newly gazetted Rubondo Island National Park, which Grzimek had long used as a dumping ground for excess zoo animals and as a retirement home for overaged lab animals. Frankfurt had airfreighted enough rhino, giraffe, and chimps to Rubondo that Markus could reasonably claim to be monitoring the ecology of ex-captive animals on a backwater in Lake Victoria—while his real mission was to worm his way inside the hierarchy at TANAPA.

Markus is barely five feet tall, and has always worn a Beatles haircut and a gold earring. He often refers to himself as a Swiss gnome—I prefer to think of him as a hobbit. But Markus possesses an infectious charm, and he only needed a few years to gain the trust of the wardens and higher-ups in TANAPA, who eventually elected him to the TANAPA board. He moved his base from Rubondo Island to the old warden's house near park headquarters in Seronera and added a second house, an office block with a generator, and a large array of solar panels. He drilled a well and always had water while the rest of us endured dust baths during the dry season. For many years, Markus also had the only functioning airplane in the park. He piloted dozens of wildebeest censuses, tracked a thousand radio-collared lions, and flew to Arusha or Dar es Salaam whenever necessary to deal with the latest political or financial crisis.

Combine Frankfurt's financial resources with the friendly persistence of a Swiss gnome and Markus Borner was able to work wonders even during the Tanzanian government's worst periods of economic austerity and despite the growing antagonism of Tanzanian voters toward wildlife conservation.

Over his fifteen years on the TANAPA board, he achieved something remarkable: he built the social capital of the park rangers, wardens, and top administrators at headquarters in Arusha. The Tanzanian National Parks became the best-run national park service in Africa—better than Kenya, better than Zimbabwe, better, even, than South Africa.

That is, until about 2006, when Mama Meghji became minister for natural resources and appointed Sheni Lalji to the TANAPA board. Construction of new buildings had been banned in all the country's parks, but Sheni quickly awarded himself lodge concessions in Gombe and Tarangire National Parks. The board of any Tanzanian organization is supposed to perform its duties pro bono, but the TANAPA board awarded itself a very large bonus when the parks raked in record revenues in 2007. The Tanzanian press broke the story. The board promised never to reward themselves again but then started calling weekly board meetings (instead of the traditional semiannual meetings) and paid out large daily "sitting allowances."

Because of all the money being siphoned off from the top, TANAPA was no longer able to raise the salaries of its rangers and wardens. Earlier this year, TANAPA projected that it could finally afford a pay raise for its field staff for the first time in years. But their revenue projections were based on an assumption of increased "bed fees" from the tourist lodges.

The lodge owners blocked the new fees at the last moment. Hotels that charged guests $250–500 per night didn't want their clients to have to pay an extra $2.50. The hotels apparently don't see any need to contribute to the management of the park.

Have I mentioned yet that the wife of ex-president Mwinyi owns the biggest safari lodge in the Serengeti—and that President Kikwete is rumored to have a 20 percent stake in the newest lodge?

Wandering around between the punch bowl and the campfire at Markus's house tonight, I run into the current warden for Serengeti tourism who tells me about the hundred or so tourist camps that have sprung up in the park this tourist season. The tented camps are mostly operated by expatriate wildlife enthusiasts who try to minimize their environmental impacts; they negotiate their fees directly with the Serengeti, and they generate about twenty times more revenue for the park than the big bricks-and-mortar tourist lodges—the lodge owners invest about the same in day-to-day management of Tanzania's protected areas as the fly-by-night hunting companies.

Alan Root is here tonight, seeming older than his seventy-five years. His speech is halting and unravels in disconnected bursts. We were next-door neighbors in the Serengeti for five years, and yet conversation feels awkward. He is far more comfortable talking to people he has just met, describing his history of animal bites: the gorilla that bit his thigh while he was shooting footage for *Gorillas in the Mist*, the hippo that bit through his calf while swimming in a crystalline spring in Tsavo, the puff adder that cost him his right index finger.

Alan was the creative genius who first used "cold light" to film close-ups of insects without cooking them, who hid cameras in tortoise shells so that he could film wildebeest hooves flying overhead, whose first wife, Joan, would provide a willing target for the venom of a spitting cobra and then wipe the spray from her eyeglasses as if it were Windex. Joan was his anchor and his fixer—both in terms of obtaining filming permits and in mending relations with whomever Alan had insulted or otherwise irritated during the course of their travels.

Alan has made his name as a storyteller, constructing engaging narratives about termites, birds nesting in baobab trees, and other animals no one else knew anything about. He has always disdained the typical tourist's obsession for large conspicuous mammals. He made stars of the unnoticed and the unloved. He once made a film called *Nothing's Happening*, in which a tourist sitting on the veranda of a safari lodge complains that "nothing's happening." Alan panned down to the guy's feet and showed what was happening with the ants, the spiders, and all the little things in the dirt, and then panned up to the trees overhead to highlight the lives of birds, bugs, and lizards.

Alan started out as the cameraman for Grzimek's Oscar-winning film version of *Serengeti Shall Not Die*. He then returned to the Serengeti in 1972 with Joan, his tortoise shell, and a hot air balloon and spent over two years shooting *The Year of the Wildebeest*. They followed the migration from the calving grounds on the short-grass plains to the western woodlands and then north to the Maasai Mara in Kenya. They filmed all the necessary milestones of parturition and predation and crocodile-infested river crossings, but the story truly came alive thanks to all the little details that only a naturalist would have known to add: the botfly whose maggot travels up the wildebeest's nostril and burrows into its brain, causing its host to walk in tiny circles until it dies; the moth whose larvae feed on the horns of dead wildebeest; the plover whose camouflaged chicks are invisible among the bones of a wildebeest skeleton.

Alan and Joan came back to live in the Serengeti in 1980 and based themselves in a research house about seventy meters from the lion house. Instead of trying to repeat the recent dazzle of *Castles of Clay*—his Oscar-nominated film on termites—Alan wanted to film an encyclopedic account of every plant and animal in the Serengeti so that future generations could see all that had been lost when the starving villagers finally overran the park. But he never found the financial support for his full-blown documentary and instead settled for a few hour-long programs (including *Islands in a Sea of Grass* about the kopjes on the plains and the peculiar birds and mammals that lived there and *The Lightning Bird* about the hammerkop—a duck-sized bird with a nest the size of a washing machine).

During his glory days, Alan never seemed to have much time to talk to anyone for very long. He was always planning his next shot—looking for the right scenery, waiting for the best evening light, figuring out the best camera angle. Each of his films took two solid years. They were epic works of art that often reported original scientific discoveries, making him more like an explorer than anything else.

One day he announced to me that he had run out of ideas. He decided to become a producer rather than a full-time filmmaker, and he hired a succession of freelance cameramen to do all the work in the Serengeti. His surrogates duly filmed the major sequences of lions hunting, nursing their cubs, and defending their territories, along with an unforgettable sequence of infanticide by an incoming male. But no one else could ever replicate the visual narrative touches that tied Alan's stories together and made his films into works of art.

I try one more time to reconnect with Alan, but he seems disinterested. Tonight is Markus's retirement party, and my encounter with Alan has left me feeling uncomfortably flat. The Whole Village Project might have been my failed equivalent of Alan's Serengeti documentary series, but, dammit, I'm still full of piss and vinegar. I've got all those camera traps to play with and plenty of other things that I still want to do.

I find Markus next to the barbecue, and we start talking about his future. This was his last day on the job. He will be flying home to Switzerland day after tomorrow. He says he'll come back from time to time, but I wonder about that. He seems most excited about the prospect of renting a house on a Greek island next spring, spending time with Sarah Cleaveland (she of the domestic dog vaccination program) in Glasgow, and doing a bit of teaching.

Markus is my oldest friend in the Serengeti. I met him in Arusha the day he set off for his very first trip to Rubondo. Our paths regularly crossed during those early years, then we became neighbors in the Serengeti, and he added immeasurably to my quality of life. We flew together, radio tracking lions. We laughed at the foibles of fellow scientists, the clumsy greed of local politicians. We ate his cordon bleu cooking, drank frozen gin, sipped Schnapps, and laughed even harder.

Markus always saw the bright side of just about everything that might have otherwise driven me to despair—there is no way I would have continued the lion project for the past thirty-four years if he hadn't lived here. I don't want to spoil the mood of fond nostalgia, but I ask his advice about the persistent corruption in the hunting industry. Despite the positive noises from the WD about lion hunting, Pasanisi has retained his monopoly in the Selous, and Sheni has more hunting blocks than ever. The two of them have stymied the necessary investments that might actually save the hunting areas.

Markus thinks it should be possible to neutralize Sheni, but he thinks that Pasanisi is too well connected. I remind him that Eric has replaced Gerard at TAHOA—and that we only need to discredit the old man back in France. But then again, Gerard Pasanisi is an ex-French paratrooper and good buddies with Jean-Marie Le Pen, the French nationalist and neo-Nazi. Maybe the French public wouldn't even care about Pasanisi's evil empire.

The barbecue team declares that the meat is ready. We load our plates and resume our conversation in two canvas chairs next to the campfire. Markus is disappointed that none of the Hollywood studios have been willing to take up Paul Tudor Jones's offer to bankroll a full-length feature film about Grzimek's relationship with President Nyerere and how the two men worked together to secure the future of the Serengeti. Nyerere had famously complained that "wildebeest don't vote," but he set aside the park nevertheless.

Markus may no longer wield the checkbook for FZS, but he is accustomed to power. Besides fiddling around with film scripts, Markus has also gotten Tudor Jones to crack open the door for him at the billionaires' club. Markus wants to meet Bill and Melinda Gates. "They are the most powerful individuals in the world when it comes to developing all these poor countries in Africa—they spend so much money on so many things. But never do they talk about conservation, about the natural world that makes all these things possible. If I can get them to at least consider putting their projects into an environmental perspective, it might help. Just shift them a little bit. That would be nice, *ja*?"

Yes, it certainly would—or maybe it would just be another trip to la-la land.

*

Meanwhile, everyone is sitting around the campfire, enjoying barbecued chicken and pork chops and throwing the bones out into the darkness. Alan Root's ten- and twelve-year-old sons have been sitting on the wicker couch next to a three-year-old boy. The three-year-old has finished his meal and is wandering around in circles, mostly lurching between the couch and the campfire, but he occasionally staggers obliviously toward the edge of the lawn where a hyena retrieves discarded bones and looks menacing enough to dispatch any small child with a single bite.

The little boy's parents bring him back to the campfire, but he is soon bored and returns to the Root boys, who have now finished their dinner. As the child resumes his aimless wanderings around the couch, the oldest Root boy spontaneously puts his hand on the child's shoulder and gently guides him away from the hyena.

This is my last night with Markus in the Serengeti. My dreams of building Savannas Forever have largely been replaced by the less noble desire just to neutralize a few people in the hunting industry. But I still have the lions to think about and a whole new project with the camera traps. I can't wait to see what patterns will emerge from the million photos per year—to discover how the cheetahs and leopards and hyenas all manage to live here together.

And out there in the darkness tonight, our cameras are silently doing their work: two hundred twenty-five quiet eyes, recording images of whatever walks by. There are hundreds of hyenas out there tonight, loping along game trails. The leopards have come down from their trees by now, slinking somewhere in the grass. Aardvarks, aardwolves, genets, civets, and all the nocturnal creatures are searching for termites and ants, rodents and ground-nesting birds.

The moon is bright at the moment, but it will set about an hour after midnight tonight. Where will all the wildebeest, zebra, buffalo, and gazelle go in the deepening darkness? No gentle hand will be there to push them safely back toward the fire, but our electronic eyes will certainly be there to watch.

Markus Borner built up TANAPA from virtually nothing during the failed days of African socialism and then watched TANAPA start to crumble again once the corrupt masters in Dar es Salaam discovered his golden goose. Tonight Markus is dreaming of shifting the priorities of a few key rich people on the other side of the planet—hoping for a quick fix. He isn't interested in hatching any more fifteen-year rehabilitation plans on the night of his retirement.

Everyone lines up in a public display of affection: a large circle has formed around the fire, and we all take turns saying a few words of appreciation or, in the case of Alan and myself, an emotional summary of how much the little hobbit has meant to us over the past thirty-four years. I want everyone to understand how impossible it would be *not* to love this man.

And I'm also thinking, like Markus, that to make a difference out here in a hurry, it is probably most important to focus on just a few key individuals. So while he's trying to get Bill Gates to shift his interests toward helping the protected areas, I wonder if it would ever be possible to convince some of the key players in the hunting industry to divert their money to some other sector of the economy.

NGORONGORO CONSERVATION AREA, 11 AUGUST 2012

Rainfall is the fundamental driver of rural livelihoods. With enough rain, people can farm for crops and cash, and Africa's expanding rural populations are placing ever greater demands on arable land. Savannas are plowed, woodlands are cleared. Crops attract elephants; bush pigs attract lions. Fences may be ugly, but separating farmers from wildlife can solve a lot of problems.

Livestock converts grass into meat, so pastoralism allows humans to thrive where the land is too dry for agriculture. But species like wildebeest and zebra move across vast landscapes in pursuit of seasonal pastures. Wildlife graze the same pastures as livestock, so separating pastoralists from wildlife would be catastrophic. Pastoralists are merely making the best of a bad job. They are poor. They are itinerant. Their lives are harsh. The Maasai are one of many pastoralist tribes in East Africa, but their red blankets and proximity to the most famous wildlife areas have granted them iconic status.

The contrast between romantic fantasies and harsh realities is clearest here in the Ngorongoro Conservation Area—a national park that allows people to live among the wildlife, provided that they adhere to their traditional lifestyle. And it is here that we foresee having to engage with people until the end of time.

Ingela Jansson drove along the eastern plains of the Serengeti at nine in the morning on 11 April 2010, radio tracking a ten-year-old female from the Mukoma Hill pride, which usually ranges inside the main tourist area in the middle of Serengeti National Park. But the March rains had been heavy that year, and the lions had followed the wildebeest migration to the short green grass forty kilometers from home. Ingela was getting a steady beep from MH37's collar as she crossed into the Ngorongoro Conservation Area, but no lions were in sight—only a large flock of vultures, feasting in the middle of the open plains.

As she drew close, she realized that the vultures were feeding on a female lion. The corpse had been speared repeatedly. The ears, tail, and one forepaw had been removed. Ingela checked the lion's whisker-spot patterns against the ID cards for the Mukoma Hill pride and recognized MH106, eighteen months old and a member of the cohort of subadults belonging to MH37 and four other mothers.

Ingela resumed tracking for MH37 and followed the signal over a rise where she saw eighteen Maasai morani walking toward the Gol Mountains. She drove in a wide arc around them and confirmed that they were carrying MH37's collar. She called a TANAPA ranger on her cell phone, then started back to retrieve MH106's body. But about a kilometer or so away, she saw a second pile of vultures and drove over to find the remains of MH37—whose ears, tail, and forepaw had also been removed.

Late that afternoon, four vehicles from the police, NCA, and TANAPA arrived and followed Ingela as she tracked MH37's collar. The signal indicated that the Maasai had climbed up Olongoyo Hill, but as Ingela drove close to the base of the hill, the beeping suddenly went quiet. She drove around to the opposite side where the signal beeped again at full strength.

The wooded hillside seemed empty until a flurry of four red *shukas* (blankets) emerged from the verdant background. Most of the police and rangers immediately gave chase. Looking to retrieve the collar, Ingela walked up the hill with four policemen. They found a morani lying beneath a bush. The police fired their AK-47s into the air, dragged the Maasai out by his hair, and started beating him with their rifle butts.

The authorities captured four morani that afternoon, but the bad news continued.

Ingela had also picked up the signal of PN91, a female from another pride that ordinarily lived well inside the national park. PN91's collar was giving a mortality signal—which indicated that the transmitter had been perfectly stationary for at least an hour. But the sun was down by the time the police rounded up the four morani that had killed MH37 and MH106.

PN91's collar still beeped the next morning, but the mortality signal had stopped, and the transmitter was broadcasting from a different hill. As Ingela drove to within a kilometer or two, the signal suddenly went dead. The transmitter must have been smashed. Ingela couldn't trace the collar, and we never saw PN91—or the rest of her pride—again.

That same week, another lion field assistant, George Gwaltu, tracked a female from a third Serengeti pride. The signal came from high up in the Gol Mountains where our lions never go. The collar must have been carried to a boma. She, too, was never seen again.

In May 2010, a dozen morani were spotted within the boundary of Serengeti National Park. On 27 June 2010, George found the corpse of a four-year-old female, SB104, who had been killed inside the park. The body was skinned, the tail and ears removed.

Dennis Ikanda had confirmed that the Gol Mountains were a hot spot for ritual lion killing, but we had reckoned that the magnitude of the problem was too small to threaten the Serengeti population. We had also assumed that most victims were a doomed excess of nomadic animals. But in 2010, four resident females were killed within a few weeks of each other in the NCA, and the Maasai came inside the national park to kill a fifth.

Accustomed to tourists and tour cars, Serengeti lions were easy to approach and spear. The Maasai were deliberately targeting Serengeti lions—they said they could tell a Serengeti lion from an NCA lion, because "the Serengeti lions don't run away."

Ingela Jansson became a nurse at eighteen, toiling intensively for four and a half of the next eight years and often working consecutive shifts to save up for travel. "I was way too restless to go to university—there was too much to see out there."

She journeyed to South America, sold jewelry on street corners in Peru. Her boyfriend was an artist; she tried making jewelry, too, but her exotic birds "ended up looking like chickens." She sold soft ice cream at a Sugar Chick kiosk in Gothenburg, and she spent a month selling Christmas trees in Manhattan where she looked so poor that people threw their spare change at her.

She was in her late twenties before she felt ready for college. After completing her undergraduate degree in biology, she spent a year and a half managing a lodge in the Selous Game Reserve. Next, she completed a Master's degree on the ecology of Swedish bears. She came back to Tanzania, toured the country on a bicycle for four months, got bored, and looked for something to do. She started work on the Serengeti lion project in 2006.

I usually hire field assistants for only two years, but I had never met anyone like Ingela. She essentially worked consecutive shifts on the lion project, tracking a phenomenal number of lions during the daytime while managing our accounts and research vehicles ("Land Rover: always sick but never dies") each night. Near the end of her first two-year contract, I asked her to stay on for another few years, but she first wanted a yearlong break with her bears in Sweden. After another year in the Serengeti, she worked on a salmon research project in Alaska.

By 2010, Ingela was terminally tired of sitting in a Land Rover. She wanted to tackle something more tangible than whisker spots and ear notches. Shortly after her trauma with the dead lions, the Maasai, and the police, I invited her to a workshop on human-lion conflict in Dar es Salaam. Bernard Kissui, Dennis Ikanda, and Hadas Kushnir were all there, as were several scientists working in Kenya and Mozambique, and a talk by an Egyptian-American graduate student, Leela Hazzah, sent Ingela's career in a new direction.

We started the ball rolling in August 2010, when I met with the top management officials at the Ngorongoro Conservation Area Authority (NCAA) to ask if they would help Ingela start a conflict-mitigation program with the Maasai in the Gol Mountains and Olduvai Gorge. But I first wanted to discuss the chronic inbreeding of the lions on the floor of the Ngorongoro Crater.

The Crater floor is a 250-square-kilometer island of ideal lion habitat surrounded by forest, desert, and livestock. The Crater lions suffered from a severe disease outbreak in 1963, and 80 percent of the population died, leaving only a dozen survivors. Several new male coalitions entered the Crater between 1964 and 1968, and the surviving females bred so successfully that one hundred twenty-four lions occupied the Crater floor by 1983. However, the recovering prides spawned such large cohorts of sons that roving gangs of six, ten, and eleven males prevented the in-migration of any more newcomers over the following two decades.

The Crater females had no choice but to mate with their brothers, uncles, fathers, and sons. By 1987, the lions were conspicuously inbred. The population fell below forty animals in 1998 and although it would occasionally recover to about sixty or seventy, numbers would suddenly drop below thirty or forty again, most likely because of periodic disease outbreaks. The NCAA refused to allow regular veterinary investigation of the Crater lions, but we were able to confirm that a die-off in 2001 was caused by the same combination of CDV and babesia that had afflicted the Serengeti lions in 1994. And the high frequency of disease die-offs over the past twenty years had provided strong circumstantial evidence that inbreeding depression was now an existential threat to the Crater lion population.

Management authorities at South Africa's Hluhluwe iMfolozi Park had recently revitalized their severely inbred population by importing lions from another reserve. No lions had lived in Hluhluwe iMfolozi until a lone male wandered into the park in 1960, then one of the park's rangers translocated two females from Kruger National Park and erected a lion-proof fence to protect the surrounding villages. The founding trio bred rapidly, but their progeny were forced to mate with siblings, parents, and cousins, and the inbred population became highly susceptible to bovine tuberculosis. Sixteen new lions were imported to the park between 1999 and 2001, some mated with the inbred residents, others with each other, but none of the outbred offspring showed any serious signs of disease.

I had attempted to bring three Olduvai lions into the Crater in 2006. We planned to build a two-acre boma on the Crater floor and hold the lions long enough to forget their way home. The NCAA gave its blessing, and Dennis Ikanda bought the poles and fencing. But then the chief ecologist refused to allow the NCAA lorry to collect the construction materials from Arusha, and that was that. Circumstances never aligned for another attempt.

So here I was today, asking if we could try again.

But this time, several of the NCAA managers were openly skeptical: What if the new lions aren't the right genetic match?

"Lions from near Olduvai Gorge are genetically connected to the Serengeti, but they're also from within the NCA, so they'd be essentially the same stock as the Ngorongoro Crater lions, only they wouldn't be their close kin."

What if the new lions carried bovine tuberculosis?

"We would make sure to test the lions for bovine tuberculosis before bringing them into the Crater."

But everyone in the room seemed worried that if anything went wrong, the ax would land on their heads, not mine.

To be honest, I was ready to watch the Crater population fade away. I had spent a lot of time here back in the days of a hundred twenty-four lions on the Crater floor. There were so many lions that tourists could expect to see dozens in a single day. Now people were lucky to see any lions at all. The future of this population was already in doubt, so let the ax fall where it may and see what happened.

I took a deep breath and changed the subject to human-lion conflicts. I gave a brief account of Ingela's adventures with the Maasai on Olongoyo Hill, tallied up the death toll of our radio-collared Serengeti lions, and then outlined Leela Hazzah's conflict-mitigation project in Amboseli.

By hiring Maasai moranis, Leela had transformed lion killers into lion guardians. Her Lion Guardian program became a conservation organization that promotes the coexistence of people and lions by teaching the moranis to use radio-tracking boxes, round up stray livestock, warn herders of the location of nearby lions, and mediate disputes between livestock owners and law enforcement officers. The Lion Guardians had become avid conservationists, not least because their traditional skills and newfound knowledge were being appreciated and rewarded. And whereas cattle-killing lions were still being speared in most other parts of the greater Amboseli ecosystem, none of the Lion Guardians' lions had been lost in the past two years.

Here in the NCA, I continued, the Maasai had exterminated most of the resident lions outside of the Crater floor. Soon, temporary migrants from the Serengeti would be the only lions left in eight thousand square kilometers of protected area. . . .

But then someone interrupted to ask, "So by killing all these lions, the Maasai have isolated the Crater lions?"

"So it seems. Back in the 1960s there must have been plenty of lions up in the highlands, so when the Crater population crashed in 1963, there were plenty of males from the outside that could come down to the Crater floor and mate with the Crater females. But there are ten times as many people in the NCA now as fifty years ago, ten times as many Maasai with spears and bomas. There are no rim-top lions left."

"So why not use the Lion Guardians to restore a transit way for lions between the Crater and the Serengeti? Then you wouldn't have to build a boma inside the Crater because fresh blood could come down on its own."

What a brilliant idea!

I immediately abandoned any further plans for a lion boma in the Ngorongoro Crater, and the NCAA granted Ingela official permission to start a Lion Guardians program in the conservation area. We raised a year's worth of research funding from an American conservation organization called Panthera, and Ingela was ready to drive off into the sunset.

But then Africa won again.

Besides conserving wildlife, the NCAA is supposed to promote the "pastoralist way of life" and to assist the economic development of the Maasai. The Ngorongoro Conservation Area Authority operates separately from TANAPA and the WD and generates thirty million dollars in tourist revenues every year.

But almost nothing goes to the eighty thousand Maasai living inside the conservation area.

Fifty years ago, the Ngorongoro Maasai preferred tending livestock and disdained working for cash, but cattle numbers were already at their carrying capacity, so today pastoralism confers only one-tenth as much wealth per capita. The younger generation of Maasai has resorted to paid employment. But the NCAA staff preferentially hire their own relatives, leaving Maasai men to travel to Arusha, Dar, and Zanzibar and look for work as night watchmen.

When Ingela asked the NCAA to organize a Pastoralist Council meeting to introduce the Lion Guardians project, the management somehow neglected to follow through—time and time again. When she first discussed the Lion Guardians with the local communities, a village elder accused her of wanting to take away their land to build a tourist lodge. When she tried to discuss her progress with the top management at the NCAA, all they wanted to know was if her research findings would help them to evict the Maasai from the conservation area for good.

As it turned out, the brilliant idea of restoring a safe passageway between the Ngorongoro Crater and the Serengeti had backfired. When talking to the Maasai, Ingela had described her goal of creating a dispersal area—and the communities assumed they would be evicted from the conservation area to make way for more lions.

The NCA had originally been established as a compromise. The Maasai lost many of their traditional grazing areas when the colonial authorities established Kenya's Amboseli and Maasai Mara Reserves and Tanganyika's Tarangire, Manyara, and Serengeti National Parks. The Maasai complained that they had been selectively punished for coexisting so successfully with wildlife. They were not prepared to surrender another inch.

Bernhard Grzimek came to Tanganyika in 1957 to ensure that the Serengeti shall not die and discovered that the eastern plains were critical to the wildebeest migration. The boundaries of the national park would have to be redrawn, and the British authorities decided to allow the eight thousand indigenous Maasai to remain in the conservation area as long as they practiced their traditional pastoralist lifestyle. However, *eighty thousand* inhabitants could not possibly survive on a pure pastoralist diet drawn from the same amount of pasture, so about ten years ago maize fields started sprouting up. The United Nations Educational, Scientific and Cultural Organization quickly alerted Tanzania that the NCA would lose its World Heritage Site status if its land were converted to agriculture. The government banned farming in the conservation area in 2010, and the NCAA promised to deliver food rations to every remote village.

The food deliveries haven't happened.

The NCAA often acts as if they wish all those impoverished stubborn people would just go away. The thirty million dollars in gate receipts mostly seems to go to the central government in Dar es Salaam, CCM faces reelection every five years, and the NCAA is more beholden to the big potatoes in Dar than to the poor people in the Crater highlands.

When Leela Hazzah visited Ingela in the NCA in 2011, very few of the local Maasai would admit to the existence of human-carnivore conflict and none of the Maasai leaders showed interest in a Lion Guardians program. Yet another lion was killed during an *Ala-mayo* (ritual lion killing) on Christmas day, 2011, and by February of 2012, Ingela was ready to pull the plug on the Lion Guardians concept. No one would tell her the truth. No one would commit to anything. She had also reached a stage in life where she needed to make some big decisions. She had no savings and no pension, and she wanted to go back to graduate school.

She was still eager to follow lions in the human-occupied areas of the NCA, and she was pleasantly surprised when her second-year research clearance unexpectedly granted her permission to attach GPS collars to six lions. The collars would show where the lions went in even the most mountainous and inaccessible parts of the area, and Ingela could use the data to locate a safe passage between the Serengeti and the Ngorongoro Crater. If the Maasai wouldn't talk to her and the NCA wouldn't mediate on her behalf, she could follow the lions anyway—and someday, maybe, someone else would be able to work with the Maasai on a Lion Guardians program. After all, Leela had needed three years to develop the necessary trust in Amboseli.

In May 2012, Ingela finally met a local Maasai who could discuss carnivore conflicts with his peers. Julius Kinini was born in the NCA and grew up in an important transit-point between the Serengeti and the Crater. The elders and morani trusted him. They gave him dozens of reports of leopards eating goats and sheep and of lions killing cattle.

Then in July, lions killed two cows near the village of Ngototo Sumbat, and the villagers asked Ingela to help with their livestock husbandry. She guided three days of open conversation, quietly coaxing them to accept the responsibility for the safety of their own livestock while moving toward a solution that would allow the lions to survive.

The ice between the Swede and the Maasai had finally been broken in at least one village. She made them understand that if the NCAA wanted to use human-lion conflicts to evict the Maasai from the NCA, the Maasai would have to prove that they could live peacefully with lions. The villagers agreed, and assured Ingela that they would work with her to develop a solution.

*

The old man in the back seat is hacking and coughing. He only has a few teeth left, he is wearing a blue *shuka* and a baseball cap, and his earlobes nearly reach his bony shoulders. The middle-aged man beside him is uncertain of his father's age—late eighties, maybe? They talk to each other in Maa. The son speaks to us in English with an American accent: "He needs to spit."

Ingela stops her Land Rover, and I reach back to roll down the passenger window. The old man leans over and hawks a perfect loogie out on the ground.

We start moving again, and the son says, "My father lived in the Serengeti before it became a national park. His boma was at the Moru Kopjes. The Maasai looked after the animals, and there were never any problems."

After the formation of the Serengeti National Park, some of the resident Maasai moved up to Loliondo, others moved to the NCA and migrated each year between Lake Ndutu (at the western end of Olduvai gorge) and Endulen (in the Ngorongoro Crater highlands). They kept their cattle a safe distance from malignant catarrh fever, the virus that is harmless to wildebeest but causes miscarriage in livestock.

The old man was unusual among traditional Maasai leaders at the time in that he wanted one of his children to become educated. He chose his second son, Godfrey, to attend a missionary school. Instruction was free, and the family only had to sell the occasional goat to cover the boy's living expenses. After Godfrey Ollemoita finished school and got a job, he devoted a sizable portion of his earnings to buying livestock for his family. Godfrey's own son has just received a first-class rating at a private school in Arusha. Godfrey beams with pride.

These days a far higher proportion of the Maasai in the NCA are going to school—some say 90 percent are now enrolled—and Ingela asks Godfrey about the quality of education.

"Not so good. The students don't come to school that regularly, and the parents are not that very much serious."

We drop off Godfrey and his father at the Ngorongoro police station. Godfrey worries that the cold air of the Crater rim will come as a shock to his father, but the old man seems unfazed and ambles excitedly to the far side of the road, eager to catch a lift home to Olduvai.

Godfrey heads the archaeological museum at Olduvai Gorge, home to many of anthropologist Louis Leakey's most famous discoveries, but the Ministry of Antiquities wants to open a modern museum at Laetoli, where Mary Leakey found the famous footprints of two Australopithecines walking side by side in soft volcanic ash, three and a half million years ago. The tracks have been under covers for thirty years, and Godfrey hopes to make the Laetoli museum a showcase.

Godfrey also has another dream. He wrote to the top brass at TANAPA last year, asking for a meeting between the park wardens and the few surviving members of his father's generation. He reckons that the park staff could learn a lot from these wise old men. There used to be plenty of rhinos in the Serengeti. The migration came through unhindered. The Maasai just tended their cattle and made no demands on the soil.

His affection for his father is so warm-hearted and sincere, and his faith in the old ways so touching, that I shake his hand for a long time, wishing him the best of luck with his new museum and his plans for a reunion at the Moru Kopjes.

But despite his nostalgia, Godfrey knows full well that everything has changed forever. Most Maasai children will now spend their days in school instead of in the fields herding cattle. And there are simply too many people in the conservation area for anyone to experience the same quality of pastoralist life that his father's generation had once enjoyed.

Two days ago, Ingela drove me along a rough dirt track that snaked along the western side of the Crater rim. Five Maasai were jammed together in the back of her Land Rover, and her roof rack was filled with heavy sacks of maize and yellow buckets of cooking oil. Ingela smiled happily as she greeted the village elders with a few words of Maa and an extensive vocabulary of Swahili. She asked after their health, their families, and the news of the day. Everyone was delighted to see her. They asked after her parents, her friends, her work. Their handshakes lasted for minutes on end.

At some point the conversation turned to wildlife, and they told her about the latest livestock attack. A leopard had caught a goat the previous day. Two nine-year-old boys had been herding a hundred forty-five goats and sheep when they heard a disturbance in the nearby brush. They ran toward the noise and chased off a leopard, but the goat was already dead.

One of the herd boys took us to the site of the attack. We walked across an open field of dried yellow grass to a twisted labyrinth of cattle trails that led through scrub and weeds that sometimes smelled like tutti-frutti, sometimes like spearmint. The boy first showed us where the village women had slaughtered the dead goat, and then he took us to the bush where the leopard had been hiding.

Ingela recorded the type of livestock that had been attacked, the size of the herd, the age class of the victim, the number and age of the herders, the species of predator, the time of the attack, the distance from the boma, and the type of vegetation.

Off to the side, the nine-year-old was practically invisible in his oversized green cotton trousers and an overlarge black hoodie. His four-foot bamboo staff was his only defense against lions, leopards, and hyenas.

Everyone knew that small children should not be sent out alone to herd livestock, but they also knew that livestock would be lost if they didn't send out enough herders. The teenagers have gone to town looking for jobs, and most kids now go to school. This boy was only here today because his school was on break.

Back at the village, Ingela consulted again with the elders as I took pictures of the children with my iPhone. I squatted down to the height of the five-year-olds and showed them their portraits. They quickly learned how to scroll from one shot to the next. After a few minutes, six or eight kids were pressing their faces all the way around the brim of my hat, hundreds of flies swarmed the backs of our heads, and the stale crowded atmosphere smelled of wood smoke and sour milk. More kids emerged from neighboring huts. They all wore tire-tread sandals. Their toenails were chipped and jagged like the toenails of old men. The livelier among them were mugging for the camera and having the time of their lives, yelling and carrying on. These, presumably, would be the ones to survive the childhood diseases, grow up, and someday wear the bouquets of colored beads and bright red cloth.

We returned to the Land Rover and drove on another few kilometers, giving three Maasai a lift to a small village called Ndepes, We learned about a spot nearby where a leopard caught a weanling goat while being tended by two young girls who ran away screaming as the leopard disappeared into a gully. On another afternoon, a leopard came repeatedly to catch weanling goats that had been left in the care of the women—who also ran away screaming. A few weeks ago, a leopard dug underneath the walls of this boma and dragged out yet another goat.

I went over to look at the boma's roughhewn wooden poles, which were two-to-three meters high. The NCAA had delivered mosquito netting to the village as protection against malaria, but the nets had instead been twisted into tight ropes that tied the poles together.

The people in Ndepes again complained that there weren't enough herders to protect their livestock from predators, but the cattle were out grazing during our visit, and there were plenty of grown men sitting around. One man lay down on the bare ground, covered his face with his orange and black *shuka*, and fell asleep. Empty plastic sachets of Kiroba Gin were scattered around on the ground. A mangy brown dog rested in the shade of the Land Rover, chewing on the hoof end of a disembodied goat leg.

On our way home for the evening, we stopped back at Ngototo Sumbat where I met the headman of the village. He talked about how the Maasai used to be allowed to live down on the Crater floor, but then there was the trouble with the rhino, and everyone had had to leave, even though his family hadn't caused any of the trouble.

After the NCAA had reneged on so many of its promises to the Maasai (all the water wells dried up, cattle troughs fell apart, cattle dips broke down), the morani protested by spearing a few rhino in the Crater in 1974. The NCAA responded by banning permanent settlements on the Crater floor, which was the first sign that the Ngorongoro Maasai might someday be evicted from their most sacred stronghold.

We passed by the remaining one hundred forty-four goats and sheep from yesterday's leopard attack—they were herded once again by only two small boys.

I looked at the headman and said, "You are an *mzee* [old man], and I'm also an *mzee*—but I still work as hard as I ever did—and, like most women in my country, my wife does the same sort of work as me. Maybe you should get your old men and your women to help with the herding, too."

Everywhere we went that day, someone handed Ingela a cell phone that needed to be recharged. She carried along a pile of phone chargers that were constantly connected to her car lighter. She gave lifts to everyone who asked. They loved to say her name: ING-ella, ING-ella. She was energized by each person she met, her face brightened, and she always had time for a chat. When she converses in Swahili, she often asks, "*Sindyo*?" [Sin-DEE-oh] "Isn't it?" She is engaging, curious, and happy to bring you aboard. The positive vibe was infectious. People lit up whenever they saw her.

We picked up a Maasai elder shortly after dark. We took him three or four kilometers and, as he got out, he thanked "Mama Simba."

"Did you know him?" I ask.

"I'm not sure; probably not, but I didn't see his face."

"But he knew who you were . . ."

She laughs. "Everyone knows the blond taxi driver in the white Land Rover. . . ."

*

Ingela lives in a cottage on the Crater rim with stone walls, running water, and a commanding view of the Crater floor. Frankfurt Zoological Society had occupied the house for several decades, and the bookshelves in the sitting room are still filled with brochures, reports, and financial accounts from their long-term rhino project. Poachers decimated the Crater rhino population by the early 1980s. Various top staff of the NCAA were arrested, others were implicated, and FZS started paying for the costs of antipoaching patrols and even translocated a few rhino into the Crater from South Africa. But funding is tight these days, and FZS wants the NCAA to start protecting their own rhinos, so Ingela now has the house to herself.

August is one of the coldest months of the year, and mornings on the Crater rim are typically cold and gray. But the weather today is warm and clear. Looking out through the sunporch windows, we can see the peaks of Olmoti and Embakai on the far side of the Crater. The soda lake on the Crater floor has dried up three months early, and the prevailing winds have kicked up delicate white spandrels of powdery dust down below.

Ingela leaves tomorrow for a month back home in Sweden. She is giving last-minute instructions to Julius so that he can coordinate her other Maasai assistants during her absence. I had first met Julius in a remote Maasai village. He was wrapped in a bouquet of bright red *shukas* and wore a set of blue, white, and yellow bead necklaces. This morning, though, he is dressed in a pale shirt and dark trousers and looks like any other Tanzanian with a desk job. He speaks enough English that his conversation with Ingela alternates fluidly between English and Swahili.

After finishing their debriefing, Ingela tells me some interesting news. Julius told her that the Maasai leaders were pushing hard for the moranis to stop killing lions. "If a Maasai kills another Maasai, he has to pay forty-nine cows to the victim's family, and the elders have just told the moranis they'll have to pay the same amount of cows for killing a lion."

Julius tells us that, after the lion killings in April 2010, every morani in the Ol Balbal Ward had to pay a sixty-dollar fine to the NCAA, and the elders all had to pay three-dollar fines. The Maasai leaders instructed their morani to quit killing lions, and they haven't conducted another *Ala-mayo* ever since.

What happened to the morani with MH37's collar?

"He was fined twelve hundred dollars. He spent two months in jail before he could raise the cash."

What about the ritual lion killing in the NCA on Christmas day. Eighteen morani were caught in the act. . . .

"They were from Loliondo not the NCA; each one had to pay a twelve hundred dollar fine. Only four have been able to pay. The other fourteen are still in jail. The elders in Loliondo also made their morani stop killing lions. The fines are too high."

I had been wondering if the new spate of lion killings in 2010 was prompted by a change in attitudes. There had been a big conference back in 2002 to stop the traditional cattle rustling between Maasai and Sukuma. Tribal leaders met with government officials and agreed to control their young men.

Without an opportunity to prove their manhood by stealing cattle, were the morani now left with only the one challenge?

No, says Julius, they don't kill lions as a replacement for cattle rustling. But the truce did make it possible for the Maasai to settle on the border of Sukuma-land where there were a lot of lions. But these new cases were retaliatory killings not *Ala-mayo*.

Had the Maasai adopted *Ala-mayo* as a new form of political protest—like with the rhinos in the 1970s? Were they protesting the World Heritage Site ban on agriculture?

No, nothing was organized, nothing was planned. Those morani just got lucky. After they managed to kill one lion, they kept finding more and more—it was just a fad.

Down on the Crater floor, we spot a group of two female lions and their five cubs at the Seneto Springs. About twenty tour cars are already bunched together on the road. We are the only vehicle with permission to drive close enough to see the whisker spots on the cubs.

We are lucky to find the lions before the Maasai herders arrive at the springs this morning. Most of the standing water is gone by now from the Crater rim, but the Maasai bring their cattle down here every day of the year, claiming that their livestock need salt from the soda lake during the wet season. But everyone knows that this is just another form of protest, reminding the NCAA that they still hold a claim to the Crater floor.

Ingela brought two of her Maasai staff down here a few weeks ago, and they were appalled that the NCA had built flush toilets so near to Seneto—they couldn't believe that the NCA had chosen to devote fresh water to the disposal of human excrement, and, surely, all that waste would pollute the springs and the lake.

Their misgivings about Western hygiene contrasted rather perfectly with the handwritten admonition for visiting pastoralists that I had seen taped above the men's urinals up at the NCAA headquarters a few days ago: "Thank you for not peeing on the floor."

After collecting ID photos of the three youngest cubs at Seneto, we drive around for an hour, looking for tour cars that are bunched together—often the best indicator of a lion or rhino in the area. Driving past warthogs and buffalos, crowned cranes, and Thomson's gazelle, we find a four-year-old male lion named MG85. Ingela points to the peculiar wound on his back, near the base of his tail.

She saw him up on the rim near Ngototo Sumbat on the eleventh of July. He had been with two females who had killed a cow the previous night. MG85 and the two females were the cattle killers that provoked the Maasai into recognizing that they either needed to live with lions or face the risk of being banished from the NCA for good. Ingela's breakthrough was brokered around the fourteenth of July.

But on the sixteenth of July, Ingela found MG85 back down on the Crater floor with a fresh wound near the base of his spine. The wound was caused by a Maasai spear.

Although Ingela is perfectly happy to be the community cell phone charger and taxi driver, I am surprised that the ex-nurse has avoided becoming an angel of mercy. "One day, the Maasai told us about this woman in her hut who was suffering from high fever. It sounded like she was dehydrated, and she was probably close to dying. The village wanted me to take her to hospital, but it would have taken me all day. I'm just one person, and I can't spend all my time being an ambulance service every time someone gets sick."

She grips the steering wheel and stares into the distance, "Why can't the NCAA do at least that one small thing for these poor Maasai?"

Ingela is confident that her field staff will be able to document enough cases of livestock depredation to sort out the ideal location for an eventual Lion Guardians program, but she still needs to find a true Tanzanian counterpart for her project—someone with a college degree who can study the problem from the perspective of a social scientist. Most highly educated Maasai have either left the NCA or are already employed. But she has just heard from one of the village leaders that his son is back from college for the summer.

The father explains that his son is at an Ol Pool—a traditional Maasai feast where the moranis and elders eat nothing but meat and entrails for whole weeks at a time.

We drive to the edge of the steep escarpment overlooking the Hadza stronghold near Lake Eyasi, hire a couple of Maasai guides, and set out walking in the light of the late afternoon. Our guides know of three likely sites for the Ol Pool, but the first two come up empty. Then we see a confluence of dusty footprints made by tire-track sandals and catch the scent of smoke and meat. Our guides ask us to wait as it would be rude for a couple of *wazungu* to walk into the middle of a traditional ceremony.

A few minutes later, we meet Saning'o Kimani, a twenty-three year-old sophomore at Montclair State in New Jersey. His English is flawless. He has an easy smile, and he seems as poised and likeable as any good-looking undergraduate, very friendly and almost sheepish that we had walked all that way to find him. Our arrival reminds him of his other life—of his college weekends in Manhattan—and he seems almost dizzy about being back home at an Ol Pool. He's not sure he can face the prospect of eating another seven goats and a cow over the next few days.

Saning'o is majoring in biology and international education. He says he wants to bring back his training to the NCA. We discuss the Lion Guardians, hoping he will decide to work with Ingela when he graduates in another three years. We tell him that we could arrange for him to start apprenticing with her next summer. Human-carnivore conflict is a whole new field of wildlife management. His research could have direct relevance to the well-being of his family and friends.

He mentions, though, that he is interested in pursuing a career in medicine. We talk frankly with him about the rather hopeless future of pastoralism in the NCA, and it occurs to me that maybe he should stick with medicine.

7 Fade Out

ZANZIBAR, 16 AUGUST 2012

In June, Susan and I spent a week poking around some of the most exotic properties in Stone Town, eating lunch and dinner in a series of refurbished palaces and homes that had been converted into sumptuous hotels and restaurants. The House of Spices had a green-walled coffee bar and restaurant on the top floor and a blue-walled courtyard that soared above a blue-tiled fountain; gold-leaf Arabic patterns adorned the blue alcoves on the upper floors. Formerly the home of a sultan, 236 Hurumzi Street was now transformed into a dream: some rooms had no walls; others were on stilts. There were bathtubs under the stars, silken canopies and wooden trellises, paintings of peacocks in the stairwell, and a tower-top restaurant high above the oldest part of town.

Tonight I am scheduled to give my tourist-operators' lion talk on the roof of the Emerson Spice with its blue-wash walls, trellised wooden balconies, hanging plants, white-tiled fountains and an exotic atmosphere straight from the Arabian nights—not to mention the rooftop restaurant with unbelievably good food.

I have come early to introduce myself. Emerson is a bulldog of a man from New York City. He wears a baseball cap with a Swahili slogan extolling the virtues of Zanzibar. His left ear looks to have been bitten by a crocodile, but it's the work of a cancer surgeon. He's a heavy smoker, so his voice is graveled, but his manner of speaking is very precise, almost fastidious, and he likes to laugh with a staccato *heh-heh-heh-heh-heh.*

Emerson had planned to spend a four-month vacation in Zanzibar but stayed on for twenty-three years. He opened the first foreign-owned business in Stone Town, converting his own home into a hotel. Eventually, he restored and designed the fantasylands of the 236, the Zanzibar Coffee House, and the Emerson Spice.

Settling in with our iced glasses of hibiscus juice, we talk about the cultural differences between the mainland and the islands, how greater political autonomy and Zanzibar's eventual independence is inevitable. I ask him about the decrepitude of so many historic houses in Stone Town and whether there is any sort of master plan to restore the place.

"If all the money spent on yet another 'study' of the rehabilitation of Stone Town had instead been spent on replacing the roofs, not one of these buildings would have fallen down." As with so many other projects in Tanzania, he says, the money was mostly spent on consultants from Sweden, Denmark, Norway, and Germany.

From the roof of Emerson Spice, we can see the giant East German monstrosity just beyond the edge of Stone Town that was built shortly after the Zanzibar Revolution. Emerson says that the German architect had originally designed a series of interlocking *L*-shaped buildings with courtyards and ground-floor shops, which were four-stories high—the same height as the buildings in Stone Town. But Zanzibar's revolutionary president, Karume, rejected the design. Karume told Erich Honecker that he wanted an exact copy of the socialist apartments he had seen in East Berlin.

Throughout our conversation, Emerson's cell phone keeps lighting up with a ringtone that sounds like Maria Callas singing an aria from Puccini. His Swahili is fluent, and his restaurant staff laugh happily as they overhear his half of a particularly animated phone conversation. Whenever the men look at him, they grin spectacularly—they all call him *Babu* ("grandfather"), and one waiter appears from nowhere to give him a spontaneous neck rub.

I congratulate him on the incredibly good dinner that Susan and I had enjoyed during our visit in June. He tells me that he and his staff have developed 150 recipes—he uses all of them over the course of the year to allow for the seasonal availability of fresh foods. He serves a fixed menu each night, and the whole meal consists of about sixteen different dishes delivered in five courses. As he describes the preparations, I remember how our entire meal had felt like a musical composition with interweaving themes and melodies.

He finishes with a flourish: "The most important spice is happiness. My staff and I have always had a terrific time preparing the food together. We've always had a lot of laughs."

The sun is setting, and Emerson is sitting alone on the roof of his hotel, wearing a creamy white fedora, a black polo shirt, and snow-white shorts. He is expecting twenty to twenty-five people, but the place is still empty, so I talk about the magic atmosphere that Susan and I felt when we first saw the interior decorations of Emerson's and the 236. He says that he still owns a half interest in the 236, but he is obviously bitter at being marginalized by the current management. When he first designed the 236, it was written up as one of the "1001 Places You Should See before You Die," and it was featured in an article in *Time* magazine. But the current management has no sense of design, he scowls; the most recent rooms are much too boxy.

Emerson is nervous that no one has shown up for my talk tonight, except, he says, that everyone shows up fifteen minutes late in Zanzibar, and we are in the last week of Ramadan so most Muslims will be staying at home, breaking their daytime fasts with their family, friends and neighbors. Two Gujarati men finally walk upstairs at 7:16, a Canadian couple appears at 7:23, and I start speaking at 7:30 with a brief break at 7:56 while the muezzins sing their amplified prayers from the neighboring mosques.

I finish my presentation, and Emerson sets two chairs toward the audience. The six of us begin a quiet discussion about lions and conservation, which unfolds smoothly enough until one of the Gujarati's asks about corruption in the wildlife sector of the mainland government.

I hadn't mentioned trophy hunting in my PowerPoint, but it's an unavoidable topic at this point. Then Emerson shakes his head and asks for the name of that Frenchmen who owns the biggest hunting company in the country.

"Gerard Pasanisi?"

"That's right, *Pasanisi*." He grins. "Pasanisi brought a party of nine hunters to my hotel in 1991. One of them kept bragging about illegally shooting a rhino. They couldn't stop talking about paying off bribes and smuggling their trophies out of the country.

"The first day they were here . . . they were unspeakably rude to my staff—treated them like dirt. I couldn't believe it; I had never seen anything like it. The next day I instructed my guys to give them the worst possible service and to walk straight past them with their noses one inch in the air and to mutter under their breath, '*Petit bourgeois*,' so quietly that the hunters couldn't be sure whether they had actually heard it.

"And it worked! The second day they were so angry at our bad service that they checked out. But they had paid in advance for three full days, so we were completely empty on the third day, and we could all take a holiday because we had already been paid!"

MINNEAPOLIS, 8 SEPTEMBER 2012

On the fourth of August, Dennis Ikanda and Bernard Kissui asked to meet with me in Arusha. They wouldn't say why—just that it was urgent. We convened at an open-air restaurant in a residential area in the middle of town. Benny was already seated by the time I arrived. Dennis joined us about a half hour later. We went through the usual ritual of familiarities: everyone's health was fine, Dennis's wife was expecting a new baby, Benny's new house in Arusha was nearly ready.

Dennis had returned to his former self. He seemed to have recovered from the stresses of writing his PhD thesis, of balancing family life with work. He still took the government's point of view in any public discussion, but his underlying humanity had reemerged, and I trusted his advice as much as anyone's. He wanted to make it clear that he and Benny were both very sad to see my name dragged through the mud. They had both been my students, and their reputations ultimately depended on my own.

While my science was widely respected within the country, I was viewed as a meddler and a troublemaker, and I had come close to being barred from Tanzania in 2011. When the one-lion-per-thousand-square-kilometer paper came out, the then-minister for natural resources tried to revoke my clearance from the Tanzania Commission on Science and Technology and to refuse my Tanzanian resident's permit. Even President Kikwete was angry with me. Trophy hunting was an important source of foreign revenue for the country, and Kikwete was convinced that I wanted to ban lion hunting at CITES.

But Dennis also reported that the forces for reform were slowly gaining power. Parliament had passed the six-year age minimum for lion trophies. He and Benny had been asked by the WD to conduct field surveys on the sizes and trends of lion populations within the hunting blocks—maybe not in the Selous, which was politically sensitive, but in as much of the rest of the country as possible. Meanwhile, all the new hunting-block holders were being held to the recommended one-lion-per-thousand-square-kilometer limit.

Dennis and Benny would be my eyes and ears, they assured me. There was too much at stake for all of us if I didn't step aside and let them take over. It was time for me to leave politics and return to science.

Up until that point, I had always assumed that any fallout from my extracurricular activities would only affect me, but the situation was inevitably more complex. Dennis and Benny were just as important for conserving Tanzania's lions as any government rules and regulations. And the big camera-trapping survey in the Serengeti was just starting to get interesting. With millions more photos of wild animals, large and small, every day and night over the next ten years, we could develop the most comprehensive understanding of any savanna ecosystem in the world.

There was nothing to be gained from pushing the WD to verify the ages of trophy lions or from talking to any more government officials about reforming Tanzania's hunting industry.

As I drove away from the restaurant that day, I felt irrelevant—and puzzled. I wondered whether Dennis was acting under orders from the WD. I wondered what else he had been told.

I had been cautioned before, but this time was different. The bad news had been delivered in person by a government employee who liked to present himself as the voice of his superiors. But I was also convinced that an elaborate facade had been concocted by the WD. Maybe they really did plan to restrict lion harvests. Maybe they really did worry about the Lacey Act and USFWS and CITES. Or maybe they had figured out how to work around meddlers like me.

The market for trophy hunting has fallen substantially since the global economic recession in 2008. Tastes have changed, hunters get older every year, and shooting large mammals doesn't hold the same mystique for the younger generation. Gerard Pasanisi has officially retired, and his son, Eric, now runs the family business in Tanzania, and it was Gerard Pasanisi who spent so much time with all the Tanzanian presidents and ministers and whose passion for secrecy gave Severre so much power at the WD. It was Gerard, in short, who was God.

The son of God may continue to bask in the glow of dad's glory, but French neocolonialism in Tanzania must surely die a natural death someday.

At Pasanisi's gala dinner in Dar es Salaam in 2006, President Kikwete stated: "There are whispers that, as it stands today, it is not so easy for Tanzanians to participate in the hunting industry. We cannot ignore the whispers. We will try to find better ways of involving them without displacing the others. I believe both nationals and nonnationals can participate and work smoothly and amicably in this industry. I do not subscribe to demagogic approach on this important matter."

I was sitting at the same table as Sheni Lalji at the time, and I was so focused on Pasanisi's wrinkled scowls that I forgot all about Sheni—but I would guess that Sheni was smiling quietly to himself as he listened to Kikwete. He was, after all, a Tanzanian citizen. Any restrictions on foreign ownership would play into his hands.

Sheni's offices were only a few blocks from the Ivory Room in Dar es Salaam. I had walked past there a dozen times the past few years, but I never felt the need to meet with him again after that encounter in 2006 where he proclaimed that our African staff at Savannas Forever would all be liars and drunkards.

I decided to pay him an unannounced visit, on the twenty-third of August. At the entrance of the building, I felt the glare of a dozen unfriendly eyes.

Where was I going?

"I'd like to meet with Sheni."

An African laborer raised his hand toward a low building. Two more laborers received me in a workshop with a small office inside. I went into the poky little warehouse office and an Asian office worker squinted at me suspiciously. I asked, "Is Sheni here?"

"No, he's traveling."

"When will he be back?"

"In a few weeks."

But I had to fly home the next day.

I'm now back in Minneapolis and won't return to Africa again until February. I've been told not to meddle in Tanzanian politics and probably shouldn't even try to talk to USFWS and CITES. But I can keep up with the Tanzanian news over the Internet, and just this morning I spotted the following story:

> TOURIST HUNTING BLOCKS: KAGASHEKI LEFT DUMBFOUNDED
> *The Guardian*, 8th September 2012
>
> The Minister for Natural Resources and Tourism, Khamis Kagasheki says he has been shocked by unfair allocation of tourist hunting blocks as he found out that one person was allocated 21 blocks.
>
> Minister Kagasheki [said] that it was incomprehensible for one person to be allocated with such a big number of blocks while other applicants complained of missing out an allocation.
>
> ". . . I was told that the said person submitted different applications some of which had his wife's name, others with children's names and yet other applications had a company name. This is unacceptable because this unfairness is so glaring and [discredits] the ministry," he stated.
>
> However he could not reveal the name of the person or the company enjoying unfair allocation of hunting blocks.
>
> "We have a number of challenges here. One . . . is this group of businessmen with limitless finances who do all in their means to corrupt the system, but I will fight this relentlessly, and the good point is that I have full support from the President who tells me I should carry on to ensure that justice is done so as to restore public confidence. . . ."

Though the reform-minded minister "could not reveal the name of the person," a brief bit of online research led to these stories:

PROBE SUBLETTING OF HUNTING BLOCKS TO URANIUM EXPLORERS—MP
The Citizen, Thursday, August 9, 2012

A commission should be formed to investigate the legality of some hunting companies to sublet their blocks to uranium exploration companies in Namtumbo District, Ruvuma Region.

. . . The Kawe legislator . . . said . . . the contract . . . involves the area around Mbarang'andu Village in Namtumbo District, which is within the Selous Game Reserve. "I understand that the village received only $10,000 while the hunting firm pocketed $6 million from the deal . . . this is not fair," she declared. . . .

"The [laws] allow a person with a hunting permit to only carry out hunting activities and not otherwise," she said.

. . . Game Frontiers of Tanzania Ltd sub-leased part of the land in Mbarang'andu Village in Namtumbo. She added that . . . should a piece of land allocated for other purposes be found to contain minerals, then ownership returns to the government . . .

GOVT ORDERS HUNTING FIRM TO GIVE EXPLANATION
The Citizen, Monday, August 13, 2012

The government has ordered Game Frontiers—a hunting firm in southern Tanzania—to issue a statement over the latter's violation of regulations.

Game Frontiers has a hunting block [bordering] Selous Game Reserve but it is reported to have leased the area to a mining firm exploring for uranium, contrary to the contract.

"The current regulations restrict a hunting firm from leasing its area to another company. We have reports that Game Frontiers leased their block to a uranium exploration firm contrary to the agreement.

"We are giving the company a chance to explain why it violated its contract. If proved to have violated regulations then the company will lose the block," said the minister for Natural Resources and Tourism, Mr Khamis Kagasheki.

Part of the Selous Game Reserve is reported to contain uranium and companies like Uranium Resources hold a portfolio of mining licences covering approximately 12,700 square kilometres. . . .

> During deliberations on the estimates of the ministry of Natural Resources and Tourism, [an opposition MP stated] that Game Frontiers leased its hunting block to uranium exploring firms. . . . But Mr Kagasheki said the company was supposed to report to the ministry and that what it did was completely wrong according to . . . existing hunting regulations.

Game Frontiers belongs to Sheni Lalji, of the Central Committee of CCM, Tanzania's ruling party. That same Sheni who is already richer than god. He who may well have been traveling when I looked for him a few weeks ago in Dar but who may well have been dealing with the uranium company.

Luke Sidewalker's dossier had linked Sheni to twenty hunting blocks, and I have no idea if the local press or Minister Kagasheki used that document to tally up Sheni's current holdings, but I'd like to think that they found it useful somehow. Sheni might have finally crossed the line; surely even he cannot lease the hunting rights from the government for a few thousand dollars and then turn around and sell the mineral rights (which he doesn't own) for six million dollars.

In the past few years, more and more Tanzanian politicians have wanted to obtain their own hunting blocks, but about a quarter of the blocks no longer support any wildlife. Sheni and Pasanisi hold a third of the surviving blocks, and, as rich as he is, Sheni could only ever afford to pay off a fraction of the bigwigs who want to join him in the hunting fraternity. With so many people left out from the pie, someone was bound to go after Sheni eventually.

Erasmus Tarimo made several genuine gestures toward reform while he was in charge of the WD, but the first chairman of the Savannas Forever board, Mzingo Pinda, has served as the prime minister of Tanzania for the past five years—and without Pinda's help Savannas Forever never would have been granted the necessary permits to conduct rural surveys, and I never would have received Tarimo's cooperation.

Pinda has been close friends with Monique Borgerhoff Mulder and her husband, Tim Caro, since before Pinda first ran for Parliament. Monique, Tim, Susan, and I visited Pinda at his home in western Tanzania in 2010. The election campaign had just started, and we all sat out under the starlit African sky less than ten meters away from the sole dirt road that runs through Mpimbwe village.

There we all were, drinking beer and soda in the dark with the second-highest-ranking official in the country, and there wasn't a bodyguard or weapon anywhere in sight. In what other country could a prime minister sit unguarded like that?

Despite the growing cynical corruption of the Tanzanian government, I have never doubted that Pinda wants to do the right thing. At our very first board meeting in Dodoma, I briefed everyone about the obstacles that Severre had thrown in our path for working in the rural areas around the hunting blocks. Pinda was still minister for local government at the time. The other two MPs on our board were both awestruck as he suggested ways to get around the problems with Severre, saying, "Let's do something right for this country for a change."

A few months after Severre was removed from the WD and Prime Minister Edward Lowassa was sacked because of his flagrantly corrupt abuse of power, Pinda was chosen as the new Prime Minister specifically because of his reputation for honesty. After his promotion, Pinda quickly brought Tarimo to heel, but although he often helped introduce Savannas Forever to high-level government officials, he was so mindful of potential conflicts of interest that he never asked any international donors to consider our services.

I mostly hear about Pinda through Monique these days, and as CCM has descended into ever worsening corruption, she says that Pinda has to spend almost all his energies dealing with internal divisions within CCM, and feels he can no longer work for the good of the country. People on the street have told me that "Pinda is too polite; he should go back home to his family. We need a strong leader—like Lowassa"—Lowassa who was almost comically corrupt but who provided a certain reflected glory to his local constituents.

So where does all this leave the wildlife? During my first twenty to thirty years in Tanzania, conservationists kept setting aside more national parks, game reserves, forest reserves, and wildlife management areas. But it is never enough because each new area should ideally be surrounded by yet another buffer zone where ever-more people are somehow expected to live with lions, leopards, and elephants. This whole fantasy is starting to fall down around our ears. In the past ten years, the elephant population in the Selous has plunged by at least 60 percent, and a quarter of all hunting blocks have been abandoned because the wildlife are all gone.

And the elder Pasanisi was right about the WMAs: instead of having to deal with just one corrupt agency in the government, the hunting companies now have to deal with the corrupt local managers of the WMAs, and those corrupt local managers don't distribute the income any more equitably to their fellow villagers than did Severre from his office in the WD.

And a lot of the international aid agencies that helped build the WMAs have added their own layer of corruption. The World Wildlife Fund just had to sack all its top Tanzanian staff for embezzling USAID funding that was meant to establish a half dozen WMAs.

The sad tragic truth of the matter is that Tanzania has too much. In an ideal world, the entire country should have been declared a world park, every square inch between the Mozambique border and the Kenyan border, from Lake Tanganyika to the Indian Ocean—this is truly the Garden of Eden. There is hardly a spot in the country that couldn't have supported healthy populations of large mammals up until the 1950 or 1960s.

But people have always lived here, too, ever since our ancestors diverged from chimpanzees, ever since we first learned to walk on two legs. The Western model of parks for animals and the rest for people was applied with vigor in the 1950s and 1960s: about a third of the country was given some sort of conservation status. But to properly manage *all* those areas, the country should be investing something like six hundred million dollars in management every year. The current figure is closer to ten million dollars. If the world really wants to protect these places, the world should help cover the costs.

The bottom line is clear: Tanzania's wildlife cannot pay its own way—the free-market economic model for conservation has failed. Trophy hunters had first crack at the challenge, but they are unwilling to raise their fees to a million dollars per trophy lion. Photo tourism can bring in a lot more people, but they have to keep their prices competitive with beach holidays and family trips to Disneyland.

And of course, if the world does decide to inject tens of millions of dollars for funding these world treasures, Lowassa, Sheni, and hundreds of others will all be available to help spend it—not to mention all the international aid agencies and the Beltway bandits. . . .

As for Ingela's project, her best hope lies in the Maasai's threat of eviction from a World Heritage Site. While the NCAA's passive-aggressive approach to Maasai management is utterly disgraceful, it appears to be a deliberate strategy for provoking the Maasai to becoming politically unsuitable tenants in the NCA. The government is eager to grant itself permission to kick all the Maasai out, and the Maasai know it. So, as Ingela says, if the Maasai can't learn to live with lions, they won't be able to live there anymore.

Meanwhile, even the best intended strategies in other parts of the country seem doomed by the sheer scale of the problem. Bernard Kissui has convinced about a hundred Maasai families around Tarangire to sell enough livestock to cover half the costs for enough chain-link fencing to reinforce their bomas. But there are nearly twelve thousand families still to go.

Bernard is a conservationist and a scientist, not a fence salesman. He can only reach a small number of families. He can only deliver so many rolls of chain-link fencing in his research vehicle, and he can only ever convince a small number of very traditional patriarchs that it's in their best interests to part with even one precious cow.

But what if someone could convince the Maasai to view their livestock as part of the cash economy? Then the costs of boma fencing would just be part of their operating expenses that could be divided between the herders, abattoirs, and dairy cooperatives.

And maybe the Maasai could learn to keep chickens while they're at it.

In western Tanzania, the Sukuma traditionally only ever killed lions in retaliation for livestock depredation, and the act of spearing a lion was considered so dangerous and terrifying that the lion killer was encouraged to travel from household to household and perform a "lion dance" to exorcise his demons—a sort of folk cure for posttraumatic stress disorder. Each household would then pay something to the dancer because he had protected all the community's livestock from future cattle killing.

But Monique and her students have recently discovered that unemployed Sukuma youth have started preemptively killing any lion they can find and then perform the lion dance as if they had provided a community service. Monique's team has convinced the traditional tribal leadership (called *Sungu-sungu*) to impose fines on bogus lion dancers in three villages near Mpimbwe, but her Tanzanian student was recently told it would be unsafe to try to abolish bogus lion dancing in many of the neighboring villages.

Professional lion killers, it seems, also have political connections.

So what to do? Tourists to East Africa love to dress up like Karen Blixen and Dennis Finch Hatton when they come out on safari—and the lodges and tented camps all oblige with black-and-white photos of *wazungu* in pith helmets, nostalgic decor, and vistas of endless plains. But the good old days were a hundred years ago. The parks have too many neighbors these days, and good fences may be the only way to make good neighbors. Without adequate management funding anywhere in the country, nowhere is being properly protected. As much as people might love Tarangire, Katavi, Ruaha, and Manyara, it would be tragic if funding were spread so thin that they dragged Serengeti and Selous out of existence, too.

Even though triage seems inevitable in the short term, there is a faint flickering light at the end of the tunnel. Malthusian predictions of doom and gloom have never been particularly accurate—in 1968, Paul Ehrlich famously predicted, in *The Population Bomb*, that four billion people would starve to death between 1980 and 1989, including sixty-five million in the United States. ("If I were a gambler, I would take even money that England will not exist in the year 2000," he famously wrote in that book.)

The "demographic transition" is the conservationist's best friend: should rural Tanzanians actually achieve a decent standard of living, universal education, and low levels of maternal mortality, typical family size could eventually drop to the same two kids per family as seen in the rest of the world. Urbanization has finally reached Africa—besides all those Maasai morani who no longer tend the cattle herds in the NCA, there has been such a rush to the cities by young people in recent years that elder care in rural Tanzania is now a serious problem. The multigenerational family is fast becoming a thing of the past.

So even in Africa, human population may well peak some day, and maybe some of the "lost parks" can eventually be restocked, the fences eventually dismantled.

Meanwhile, I will stay out of Tanzanian politics. The Hopcraft effect, Peyton's manes, Karyl's harvest model, Sarah's vaccination campaign, Linda's coinfections, and Anna's real estate map were all such hard acts to follow that I'll confess to having felt the need to reenergize myself by breaking out of the Serengeti the past few years, playing around with Bennie's cattle killers and Dennis and Hadas's man-eaters for a while—and I really did want to discover if the Whole Village Project could treat foreign aid projects as large-scale experiments in biocomplexity.

I'm ready to cocoon myself once more inside the comfort of the Serengeti.

I have always concentrated on the ways that lions live together—I was most interested in how lions reared their cubs, the extent to which they cooperated when they hunted, the forces that drove them to gang warfare. Even after the big disease die-offs in 1994 and 2001, my graduate student, Meggan Craft, spent most of her time studying how an airborne pathogen like CDV might persist in the population by traveling from one lion to another according to patterns of social interaction and interpride encounters. But the Serengeti is home to a wide range of carnivores that can all carry the same disease—and Meggan found that lions were only one piece in a complex multispecies network and never could have sustained the outbreak by themselves. The real villains were the intermingling domestic dogs, jackals and hyenas: lions were just spillover hosts.

Ali Swanson joined the project in 2008, wanting to know how lions, leopards, hyenas, and cheetahs all manage to live together in the Serengeti. Maybe certain stretches of lion real estate are less valuable if they attract too many undesirables (e.g., hyenas or leopards). Maybe lions thrive better in neighborhoods where they can reliably scavenge from cheetahs. Lions go out of their way to kill leopards, cheetahs, and hyenas—how do these smaller species manage to cope?

At first we thought about attaching dozens of radio collars on these three species and following them around like the lions, but while my mind was wandering one day on the eighteenth floor of a Manhattan skyscraper at an advisory board meeting for Panthera, I suddenly realized that it would make a lot more sense for Ali to set out a ridiculous number of camera traps and to look at how the four species divvied up the central landscape of the park.

Statistically, the technique has been a great success. Conservationists had worried that lions could drive cheetahs to localized extinction, but the Serengeti cheetahs have managed to thrive despite the tripling in the lion population over the past forty years—and Ali has already found that the cheetahs are able to stay safe by waiting at least twelve hours before venturing to the same site where a lion had rested.

The real joy comes from the individual photographs themselves. One flash photograph lights up a lion with her mouth and forepaws gripped tightly around a zebra's neck. A photo at sunset portrays a bouquet of rufous ears and short black horns from a small herd of eland. A night shot shows a giraffe's crotch crowded with roosting oxpeckers. On another night, a male porcupine carefully aims his phallus beneath his lady's quills. And I never would have guessed that a couple of bat-eared foxes would ever have the courage to chase off an aardwolf.

There's always something new to discover. It never gets old.

ROCHESTER, NEW YORK, 16 SEPTEMBER 2012

"Can I *touch* him?" I am astonished, amazed to be this close to a *Ceratotherium simum*, more popularly known as the white rhino.

This one is a male, six years old and five thousand pounds, and he is as gentle as a lamb, only a lot less flimsy. He eats bananas from my hand. He has no front teeth, and his lips are about a foot thick with a small opening to whatever that area is back there where he chews his food with his massive molars.

I can't take my eyes off the creases, folds, and rugged texture of his skin. This white rhino really is white, and his skin looks like it belongs to another world—a world of dinosaurs and giant club mosses. But touching that skin is irresistible. His shoulders and back feel like asphalt, the bumps and ridges on his sides are rough as sandpaper, but just behind his ears, on his upper lip, in the folds of his creased armor, his skin is smooth and supple as the softest leather.

And he loves being touched!

The white rhino is a solitary animal. Once they have been weaned, these animals never touch each other, except during sex. The rhino keeper at the Seneca Zoo tells me about a certain spot on a white rhino's chest and another somewhere near its groin: "Touch one there, and they immediately fall asleep. They doze off so fast you have to be careful not to get your arm stuck underneath."

Who knew?

8 Exile

WASHINGTON, DC, 26 JUNE 2013

Today was a big day: the U.S. Fish and Wildlife Service held a formal workshop to discuss the potential listing of the lion on the Endangered Species Act (ESA). The stakes were high. A "threatened" listing would only require additional oversight into current hunting practices, but an "endangered" listing would ban imports of lion trophies into the United States and create momentum for an eventual appendix 1 listing at CITES, resulting in a worldwide ban on lion hunting.

The room was jammed. Safari Club International and various animal welfare groups had each sent half a dozen representatives. John Jackson and Dennis Ikanda were both in attendance, and everyone was awkwardly polite to each other.

Anti-hunting platitudes flew from the left side of the room, pro-hunting platitudes from the right. It was like a small-scale version of the CITES meeting in 2004, only this time I felt compelled to fight against both sides.

But I had to be careful. My Tanzanian research clearance had recently been revoked, and my appeal was still under review. John Jackson and Dennis Ikanda would nail me if I even hinted at the word "corruption." But I had to speak out—respectfully, levelly—against so many blatant misrepresentations that I was dead on my feet by the time it was all over.

I left the conference room in a trance, my brain addled.

Everything had started falling apart seven months ago, on the twenty-seventh of November 2012. The U.S. Fish and Wildlife Service had requested comments on the petition to list the African lion on the ESA. An endangered listing would prohibit American hunters from bringing home their lion trophies. Americans comprised 60 percent of the lion-hunting market.

I e-mailed USFWS, asking to meet the next time I came to DC. I provided a few key details about corrupt practices and named a few key names, but I made it clear that it would be risky for me to write out everything I knew about all the corruption in the Tanzanian hunting industry. I should have written "confidential" on the subject line, but I didn't, and the recipient posted my e-mail on a public notice board on the USFWS website the following day.

When I learned that my e-mail message had gone public, I felt like I had been hit on the back of the head. I was in my living room in Minneapolis, and I could barely see straight. To calm down, I convinced myself it was better this way. If lions were going to be considered for listing, the authorities would have to decide on the basis of all the realities in Africa, and corruption was fundamental to understanding those realities.

Three days later, NSF declined my research proposal to extend the lion project. I had received twenty-nine consecutive years of NSF support up until then, and I had often changed the subject of my research (from cooperative behavior to the lion's mane to the ecology of disease to biocomplexity, et cetera), but the camera-trap study brought me into the field of community ecology, a complex, rigorous field with an associated academic literature that was so extensive and detailed that our inexperience with food webs and spatial ecology was too glaring for the review panel at NSF to ignore.

The next round at NSF would be on 14 January 2013. There was still time to recruit the right people to join Ali, Margaret, and me to overcome the referees' criticisms. If we were lucky and got the grant the following round, there would only be a three-month break in funding—surely I could find enough money to bridge the gap.

On the eleventh of December, Ali and Margaret premiered a brilliant new way to deal with the million-camera-trap-photos-a-year problem. They had teamed up with a citizen science program called Zooniverse that recruits thousands of volunteers to help classify the galaxies captured by the deep-space photography of the Hubble Space Telescope. Zooniverse is based in the astrophysics department at Oxford, with partners at the Adler Planetarium in Chicago. They had been looking for a nonastronomy project to diversify their portfolio; Ali and Margaret applied, and "Snapshot Serengeti" went online a few months later.

After a quick tutorial in how to distinguish a wildebeest from a buffalo, a cheetah from a leopard, a hartebeest from a topi, online visitors could view a random selection from the first two million camera-trap photos from the Serengeti.

Thousands of people flocked to the site from around the world, and many became hopelessly addicted. At first, you might get a picture that had been triggered by the grass moving in the background, so you entered "nothing." The next photo might show a few wildebeest in the middle distance, followed by more empty frames of grass stalks or tree branches waving in the wind, more wildebeest, wildebeest, nothing, nothing, nothing, then WHAM!

A leopard as sexy as silk lingerie! How cool is that?

Volunteers complained that they couldn't quit. They stayed up all night, classifying animals, eager to see what popped up next.

If the first ten people to classify a particular photo all agreed about which species had been captured, we accepted their judgment, and the image was "retired." If there was any disagreement among the first ten judges, the system showed the same picture to twenty more people—and we accepted the plurality vote.

So two million pictures, each viewed ten to thirty times, and do you know how long it took our Zooniverse volunteers to work through the whole pile?

Eleven days.

By the first of January, Margaret and Ali had produced a grid of pie charts illustrating how many animals passed by each trap each month. A young professor at Wake Forest, Mike Anderson, overlaid the pie-chart grid onto satellite data showing the green growth of the grasslands in the Serengeti each month.

And voila! Looking at the late dry season in two consecutive years, for example, we could see how the wildebeest stayed away from our part of the Serengeti during the brown months of September through November in 2010, but overran the grid when the rains broke in early November 2011. In contrast, the nonmigratory hartebeest remained scattered around the grid in both years.

We had solved the primary logistical problem of handling such a massive accumulation of data. The project wasn't just feasible; it was a thrill! By working with Mike Anderson, a botanist, soil scientist, and member of the original Serengeti biocomplexity project, we could investigate how geographical variations in soil chemistry across the one thousand–square-kilometer camera-trap grid ultimately passed upward through the vegetation to affect the distribution of the herbivores—and from the herbivores up to the carnivores. We also recruited a second new partner, James Forrester, whose sophisticated skills in spatial statistics would solidly anchor our eventual portrait of the Serengeti food web.

We finished our preproposal just in time. If NSF liked our four-page effort, they would invite us to submit a full proposal in August. The 2013 document was far stronger than 2012's, and we had been invited back in 2012, so life seemed good. I just needed to fill that three-month funding gap at the end of 2013.

I wrote to Frankfurt Zoological Society and asked if they could help out. My current funding ended in September 2013; if the new NSF proposal were successful, funding would resume in January 2014. The Serengeti lion project wasn't strictly a conservation project, but it was an iconic study that was closely allied with FZS, and, with Snapshot Serengeti, the camera-trap photos were a brilliant way to engage thousands of people in Serengeti wildlife.

The Frankfurt Zoological Society assigned Peyton West to help us out. Peyton had been working at the American Association for the Advancement of Science in Washington, DC, for the past few years, and she wanted to reconnect with African wildlife. Markus asked her to head the new fund-raising wing of FZS that had just been launched in the United States.

I had seen Peyton and her family every few months for the past couple of years. Her husband had helped me navigate the halls of Congress back in the days of the Whole Village Project. The U.S. wing of FZS was well connected. Raising enough money to extend the lion project for three months should be relatively painless.

Everything was bound to be OK.

In late January, Zambia banned the hunting of lions and leopards for the 2013 season. John Jackson immediately blamed me, saying that the Zambian government had quoted my research findings from the Tanzanian hunting blocks. In fact, the hunting ban was largely an attempt to fight corruption in their block allocation system. The Zambian equivalent of Sheni Lalji, an Asian named Rashid, was gobbling up most of the blocks in the country. The American government had a quiet interest in Rashid: he was believed to have links with al-Qaeda.

Meanwhile, a backlash was developing against the stonewalling of the Tanzanian hunting industry. Several American hunters were concerned that the lion might be listed on the Endangered Species Act unless the hunting community recognized the consequences of past practices and embraced the recent reforms. They had attended various Safari Club International conventions, they had seen the underaged "trophy" lions, and they knew that there had been very little oversight in the African countries, that lion offtakes had been virtually unregulated. The hunting community needed to embrace the age minimum publicly.

Thus arose the definition of a "huntable lion" as a male old enough to be harvested safely without the risk of infanticide. John Jackson and his cronies opposed the idea, but they were outnumbered by the reformists.

Then the head of Dallas Safari Club called one Sunday, and we discussed the logistics of trophy monitoring should the lion be listed as endangered. We decided that if the age of the harvested lion couldn't be verified upon export, the trophy could be inspected after arrival in the United States. Most American hunters are eager to show off their trophies, thus they submit measurements and photographs to groups like the Dallas Safari Club so as to qualify for the record books. What about setting up an age-verification system in Dallas?

I went to Washington a few days later and met with the staff at USFWS who would make the decision on the lion listing. The animal welfare community and the hunting industry were both lobbying hard, and USFWS dreaded the coming battle.

The absence of age verification of Tanzanian lions was a problem. The U.S. Fish and Wildlife Service lacked the resources to inspect every trophy imported into the United States and sounded relieved at the idea of self-policing by the American hunting organizations.

A few days later, I talked to Steve Chancellor, asking if he could get SCI to adopt a policy of lion inspections similar to that of the Dallas Safari Club. He wasn't interested. He said that most of the lions left in Africa lived in hunting areas, that hunters gave value to lions, and that the listing would lead to their immediate extinction. Lions were only in real danger in West Africa, so lions from those countries might have to be listed, but the rest of Africa was fine. He had just met with the Zambian minister of wildlife—she was on the verge of changing her mind about the lion ban. He had just helped SCI raise a million dollars to fight the lion listing in court.

In March 2013, the paper "Conserving Large Carnivores: Dollars and Fence" finally came out. The story was picked up by *Science* magazine, the *New York Times*, and the British press. Some of the headlines were a bit over the top: "To Save the King of the Jungle, a Call to Pen Him In." I had originally written the paper to determine the true costs of lion conservation but came to discover that fences could even be more cost-effective than constantly fighting poachers and mitigating wildlife conflicts in an unfenced park. I felt that fencing was less important than the magnitude of the financial challenge, but it turned out to provide the sound bite that caught the ear of the press.

I wrote an op-ed for the *Los Angeles Times* that directly addressed the fencing issue. As background, I outlined the history of wildlife management in South Africa where wildlife had largely been eliminated in the nineteenth century. In the 1990s, rural land owners sought permission from local people to bring back dangerous animals after assuring everyone that the elephants and lions would be safely fenced inside—an approach that contrasted dramatically with the laissez-faire traditions of East Africa, where wildlife were only now being eliminated by burgeoning population growth, and conservationists clung to their memories of untrammeled landscapes.

I pointed out that more wild lions live today in South Africa's fenced parks and conservancies than existed in the country a century ago, yet no one in South Africa complains of livestock losses—no one worries about man-eating lions. In considering the rest of Africa, I focused on the Selous—an unfenced nonmigratory ecosystem the size of Switzerland, with a management budget of five million dollars a year that would need over a hundred million dollars per year to safely sustain even half its potential lions. Then again, fencing the Selous would cost something like thirty million dollars, and a twenty-eight-million-dollar annual budget might subsequently be enough to secure 80 percent of its lion population. This would still require a large budget increase, but not as astronomical as trying to sustain an unfenced Selous.

The donor community gives Africa billions of dollars every year for human health and economic development. Ecotourism directly contributes to economic development. Why not get the World Bank to fund infrastructure projects that would maintain a wildlife area the size of Switzerland? Otherwise, lion populations will be fragmented into an archipelago of tiny parks with little hope for long-term viability.

I ended my op-ed piece with: "Conservationists have failed to save elephants and tigers, and lions won't fare any better unless everyone contributes. If the world really wants to conserve iconic wildlife for the next thousand years, we need a latter-day Marshall Plan that integrates the true costs of park management into the economic priorities of international development agencies."

When Africa's protected-area networks were first developed in the 1960s, wildlife was expected to pay its way. The data are now in: wildlife cannot pay its way through photo tourism and sport hunting. The world loves African wildlife, but not everyone can afford to go to Africa to see it in person. Is the world willing to see their tax dollars used to protect these last bastions of nature for the next thousand years?

*

About the same time, the hunters who had developed the definition of the "huntable lion" sent an e-mail, asking if I would make a public statement opposing the potential listing of the lion on the ESA.

I refused.

The WD still hadn't established any sort of verification system for their age assessments of lion trophies. In fact, none of the Tanzanian hunting companies had received any feedback on the ages of their lion trophies for the past three years.

I insisted that I would publicly support trophy hunting only if the recent reforms were shown to be real. At this point, John Jackson was with the Tanzanian delegation at the 2013 CITES convention in Bangkok. He wrote that the age assessments were indeed being verified. I contacted the scientists whom he claimed were helping the WD. They had not received anything from the WD. If ages were being estimated by anyone, the work was being done in-house, behind locked doors.

I reported back to the hunters, and Jackson accused me of trying to shut down lion hunting. I pointed out that he and Eric Pasanisi had been instrumental in keeping the age assessments secret. The WD had such a bad reputation for corruption that they needed to be above suspicion with the age estimates. Otherwise, no one would believe them.

Our e-mails bounced back and forth, and they were still being cc'ed to a dozen different hunters. Jackson started saying that I should be denied permission to return to Tanzania and that I should have been deported years ago.

I fought back, refusing to give him the final word. The exchange went on for weeks. One of the e-mail bystanders invited Dennis Ikanda into the conversation.

A few days later, Dennis chimed in, sounding like the perfect echo of John Jackson. Beyond my other crimes, I was disrespectful of Tanzania: I was giving the country a bad name. Here was the same Dennis that had been so vicious toward Hadas back in the man-eating days, the same Dennis who accused me of racism while writing his thesis in Norway.

I returned to Tanzania in the middle of March. The recent e-mail exchange had made me sufficiently nervous that I wanted to try to soothe the hunters. Our dollars-and-fence paper had only included a few examples of where trophy hunters had successfully enhanced lion populations. Lions in three fenced hunting reserves in Zimbabwe had prospered as successfully as any other fenced population, but all the other hunting sites had suffered from overconsumption—so hunting had emerged as one of the most harmful factors affecting the unfenced reserves.

If the case could ever made for keeping lions off the ESA, we would need hard data of positive impacts from as many places as possible. But none of the Tanzanian hunters had conducted any sort of lion monitoring.

I issued an appeal, asking if any of the operators could provide any sort of data showing a positive effect of lion hunting on lion conservation. Only one company volunteered. Tanzania Game Trackers had noted whenever lions appeared at their baits for the past ten years. I ploughed through their files as soon as I got back to Arusha. Sure enough, the proportion of baits that attracted lions had actually increased over the same time period that lion-hunting offtakes had dropped in the rest of the country. Consistent with our dollars-and-fence paper, TGT had invested about a thousand dollars per square kilometer in antipoaching and other management activities each year. Other than Paul Tudor Jones (who never hunted his blocks), none of the other Tanzanian operators could ever have made that scale of investment.

Sadly, the rest of the industry was giving them so much trouble that TGT requested I keep the good news to myself. The bad guys didn't want to be shown up in comparison.

And speaking of the rest of the industry, what had happened to Sheni since being exposed for selling "his" mineral rights to the uranium miners? The official story had changed. He hadn't sold the mineral rights after all (though Rolf Baldus had managed to obtain a copy of the original mineral-rights contract). Sheni had merely been compensated the six million dollars for the "annoyance that the mining activities would cause to his animals."

In April, I met with the newest director of wildlife, Professor Alex Songorwa, in Dar es Salaam. Songorwa had left the WD in the 1990s, resigning after a serious argument with Severre. He enrolled in a PhD program in New Zealand and wrote his thesis on the relative merits of community conservation programs versus "protectionist" policies of wildlife management. He then taught at Sokoine University in Morogoro, before being conscripted back to the WD when the new wildlife minister, Khamis Kagasheki, started cleaning up after last year's live-animal smuggling scandal with Qatar.

With his rocky history during Severre's reign, Songorwa seemed uniquely qualified to reform the WD. But two of his closest friends told me that he was extremely uncomfortable in the job, hinting strongly that Songorwa was reduced to mere window dressing while business continued as usual. Indeed, Songorwa wrote an op-ed piece for the *New York Times* in March, claiming that lion numbers across Tanzania were so healthy that the species needed no extra protection. Besides, lions were too economically important to the hunting industry to allow them to be listed on the ESA: the loss of hunting revenues "would threaten the country's capacity to protect its wild lions."

A week later, the *Minneapolis Star-Tribune* had run a "counterpoint commentary" titled "Lion Hunting in Tanzania: Getting away with Murder?" by an incoming faculty member at the University of Minnesota named Marc Bellemare. The commentary noted that Tanzania is in the bottom half of the world's corruption rankings and that Tanzania's minister for natural resources had recently warned the Tanzanian hunting outfitters "that corruption usually began with wealthy hunters bribing officials so that they would turn a blind eye to illegal behavior." Instead of lobbying against the lion's listing as an endangered species, Bellemare recommended that Tanzania should instead reform its corrupt, dysfunctional institutions, which were a major cause of the country's underdevelopment. He concluded with: "Failing to put the African lion on the endangered species list would enable both those who turn a blind eye and those who get away with murder."

A few days later, I was of course blamed by John Jackson et al. for inciting this inflammatory rhetoric, coming as it did from my hometown paper. But I had never met Bellemare. His recent arrival was purely coincidental. The situation reminded me of when Kerri Miller, a prominent personality on Minnesota Public Radio, came over to Tanzania in 2007 to do an hour-long report on the Whole Village Project. Being the newshound that she was—and given the virtual absence of any real progress on the Whole Village Project by that point—she became immersed in our problems with the Wildlife Division.

And by some fluke of fate she ran into President Kikwete and his entourage one day in Arusha, whereupon she asked what he thought of the recent allegations against Severre.

Kikwete glared at her before snapping, "The director of wildlife has my full support."

Of course with the limited number of foreign correspondents in the country at any one time—and the Tanzanian security force's training with East Germany's secret police—everyone in the government knew that Kerri was on assignment with me, and half the government was already convinced that I had compiled Luke Sidewalker's dossier against Severre.

Needless to say, relations with Kerri became strained after I told her that she had likely just gotten me killed.

When I met Songorwa in his modest white-walled office, he was visibly uncomfortable. He sat so stiffly and spoke so softly that he almost seemed shackled. I had come to see him to discuss the dollars-and-fence paper, and, as outlined in my op-ed piece for the *Los Angeles Times*, I made the case that since large-scale conservation was so incredibly expensive, it might be a good idea to try to get help from the World Bank. Tanzania could never afford to cover the necessary management costs for an unfenced Selous, but the construction costs for a fence around the Selous were well within the reach of the bank—and, besides, it was time that the big donors recognized the enormous value of Tanzania's wildlife to its economic development.

But he was skeptical. He didn't see how he could get the government to ask the World Bank to redirect money from starving children to wildlife.

Our conversation was halting, awkward, but I had at least been able to talk to him about funding prospects for wildlife management. And I had hoped to establish a positive relationship that was untainted by my "meddling in lion hunting."

But as I stood to leave, he made it clear that he still had something to say.

"Forgive me for what I am about to say. I am very, very sorry for bringing this up," he said so softly that I had to strain to hear him. "Why have you been saying such bad things about Tanzania?"

Sitting back down, I knew he was talking about my e-mail to USFWS. I took a deep breath then narrated my long history with his predecessors and reviewed how bad things had been back in the days of Ndolanga and Severre—the stories of corruption that had been in all the local newspapers. I told him how I had worked so well with Tarimo and how I had once hoped to work with Serakike on the age-verification system.

He barely listened. In fact, he barely seemed to be there. He knew that I knew that he knew everything I was saying was true. He looked like a man who wanted help but could never ask.

About this time, Luke Sidewalker met with Source D, who had recently been to Mloka, just outside the eastern border of the Selous. Source D had gained the confidence of Mr. Big in the local ivory smuggling ring. Mr. Big liked to brag about all the huge houses he now owned, all paid for by ivory. He only spent part of his time in Mloka, the rest of the time he was in Lindi and Kilwa. Ivory smuggling was easy. The big shots in Dar would send brand new cars to Mloka. He had been given all the necessary tools to remove and replace the door panels. The ivory was broken into chunks small enough to fit inside. With the right equipment, the ivory-stuffed vehicles looked good as new.

So, Luke asked, who are the big potatoes that run the whole operation?

Source D mentioned an Asian family who owned several large shopping centers and hotels in Dar es Salaam, then he turned pale, lowered his voice and barely whispered the name of the oldest son of the highest-ranking elected official in the country.

A few days later, Henry Brink asked to see me. He had met with Dennis Ikanda at a workshop in Morogoro. Dennis was furious over the comments I had made in my messages to John Jackson. I was destroying Tanzania's reputation. I should be kicked out of Tanzania. Dennis was on the rampage, and, according to Dennis, Dennis was the only person that the Tanzanian government could trust about anything to do with lions.

The next day, I met with Dennis in Arusha. He ranted for over an hour. He told me I was an American spy—my funding from NSF was just a cover story, I was actually employed by the CIA, the State Department, USFWS. I had ruined Tanzania's reputation. America was trying to destabilize his country.

There were forces in the government that had wanted to get rid of me for years. The Tanzanian Wildlife Research Institute had been too lax. They had continued to renew my research clearance despite all these complaints.

But that was all about to change.

Ingela had returned to Sweden for a month-long break, and I had promised to monitor the Ngorongoro Crater lions during her absence. When I arrived in the NCA on the twelfth of April, the first order of business was to inspect the disembodied heads of two male lions that had been killed during a takeover by a quartet of young adult males on the Crater floor the previous week. The NCA vets had frozen the heads, and I wanted to identify them from their ear notches and whisker spots.

Waiting for the orange blobs to thaw in the warm sunshine, I wondered if I might be next in line for decapitation.

The lions turned out to be MG69 (nine and a half years old) and LA87 (six years old). They had been born in separate prides on the Crater floor; MG69 had teamed up with LA87 shortly after MG69's brother died, and the new pair managed to control a small Crater pride for the next few years. Their orphaned offspring were now about a year old and unlikely to survive.

I spent the next four days commuting down from the rim to the Crater floor, each day thinking about this possibly being my last trip to Eden. Some days were so rainy that the roads ran like riverbeds. Other days were sunny and green, with the wide swathes of green, green grass streaked with acres and acres of yellow and purple wildflowers. Thousands of wildebeest, zebra, and buffalo, long-tailed birds in their breeding plumage, everything looking fat and sleek: there was no place on earth this lush, this vibrant.

The first day in the Crater I found members of the incoming quartet that had killed MG69 and LA87. MG90, MG91, MG93, and MG95 (all about four and a half years old) had been born in the same pride as MG69. The same female from the MG pride, MG46, was the mother of MG90 and MG91, the great-grandmother of MG93 and MG95, and the grandmother of MG69. The new quartet's grandfathers were MG69's fathers; their fathers were the older half-siblings of LA87.

Cousins murdering cousins just like the Borgias—but at least the royal family of the Crater had the courage to kill each other face to face.

The second day I found LK67, informally known as Dennis, who had been born back in the days when Ikanda still worked here. Dennis the lion was now over fourteen years old, the oldest male by far on the Crater floor. He was the last surviving member of the quartet that had sired MG69, LA87, and the mothers of MG90, MG91, MG93, and MG95. Though his orange mane was still long and full, it was coarse and ratty. Male lions can seem sluggish at the best of times, but Dennis moved about as if in a trance. He looked like he might drop dead any minute.

I was in the middle of paradise, with animals I had known for years, but my mind kept turning to my uncertain future. Would this be my last tour of Eden? What would I do instead? Where else could I go? I hadn't wanted to contemplate retirement for at least another dozen years, and I had little interest in starting a new long-term study anywhere else.

I did still have a huge stack of lion data back home in Minneapolis which could keep me and my students busy with papers for at least a few years. Then what?

This chapter of my life was possibly ending, and why?

For telling the truth?

For losing patience?

Or for being so naive to think I could actually make a difference out here?

A few days later I bumped and rattled my way along the dusty road to Lake Ndutu on the border of the Serengeti and NCA. The lake is located at the western end of Olduvai Gorge. The wildebeest migration lingers here each wet season, and all the big cat species are attracted to the permanent water and associated woodlands—a true oasis in a vast expanse of short-grass plains.

Ndutu Lodge was built in the 1970s by a larger-than-life ex-trophy hunter who was later forced to leave Tanzania because he had decorated the grounds of his lodge with the skulls of wildebeest that had drowned in the lake. He had been branded as a poacher, apparently, because a well-connected Asian family from Dar es Salaam wanted to own a lodge in the Serengeti. But they were unable to make a profit, so Ndutu Lodge passed back into European hands a few years later.

I shared my research-clearance woes with a Dutch friend who had managed the lodge for the past twenty years. She had held onto her business despite the grinding corruption and continual takeover threats. She had stayed safe by remaining under the radar. She worried sometimes that her silence had implied acquiescence, but she had at least nurtured her two dozen employees and kept the lodge out of the hands of speculators.

That afternoon, I showered in the alkaline water brought up from a well near Lake Ndutu. The water is so slippery that you can never be sure if you've washed off all the soap, but you still feel clean by the time you're dry.

Genet cats climb the rafters of the open-air dining hall each evening, and guests sit around the campfire listening to hyenas whoop and lions roar. I sat alone at dinner that night not wanting to talk to anyone. I just wanted to keep my troubles to myself and watch the genets.

The next morning at breakfast, a black-bearded thirty-something from India introduced himself and handed me a book called *Exotic Aliens* that had been written by his uncle.

Once upon a time, lions were one of the most widely distributed mammalian species in the world. The Assyrians drew lions from first-hand experience in ancient Mesopotamia, lions were familiar to the Egyptian pharaohs, the Old Testament described lions near Jerusalem, and Alexander the Great killed a lion in Greece. But lion portraits also appeared in countries where they never lived. Lions adorned the medieval crests of England and northern Europe and the art of dynastic China. Just because a nation's cultural tradition celebrated the lion doesn't mean it was a native species.

Today's distribution of *Panthera leo* is restricted to the continent of Africa—with the curious exception of the lions of Gir Forest in the Indian state of Gujarat. For nearly two centuries, biologists have labeled Gir lions as "Asiatic lions" and designated them as a unique subspecies, *Panthera leo persica*. Assays from the 1980s supported this classification, based on the Gir lions' exceptionally low variation in blood enzymes—as expected from a long history of inbreeding in a small population. But they show no obvious morphological distinctions from African lions, and recent DNA tests cluster them together with North African lions, though they are genetically distinct from eastern and southern African lions.

I attended a workshop in Ahmedabad in 1995 to help identify a second home for the Gir lions. Because canine distemper is also common in Asia, the Indian wildlife authorities were worried that an epidemic as bad as the 1994 Serengeti outbreak could wipe out the remaining Asian lions. Their goal was to divide their eggs into more than one basket.

After the workshop, my Indian hosts took me to the Gir forest. The Gir lions were famous for being so tame that tourists routinely watched them on foot. Up until the 1980s, tour guides led goats into the forest and tethered them to stakes. The lions patiently waited, not feasting until the guides moved away. Compared to wild African lions, this forbearance is inconceivable. My biggest frustration in the Serengeti was always being imprisoned in my metal box while the animals roamed free, and I was thrilled to finally be allowed to walk with wild lions.

We drove around Gir until we found a cattle herder who pointed off to the side of the road, saying he'd just seen a mating pair. We left the car, and my hosts warned me not to walk too close: the male could be unpredictable while mating. But I couldn't resist. I ducked beneath low branches, stepped over thick vines and sat on the ground about three meters away. The female paid no mind, but the male reacted jealously. Then she rolled on her front and started the low growling that invites the male to mount.

It was an intense moment. On the one hand, I was transported by the sensation of her low grumble vibrating through my fundament. But I also felt stupid for being so close at the impending moment of climax—when she would roll over on her back and swat him in the face, whereupon he might charge at a third party. I reckoned I had a half chance of trouble. If she rolled counterclockwise, her swat would shove him toward me. But I was lucky. She rolled clockwise and pushed him in the opposite direction. He merely snarled like an irritable house cat from ten feet away.

After the day was over, we went to see an "African village," which was home to the Sidi people whom the local Nawab had brought to Gir to look after his lions in the nineteenth century. "Lions come from Africa," my hosts explained. "There are still lots of lions in Africa. Therefore, the Nawab figured that Africans must know how to look after lions."

Another host blurted out: "The Nawab also brought over some lion cubs—"

But before he could finish, the other Indians interrupted him and changed the subject. Everyone acted as if someone had let the cat out of the bag, and I left India with the suspicion that the "Indian" lions might not be so Asian after all.

The authors of *Exotic Aliens* made the case that the Gir lions had been brought to India as part of an international trade in wild animals that was well established by ancient times. Westerners have known forever that Romans imported lions from North Africa to the Coliseum to battle gladiators and to consume the occasional Christian. But Eastern rulers like Alexander the Great gained considerable prestige from being courageous enough to kill a lion. Thus the Persians and Indians practiced an early form of canned hunting by importing lions for the glory of their shahs, rajahs, and emperors. For centuries, these brave heroes (and their wives and children) confronted their royal quarries in hunting parks and menageries: tigers, lions, cheetahs, and leopards were corralled and sometimes even drugged with opium so as to be speared, stabbed, or shot in comfort and safety. It would not have been impossible, therefore, for the royal parks to have been restocked with African lions for centuries.

Gujarat had long been a landing point on India's west coast and was thus the most likely portal for importation. The nineteenth-century rulers in Gujarat bred lions, and Gir was their private hunting park. The Nawab probably wasn't still importing lions from Africa by the time he brought over the Sidi. The ancestors of the Gir lions could conceivably have entered the forest after escaping from earlier menageries during periods of political upheaval.

In March 2013, the Indian Supreme Court ruled that Gujarat could no longer monopolize its lions. Distraught Gujaratis went on a hunger strike protesting the decision; a dozen people died. The Gir lions still hold political significance today, even if they are semidomesticated descendants from ancient menageries of African lions. Lions have persisted in Gir Forest for at least two hundred years in a country that is now home to a billion people. And the Gir lions may be living relics from a time when fierce creatures were brought to India, tamed, and drugged, so that royal hunters could perform repeated feats of "courage."

Reading about the Asian lions lifted my spirits until the beginning of May, when I returned to Arusha and received an official e-mail from TAWIRI informing me that my research clearance had been withheld because of my efforts to "sabotage the Tanzanian hunting industry."

It could have been worse. Ingela and my other lion project staff had all been renewed, so at least TAWIRI wasn't planning to close down the whole lion project, and TAWIRI had also requested a reply to the charges that had been made against me.

According to the letter, I was guilty of making disparaging comments about the hunting industry in my e-mails to USFWS in November and to John Jackson in March. The Tanzanian Wildlife Research Institute listed the same set of quotes that Dennis Ikanda had recited to me the previous month.

I spent the next twenty-four hours composing a response. I started out by reminding TAWIRI that I had served as an official member of the Tanzanian delegation to CITES in 2004 and helped convince the Kenyans to drop their proposal to list the lion on appendix 1 and that the surveys I had supervised in the man-eating areas in southern Tanzania had illustrated that lions were still widespread outside the protected areas.

In my e-mail to USFWS, I had reported that despite the WD's long history of scandal, they had made many excellent reforms to ensure sustainability of lion hunting in Tanzania. The most important step left was to ensure the transparency of the WD's age-assessment system. When I later met with USFWS, they told me that they had already informed the Tanzanians of their concerns about these precise issues at a meeting in Botswana in 2012.

I had also communicated my concerns to several international hunting organizations, and we had come up with an alternative approach. If age assessments could not be verified in the range states, Dallas Safari Club would develop their own age-verification system, restricting entries in their record books to lions that could be confirmed to be six years or older. Just last week, the International Council for Game and Wildlife Conservation, a European hunting organization, announced a similar policy. The U.S. Fish and Wildlife Service had told me that this strategy might help to prevent lions from being listed as endangered.

My offending e-mails had merely highlighted the most recent examples of scandal in the Wildlife Division, all of which had been well documented in the Tanzanian and international press. To prove the point, I went online and found direct quotes of the current minister of natural resources, Khamis Kagasheki, complaining, "of businessmen with limitless finances who do all in their means to corrupt the system" (*Guardian*, September 2012) and ". . . of course, there's been corruption" (*Morning Edition*, NPR, October 2012).

As for my specific claim to John Jackson that recent recipients of new hunting blocks were well connected to the Tanzanian government and to the ruling party, I provided a clipping from June 2012 where the Tanzanian MP criticized the "corrupt environment" in the ministry that had recently awarded hunting blocks to Sheni Lalji and other rich politicians and businessmen.

In defending my statement to USFWS about President Kikwete having considered revoking my research clearance for "meddling in lion hunting," I revealed that I had received that particular information from Dr. Dennis Ikanda in August 2012 and that he had in fact made similar statements again in April 2013.

Saving the most important charge for last, I explained why I had written to USFWS that "reforms for lion hunting in Tanzania are being vigorously opposed by the corrupt hunting companies that operate with apparent impunity." I didn't mince words. Eric Pasanisi, Sheni Lalji, and John Jackson had not only worked to prevent the WD's age assessments from being made public, they had also tried to get the Tanzanian Parliament to eliminate the age minimum. Lalji had a well-documented history of dubious behavior, the Pasanisis had maintained far more than the legal maximum number of hunting blocks for decades, and Jackson was well-known in the United States as a combative, adversarial advocate who had represented several Tanzanian hunting companies with poor ethical reputations.

I had wanted to emphasize to USFWS that while the WD had indeed made many important steps toward meaningful reform, they were under considerable pressure from several influential representatives of the industry. I had only wanted to meet with USFWS to discuss how to prevent lions from being taken off quota so that the Tanzanian nation could benefit from lion hunting in the long term despite efforts by several businessmen to maximize their own short-term profits.

My letter ended with: "I have only ever tried to help Tanzania to maintain lions on quota despite the threat of international sanctions. My efforts may have run counter to the short-term interests of a few individuals, but the WD's adoption of my prior recommendations have only improved Tanzania's reputation. Tanzania is now known as the only country in Africa with a legal age-minimum for lion trophies. . . . Thus I hope you will consider my overall record and my sincere desire to contribute to long-term lion conservation throughout Tanzania."

I flew to the Serengeti the following day. The twelve-seater passed only a thousand feet above the gray summit of Mount Lengai, still steaming after its eruptions of a few years ago. Farther north was Lake Natron, a vast mirror reflecting dull cottony clouds that were themselves tinted pink by countless nesting flamingos. The pale green escarpment of the Great Rift Valley rose up below to reveal the details of every bush and rocky outcrop.

We circled around an afternoon thunderhead in the middle of the park, and bullet-sized raindrops pounded the windscreen for a few moments above a large black herd of buffalo before we broke clear of the weather and coasted low over the Seronera River, landing on the gravel airstrip in brilliant sunshine.

That night was Ali's farewell at the lion house. Twenty people showed up for the party: balloon pilots, FZS staff, field assistants from the vegetation, cheetah and hyena projects, and domestic staff from neighboring households. The evening was bittersweet. Ali is one of the most effusive, vivacious, and lively people ever to work on the lion project. In many ways her enthusiasm reminds me of Peyton's, Karyl's, and a few other rare souls, but Ali is virtually unique in combining a love of the Serengeti with a full-scale joy for complex data analysis.

Mentoring graduate students is almost the same as adopting them, except that adoptive relationships are expected to last a lifetime.

In the week following Ali's departure, I gained a much deeper appreciation for my Serengeti staff. Stan Mwampeta had joined the project a few years earlier, but just a few days after arriving, he watched a group of lions attack and kill a female leopard. After two of the lionesses blocked her escape route up a nearby tree, the third lioness leapt forward, bit the leopard on the throat and strangled her.

After the leopard was clearly dead, the lions abandoned the carcass, making no attempt to feed. As when killing cheetahs and spotted hyenas, lions don't catch leopards for food—they view them as enemies. But how do leopards, cheetahs, and hyenas all manage to endure in the midst of lions, when lions take every opportunity to kill them? And why are lions so beastly towards these species?

Killing leopards makes sense: leopards frequently kill lion cubs, so lions do well by ridding their nursery rooms of these mortal threats.

Lions and hyenas eat pretty much the same prey, and hyenas also kill lion cubs and even the occasional adult, so killing hyenas minimizes the competition at the dining room and increases security throughout the household.

It is very odd, though, that lions are so vicious toward cheetahs. Cheetahs don't ambush unsuspecting prey next to water holes as lions do. Cheetahs catch fleet-footed prey out in the open where a lion would never be successful, and a single lion can easily displace a cheetah from its meal. So surely lions ought to domesticate cheetahs, each pride keeping its own pet cheetah—kind of like Arabs and their falcons.

In order to better understand how all of these species coexist, we aimed to use the camera traps to discover if leopards largely avoided lions by mostly occupying the no-man's-land between neighboring lion prides or by generally venturing forth when most of the lions are asleep. Or maybe the leopards' spots render them invisible even to the lions' keen night vision.

Meanwhile, Ali had already investigated the effects of lions on cheetahs. Tim Caro started the long-term cheetah project in the Serengeti in 1981. A few years later, he radio collared a dozen cheetahs in the lion study area and tracked them over a five-year period. One of Tim's students observed several cases of lions killing cheetah cubs. Lions were also implicated in several other cases where cheetah cubs simply disappeared. A few years later, Tim's successor found that certain female cheetahs were much more likely to flee when they heard recorded lion roars and that the flightiest females had the highest cub survival.

Around the same time, studies in other parts of the world revealed how conflicts could influence interacting species in other systems. Wolves, for example, can virtually eliminate coyotes from large areas. In Africa, the most sensitive species is the African wild dog, *Lycaon pictus*. Wild dogs chase their prey for miles before finally running them down, an exhausting hunting strategy that requires an enormous amount of energy. Should the pack be chased off their kill by hyenas or lions, the dogs may run a negative energy balance and literally starve from their own efforts. Dogs also keep their pups at the same den for weeks at a time, dogs chatter to each other, and they are visible from a mile away. Consequently, lions can easily find and harass them.

So when the Serengeti lion population grew for several decades, the wild dogs struggled in their wake. By the late 1980s the wild dogs only persisted in the gaps between lion territories. As the lion population continued to grow, they essentially pushed the last few dogs out of the Serengeti National Park in late 1992, and the wild dogs have been limited to the NCA and Loliondo Game Controlled Area ever since.

In contrast, lions mostly affect cheetahs by killing their cubs, but a cheetah female often has six cubs per litter, she gives birth repeatedly over her lifetime, and she only needs to raise one daughter and one son (on average) for the population to remain stable. Despite the numerous observations of lions killing cheetah cubs, lions don't kill enough to keep the population down: cheetah numbers in the Serengeti have held constant for the past thirty years despite the threefold increase in lions.

How did cheetahs manage to persist so well? Ali compared lion and cheetah ranging patterns over Tim's five-year span of radio tracking and found that the two species used the same stretches of river during the dry season and similar areas in the wet season. Unlike the wild dogs, the cheetahs remained in the middle of the park when the lion population tripled. Cheetahs and wild dogs are both alert and wary, but wild dogs expend enormous energy on virtually every meal, and those long chases attract a lot of attention from greedy scavengers. Cheetahs, in contrast, are the fastest land mammal, they catch dinner in a hurry, and they can finish most of their meals before being pushed off by lions or hyenas.

My other Serengeti assistant, Daniel Rosengren, had been Ingela's cycling companion in earlier days. Daniel's first big trip was from Norway to Africa by way of Cairo. Ingela joined him in Nairobi, and they sometimes followed roads, sometimes footpaths and cattle trails. Then Ingela departed for a job in the Selous. A few days later, Daniel continued on alone from Dar es Salaam and pedaled south through a construction zone south of Rufiji. As he passed by the road crew, one of the workers leapt forward and hacked his face with a panga (machete), cutting right through his jaw. Daniel managed to keep his balance and bike down the road as the blood poured down his chest and half his jaw hung below his mouth.

If he didn't find help, he would bleed to death. He cycled on for another few kilometers, where he met two men who were carrying firewood on their bikes. They flagged down an oncoming truck that carried him back past his assailants and dropped him off at the police station of the nearby village. A medic stitched his face back together and helped him catch a crowded public bus where he sat on the packed floor with his suppurating wound dripping on his neighbor's shoulder. He made it to a large public hospital in Dar es Salaam that lay idle because of a work stoppage. A policewoman convinced the striking surgeons to wire Daniel's jaw back together. He flew off to Europe the next day.

Once he recovered, he returned to Dar, reclaimed his bike from the Swedish Embassy, and finished his trip through southern Tanzania and Mozambique to Cape Town.

The male cheetah walks along the short-grass plains toward some distant goal. He is a particularly handsome animal with pale fur and an unusual dotting of small black spots around his larger spots. Daniel drives in a wide circle to position our car in his path, and I place a toy lion cub on the ground out of view on the opposite side.

The cheetah is so preoccupied that he walks to within five meters of the toy without noticing. Daniel imitates a meowing lion cub. The cheetah stops, glances toward us and finally sees the toy. He looks around cautiously then approaches it from behind. He looks around again then swats, lunges, and bites the toy in the back of the head.

He hurriedly runs off at a right angle to his original route with the cub clamped between his teeth, but he soon turns back toward his original destination, occasionally looking back as if fearing the sight of the toy's mother.

After a few hundred meters, he drops the toy on the ground and licks its face and head. Unlike leopards, which kill cubs for food, he makes no attempt to eat it, but he wants to make sure it's dead, so he picks it up and carries it another few hundred meters. But he is still nervous of an attack by the bereaved mother, so he looks back repeatedly toward the scene of his crime.

No one has ever seen a cheetah kill a real lion cub. Cheetahs are scarce and cubs as small as our toy usually remain well hidden, but this encounter gives a good idea why lions hate cheetahs so much.

But why would cheetahs kill lion cubs if not for food? Lions are habitual scavengers of cheetahs, so by reducing the local lion population, our killer here may lose less food in the long run. Or maybe cheetahs hate lions because lions hate cheetahs, each locked in an eternal cycle of mutual recrimination: one killing the other because the other kills them . . .

In answering why lions are social, Anna Mosser found that reproductive success not only depended on the size of a female's own pride but also on the size of her neighbors' prides: females in relatively small prides suffer higher mortality and lower cub recruitment. She also discovered something unexpected. Females suffered higher mortality when neighboring prides were controlled by larger male coalitions—and the effect of larger coalitions of neighboring males was much stronger than from larger prides of neighboring females.

Male chimpanzees enlarge their territories by killing the females of neighboring communities. This frees up more fruit trees and other resources, and essentially removes competition. Male lions had occasionally been seen to kill females of neighboring prides—sometimes in attempted takeovers when mothers defended their cubs, sometimes for no apparent reason. But even though we had never seen any clear example of males killing neighboring females on behalf of their wives, the statistical impact of neighboring males was so strong we were convinced that lions must be fundamentally similar to chimps.

Then a few months ago, Daniel saw something extraordinary. Two females of the Barafu pride were resting only fifty meters away from five females and two males from the Kibumbu pride. The Kibumbu's resident coalition was known as the Killers because they had twice been seen near dead females of neighboring prides and were presumed to have been the murderers. The Kibumbu females started chasing the Barafu females. The Killers followed their wives at first but then took over the hunt and pursued the Barafu females for several kilometers. Once the Kibumbu females saw their husbands in action, they stopped and watched events unfold. The Barafu females split up, while the males focused their attention on BF18. Exhausted, she slowed down. The two males caught up and bit her flanks and spine. She collapsed, paralyzed from the waist down.

Their work done, the Killers sat nearby; there was no need to finish her off.

Unable to move, unable to eat, BF18 died ten days later.

The Killers had done the dirty work. In the following months, the Kibumbu females expanded their range eastward into what was once the Barafu's exclusive territory.

Many years earlier, I had similarly seen two pairs of females from the neighboring Seronera and Shambles prides resting uneasily about a hundred meters apart. They were unhappy to be so close, but the score was tied, two to two.

A resident male from the Seronera pride appeared, and, as he joined the two Seronera females, they immediately got up, and the three of them started running toward the Shambles pair. Now outnumbered two to three, the Shambles females started to flee. The Seronera females continued for a few hundred meters but then stopped when their husband passed them by. But as the husband caught up with the Shambles pair, they realized he was *their* husband, too, so they relaxed, stopped, greeted their mate, then turned around and went back toward the Seronera females with husband dutifully in tow—now outnumbering the Seronera's three to two.

The situation reversed itself twice more, each pair chasing the other whenever hubby was on their side—until hubby, exhausted by his two-timing, finally refused to budge.

I had always thought this story quite innocent and amusing, assuming that the females were just trying to chase each other away amid all the fun and games. But now I realize that the Seronera and Shambles female had wanted their neighbors dead, and they wanted their man to do their dirty work for them.

The week after Ali's farewell party, I went out every evening to absorb the Serengeti in case this would be my last wet season in the Serengeti. My current research clearance was due to expire at the end of August, so I would at least be able to return for the coming dry season. But the wet season always held the most magic for me. The birds were in their breeding plumage: whydahs, widowbirds, and bishops. The days were so clear that all sense of distance had vanished.

For some reason, the lions kept themselves up trees practically every day that week. One female named Snork has a deep scar on her snout running halfway up from her nostril to her eye. The open passage doesn't seem to bother her much, but it gives her a goofy look. The rains were winding down a bit early, and the grass around Seronera was the perfect height for stalking prey, but the wildebeest migration was only just starting to bunch together and point their way back toward Snork's familial territory. She couldn't sleep properly while perched on the horizontal branches of the sausage tree twenty feet above the ground. Between fitful naps, she stood up and peered off to the south and east looking for signs of returning wildebeest. So maybe she was up the tree for the view, or maybe all the lions in the area had decided it was safer to stay up trees now that so many hostile elephants had fled the ivory poachers at the periphery and found refuge in the middle of the park.

I had to head back to Arusha to start the journey home—Catherine was getting married on the twenty-sixth of May—but I wanted to check my e-mail one more time before disappearing out on the plains.

A message from NSF from the sixteenth of May: our 2013 preproposal had been rejected.

I had secured continuous NSF funding for the lion project since 1978, but the economic hardships caused by the financial crisis of 2008 and the Tea Party election in 2010 (which killed any hopes for the la-la land version of the Whole Village Project) also affected the National Science Foundation.

The National Science Foundation would only be able to fund 5 percent of proposals in 2013. The ecology review panel gave various reasons for knocking us out, but their criticisms really boiled down to the fact that what we were doing was too unfamiliar—too speculative. No one had ever used a grid of camera traps to look at fine-scale variations in animal abundance in space and time before, and no ecologists had ever relied on citizen scientists to classify their primary data set.

We were, in short, too high risk.

But I wasn't about to give up a forty-seven-year project without a fight. We would try again at NSF the next time around, but now I faced a gap of at least fifteen months. I had never needed to beat the bushes for funding before. I had received NSF support for so long that I barely thought about alternatives. Those two years' funding from the MGM Grand were just gravy. Panthera's funds were for conservation projects such as how to protect lions from angry pastoralists and how to deal with man-eaters.

But the long-term lion study is pure science. Why do lions live in groups? Why do they have manes? How have they coped with disease? This was where we would discover how lions coexist with hyenas and leopards, how predators follow their prey.

Seeking support from a conservation organization for basic science would be difficult at the best of times, but I had made myself into a pariah among the conservationists by tossing around the *f* word of conservation: fencing.

Whereas Frankfurt Zoological Society initially authorized Peyton to raise money to bridge a presumed three-month gap, they rescinded their offer in June, shunning me because of fencing. Their donors wanted visions of virgin wilderness: fences were *verboten*. Some of their board members were trophy hunters. My reputation could hardly have been less suitable.

Fortunately, there is a new model for supporting research: crowd funding. Put a good idea on line, make a video to explain what you want to do, and post a fund-raising goal. Sites like Kickstarter and Indiegogo have helped finance projects ranging from scientific exploration to new-product development to film production.

Snapshot Serengeti had attracted over a hundred thousand users over the previous six months—twenty thousand people had subscribed to Zooniverse's weekly newsletter. And *National Geographic* had been working on a story about the Serengeti lions for the past two years. The magazine article would come out next month, and the magazine had agreed to link the story's website to our Indiegogo campaign. Out of four and a half million *National Geographic* magazine subscribers and a hundred thousand Snapshot Serengeti users, we would find out soon enough if the project could be extended through a public appeal.

Today was the day when USFWS held its workshop to begin formal review of the African lion for the ESA. At the start of the session, USFWS insisted that the meeting was not concerned with the pros and cons of trophy hunting, but there are five criteria for listing a species as threatened or endangered: range contraction, population decline, susceptibility to disease, demand for body parts, and overconsumption. With "overconsumption" on the list, hunting inevitably dominated the discussion.

The first speaker was from the Humane Society, one of the copetitioners for the lion listing, and she gave evidence that lions qualified for listing by all five criteria. Next, the head of Safari Club International asserted that sport hunters did more than anyone to conserve lions: without lions on quota, the lion would be lost. The International Foundation for Animal Welfare then blamed sport hunting for the lion's decline.

Both sides overstated their case. The antis claimed that lions were so susceptible to FIV, CDV and bovine tuberculosis that range reduction, overhunting, and demand for body parts would lead to inevitable extinction. They implied that a thousand wild lions a year were being shot as trophies each year, when in fact over 80 percent of these animals were from canned hunts, and all the lion bones came from South African lion farms.

Safari Club International had bet the bank on Tanzania. Lions were given value to local people by sport hunting in Tanzania, and without hunting they would disappear like in Kenya. Half the lions in Africa were in Tanzania, and SCI's surveys had shown that Tanzania's lions were thriving. To be fair, their "survey" had done a very thorough job of traveling around the country and interviewing people. But they merely asked whether respondents had heard or seen a lion any time in the past ten years. Mapping out the responses, multiplying by some number, the total became magically precise: 16,800 lions in 2013!

And, since earlier guesses had been lower, Tanzania's lion population was actually growing.

Having been foolish enough to throw out numbers for those earlier guesstimates, I politely insisted that the SCI survey was incapable of producing a solid number and that these numbers couldn't be used to indicate population trends one way or the other—especially since my own guesses had been pulled out of a hat. Meanwhile, John Jackson was off to one side insisting that guesses like mine were "the best possible science, the best possible science."

At the coffee break, I went over to shake hands with Dennis Ikanda. It was the first time I had seen him since my research clearance had been withheld. We exchanged pleasantries, Tanzanian style, and then I asked what was happening with my appeal. He said he didn't know.

We remained smiling and polite, and I decided not to tell him the news that Dennis the Ngorongoro lion had last been seen alive, but badly wounded, on 22 May and that the next day all that was left was a puddle of blood and a swatch of mane.

I also didn't ask him how he could afford to send his two older children to private school in Arusha—how he could suddenly afford to send his wife to college in Kenya and own a fleet of cars—all on a middling government salary at TAWIRI.

A vague light sparked in my memory—recollections of how Ikanda disappeared for whole weeks at a time for god-only-knew what reason back when he was an otherwise steady research assistant in the Ngorongoro Crater. How I had once paid his airfare to a conference in the United Kingdom, but he had failed to go, and how it had taken him nearly a year to pay me back. How we had once spotted him money for a new steering box but he never bought it. How a German film team almost took him and his brother to court for stealing an expensive piece of camera equipment. How his Tanzanian staff in the man-eating area was so perplexed by his spendthrift ways.

And here he was today, SCI's man from Tanzania: a full-fledged member of the hunting fraternity—ready to report whatever I had to say back to the WD and to whoever else was supporting him and his lavish lifestyle.

After the break came the first real data. A team from Duke University had mapped the extent of remaining habitat that might conceivably be home to lions, given agricultural activity and human population densities in the surviving savanna across Africa. Although much of West and Central Africa might look suitable from a coarse-grained perspective, close-ups on Google Earth showed small farms and huts scattered throughout. Only 26 percent of African savannas could still conceivably host lions, and most of the remaining habitat was in eastern and southern Africa. But there were still a number of "strongholds" where lions somehow managed to thrive, and by taking everyone's rough guesses of overall lion population size and adjusting for the new range restrictions, they estimated a total of thirty-two thousand lions for all of Africa.

Then it was my turn, and I started with disease, as I had done so long ago in Bangkok.

Canine distemper virus was only a problem in combination with babesia, and such coinfections only happened about once every fifty years. Even with 30–40 percent mortality, the population could bounce back in a few years.

Lions got sick from eating TB-infected buffalo; dozens of lions had died lingering deaths in Kruger and Hluhluwe iMfolozi Parks in South Africa. But bovine tuberculosis was only likely to affect the size of outbred populations by about 6 percent. No one would even notice if a large population was "only" at 94 percent of its potential.

As for FIV, new data were available from Botswana and from captive lions showing that the virus registered with the host's immune system, but the effects were not life threatening. We had recently found three different strains of FIV in the Serengeti: A, B, and C. The A and C strains were associated with higher mortality in older animals during the 1994 CDV outbreak, but the overall effects were too small to threaten the overall population.

Bottom line? In my view, disease should be stricken from the criteria for listing the lion.

Next, a brief overview of human-lion conflict: Kissui's data showing that about one Tarangire lion died for every livestock attack then the latest figures on man-eating from southern Tanzania. The problem had peaked between 1998 and 2004; the man-eaters had largely been poisoned out of existence shortly thereafter.

What I really wanted to say was that the Tarangire lions were mostly being speared and poisoned in hunting blocks and that any incentives for living with lions came from the national park and any conflict-mitigation projects were implemented by conservationists, not by the WD or the hunting companies. I also wanted to say how conservationists didn't like hearing about man-eating because it might affect their donations from lion lovers.

But I kept quiet and moved on to the next slide, which focused on snares. Nearly 20 percent of lions were found with snares around their necks in Luangwa Valley, Zambia. Snaring was the primary cause of lion mortality in Niassa Reserve in Mozambique.

But I didn't say that all these snares had been set out in hunting blocks that had somehow failed to give wildlife sufficient economic value, that one of the only hunting companies in Tanzania to invest in antipoaching was at risk of losing its blocks to corrupt competitors.

Then on to sport hunting, starting with the dichotomy between wild lions and canned hunting and reiterating that the lion farmers were so far meeting the Chinese demand for lion bones, then the usual spiel on infanticide, age minimums, and the evidence for overharvesting from all across Africa and across the hunting blocks in Tanzania.

Then a graph showing the most recent countrywide hunting-offtake data from Tanzania and how the trend was pretty much a perfectly straight decline between 2000 and 2011. Though the hunters claimed that the recent low numbers had resulted from strict observance of the age minimum, age verification of the trophy animals was essential since hunters might have become more selective or the downward trend might merely reflect the continued consequences of overhunting and unimpeded habitat loss.

But I didn't say anything about *why* trophy verification hadn't taken place. I said nothing about the continued loss of hunting blocks because the wildlife was gone. Human population was growing fast and most of the hunters did nothing to fight poaching and habitat conversion.

Then the dollars-and-fence paper: first, half the unfenced populations were likely to fall to near extinction in the next twenty to thirty years.

Trends were strongly downward even in the Duke study's "strongholds." Second, the enormous expense of looking after these unfenced reserves—where would the money come from?

I got a lot of questions, but the only one I can remember clearly came from SCI: "Is the lion likely to go extinct any time soon?"

"No, a number of lion populations are safely fenced."

Then the Born Free representative made the case that trophy hunting wouldn't be missed in Africa even if it disappeared tomorrow because it contributed so little money compared to phototourism. But he neglected to say that trophy hunting could generate money in remote areas that photographic tourists would never visit.

Then John Jackson once again placed everything on Tanzania, pointing out that hunting blocks comprised 70 percent of the country's lion habitat, before talking in broad terms about the value of "sustainable use" and hence of sport hunting across the world. I merely commented that those huge areas of Tanzanian hunting blocks presented an enormous financial challenge, if our estimated requirements of two thousand dollars per square kilometer were remotely accurate.

To be listed, the lion only had to be at risk from three of the five criteria. It seemed clear that the lion's range had clearly shrunk and that a significant proportion of populations were in decline. But lions were not seriously threatened by disease and no one had provided any evidence of an impact of the bone trade on wild lions. So I felt that the score was tied two to two.

The tiebreaker was "overconsumption," and lions had clearly been overhunted in the past, but claims could be made that new rules would prevent overconsumption from this date forward—making the score either three to two or two to three depending on what you wanted to believe about the enforcement of well-intended offtake policies.

After everyone was finished, USFWS asked for summary comments. I used my final statement to say that no matter what happened with the listing, lion conservation would cost a fortune. New economic models must be developed to find the funds since no one was raising enough cash at the moment. Although a lot of lion areas would likely be lost in the coming years, the lion had an advantage compared to species like tigers: there were still a few huge areas left in Africa that could easily conserve the species in perpetuity, and resources should focus on these big areas. It might be impossible to save every last lion, but some way should be found to save those places that are the size of Switzerland.

*

Over the past few years, part of me has been like a cheetah, operating at high speed and willing to live with danger, but part of me has felt like a wild dog, burdened by the prospect of more exertion, ready to flee to no-man's-land.

After the meeting tonight, I was too drained to make pleasant with John Jackson and Dennis Ikanda, too cautious to talk with the antis. I needed to unwind, and I hated myself for staying so quiet about the things that I most wanted to say.

I have never felt more alone.

SERENGETI, 8 JULY 2013

The morning is clear and bleached, the yellow plains a carpet of pale straw dotted by an occasional green tree. I am radio tracking the Kibumbu and Barafu prides, but the only sound on my headphones is a constant white hiss of static. The land is dry. The wildebeest migrated north months ago. Only a few hartebeests, warthogs, and gazelles still linger in this part of the park. The lions must be somewhere along the Ngare Nanyuki River, but the network of informal dirt tracks has changed so much the past twenty years that I'm not sure where this one leads.

But then the road snakes north and east before curving down to a stretch of woodlands at the confluence of the Kibumbu and Ngare Nanyuki Rivers. A mixed herd of Thomson's and Grant's gazelles graze in an open patch of grassland. The ground beneath the tires is now black—a wildfire spread through here a few days ago. The vegetation has been burned to stubble, the smaller trees singed, their leaves toasted coppery brown.

A hundred meters farther, the grass is intact. A tented camp has been staked out beside an attractive stretch of the Ngare Nanyuki. Still no beeps. The track ends by the camp's dining tent. I continue cross-country parallel with the river, trying not to leave tire tracks on the camp lawn.

A pile of lions lounges on the riverbank a few hundred meters upstream. They seem jumpy, nervous of the car; they begin to move rapidly away—five subadults and an adult female. The three young males are the most eager to flee. The two young females also seem shy, but one of them is blasé enough to catch a marsh mongoose that she carries along, one small mongoose leg sticking out of her mouth. The adult female doesn't seem shy so much as attentive to her brood, sensitive to their fears.

I can't get close enough to identify them yet, but they fit the classic profile of recently evicted subadults chaperoned by a mom.

The morning passes, waiting for the lions to get too hot to move any farther. The landscape here is stunning. The Ngare Nanyuki meanders through a long, broad meadow surrounded by umbrella acacia trees. But the grass is so tall and the steep-sided channel so narrow that I'll only be able to photograph the lions' faces if they remain on this side of the river.

After a few more kilometers, the lions finally start to relax, acting as if they have moved closer to safety—either nearer to home or farther from the incoming males. But they are still jumpy enough to stay just out of range of my camera—except for Mom, who finally poses long enough to be captured for posterity.

She then follows the five subadults into a network of interwoven channels that render them inaccessible. They pause for a few minutes, then pass over to the opposite side of the river.

I had photos of the mom. If the group belongs to the study population, we'll figure out the subadults from their age and sex: three males and two females, all about twenty months old.

I continue up my side of the river, circling around several dry tributaries almost as deep as the Ngare Nanyuki. The white noise hisses on my headphones, still no sign of the collared lions. After maybe another kilometer or two, I spot another uncollared female lion. She seems agitated, too, but soon settles after finding a shady spot to sleep through the day.

The meander of the Ngare Nanyuki is wider and more winding than I remembered. There is still no way to cross over to the subadults. Searching for the Kibumbus and Barafus will be more productive. I should drive somewhere with better conditions for radio tracking.

I head away from the meander as best I can when I reach a slightly greenish patch that looks like the sort of low brush that grows next to a pig hole or hyena den when—THUMP! The front end of the car falls into a steep-sided gully four feet deep and six feet across.

The car is pointing downward at a thirty-degree angle. The front wheels are resting on dried mud at the bottom of the gully; the rear wheels are still up on the grassy ground of terra firma. I try to reverse; the wheels spin: no traction in front or back. Going forward, the car only moves six inches before the bumper bangs against the vertical wall of the opposite bank.

It is now eleven o'clock. The sun is bearing down. There is no cell phone coverage in this part of the park. Stan is working in another area, also without cell coverage, Daniel is on leave in Sweden. I have no radio. There are no roads out here, no tourists.

I have 1.5 liters of water, and it is already hot.

The most important tool in these situations is sometimes called a Tanganyika jack or more commonly a high-lift jack. The contraption is about four feet tall with a six-by-eight-inch platform at the end of a heavy metal bar perforated with a series of half-inch holes all the way along—kind of like a narrow ladder. A second component moves up and down the ladder by ratcheting alternating metal pegs in the series of holes. The ratcheting movement requires an up-and-down motion with a three-foot metal handle.

On level ground, the jaw on the ratcheting component fits below a metal bracket underneath the bumper of the Land Rover. The handle is then levered down to a near horizontal position, thus lifting one of the metal pegs up to the next hole in the ladder. Jack the car high enough, and it is easy to change tires and get out of most animal holes.

However, the situation here is a bit more challenging.

The bracket beneath the bumper is almost touching the gully bottom. The bumper is so close to the wall of the gully that there is no space for the jack, let alone for a horizontal handle.

The next most important tool in these situations is a shovel.

It doesn't take long to remove enough dirt beneath the bracket to position the jack. But it's quite a bit harder to dig away enough of the gully wall to lower the jack handle.

The third most important tool is a pickax.

The ground is soft enough to hack out a trench on the opposite wall of the gully, but the space is so tight that I have to work awkwardly, half bent over. It is stiflingly hot down here, my shirt is soaked with sweat, my heart pounding. I have to stop every few minutes to catch my breath and try not to drink all my water at once.

The topsoil is bound together with roots from the grass, while the dirt deeper down is packed but crumbly. When the trench reaches the floor of the gully, though, the soil is so hard that the pickax can barely break through. The working space gets tighter, and progress slows.

Half my water is gone. There is barely a cloud in the sky. It is already one o'clock.

The high-lift jack leans forward slightly, pushed by the angle of the car. The trench will need to go deeper than the gully floor.

The jack handle is a heavy pipe, and I use it like a pile driver, breaking up the soil, then using the narrow tip of the pick to poke holes in the ground. I clear the loose dirt with my hands.

I replace the jack handle in its holder, press it down to the new floor of the trench, and—CLICK—the car goes up a notch!

But after raising the handle back up to the vertical and pushing it down again, it no longer aligns with the trench—everything has rotated.

Take off the handle again; widen the trench with shovel and pick, clear out the loose dirt.

Bring down the handle again. Not deep enough.

Again with the pick, again with the fingers, another inch scooped up from the bottom.

Handle up, handle down: CLICK.

Up, down: CLICK.

Up, down. Nothing.

Everything has rotated again.

Widen the trench again—even more this time, the rotation is getting worse.

Handle up, handle down: CLICK.

Up, down: CLICK.

Up, down: CLICK.

The front left wheel is off the ground!

But then everything rotates again.

The various torsions of the tilted car have destabilized everything so badly that the front of the car wants to shift sideways. I had planned to lift the front left wheel high enough to push a spare tire underneath, but the jack is too wobbly.

Everything I have done in the past three hours has only made things worse. For every click, the car teeters more precariously. Every bit of work makes things worse. The more I keep trying, the worse it will get.

A car stuck in a gully in the middle of nowhere, everything falling apart thanks to the half-assed efforts of some moron who doesn't know what he's doing: the perfect description of the past few years. The futility of working against a hopeless situation, the energy devoted to digging myself deeper and deeper into a hole of no significance whatsoever.

I am out of breath and out of water, but I am not out of time. I took a GPS coordinate when I first saw those shy subadults, which were close to the tented camp. I'm about four or five miles away. It is only two in the afternoon. I'm only wearing a polo shirt, shorts, and sandals, but I have a wide-brimmed hat, and there is no other option.

I head toward the afternoon sun. The grass reaches above my knees. The backlighting and thick vegetation make it hard to see any holes, rocks, or reptiles. There used to be an old road somewhere off to the left, but it may have disappeared by now, and, anyway, it may run too far to the south—the tented camp is northwest from here.

The grass has gone to seed. Most of the seed stems are pale yellow, but some are salmon pink. A few low weeds are a faded green, but they have needle-like spines that break off in the sides of my feet. I try to avoid the greenish patches and keep a steady rapid pace through the yellows and pinks, firmly focused on each step at a time, one after another. My cell phone picks up a faint signal at the first low ridge top. I send a text to Stan then resume my march.

I make a wide circle around a thorny thicket to my right. Thirst dominates everything. The futile hours of digging expended far more than the

1.5 liters in my water bottle, and the sun is still blazing above the dry open plains. Up ahead, two bull buffalos watch my approach for a few moments before snorting at me and running off.

I reach the faint traces of a game trail. The narrow grass-lined path is smooth, so I can at least stop worrying about hidden rocks and holes. But in a bare patch at the top of the next ridge, I spot a half dozen ticks clinging to each of my legs.

At least they haven't yet sunk their heads into my flesh. People can get babesia, too.

A half mile away or so—is that a building? I'm not halfway to the tented camp yet. Nobody else ever comes out here—but wouldn't it be nice. . . .

I stare at the ground ahead, avoiding the thorny weeds, stopping at patches of bare earth to check for ticks, focusing one step after another after another. I look up again and the building still seems to be there. It is remarkably well camouflaged—the same height and color as the surrounding acacia trees. But it also has weird dimensions: taller than it is wide. Maybe it is a mirage? But mirages shimmer horizontally in the distance, they are never three dimensional.

I concentrate on the ground ahead, make toward the trees, take one step at a time, ignore the thirst and heat.

The next time I look up, I'm in the woodlands, but the "building" is gone—it was nothing but a trick of the eye. I have been walking for forty-five minutes, my mouth is dry, and the sun is lower, but the air is as hot as ever.

I veer back toward the open plains, keeping as much height as possible in case Stan got my text. Walk, walk, keep walking, stop under that lone tree up ahead, and get a GPS reading. The tented camp must be near the Ngare Nanyuki. I walk past scattered trees, stop on bare earth to pull off yet more ticks.

After another half hour, I come upon the twin ruts of a dirt road. If I follow it toward the Ngare Nanyuki, maybe it will curve to the tented camp. But half a mile along, the track ends at a picnic site at the edge of the river. The grass here is taller and thicker than back near the ridge tops, so I turn around and retrace the last half-mile.

A mile-long miscalculation, but at least there is no fear of hidden snakes while walking on bare ground, and there is less chance of picking up more ticks.

The track makes a wide curve before carrying on to the west, straight into the sun, which strikes my face below the brim of my hat. I am so dehydrated I can barely sweat.

Then another twenty minutes, and there are the sand-colored tents of the luxury camp. This is no hallucination—the GPS says it is so.

The road would take a long circuitous route down to the camp, but I am now back to the site of the grass fire, so I can take a tick-free shortcut across charred open ground. The way may be clear, but it is hotter than ever. My mouth is so dry I can barely swallow.

The camp attendants are busy around the supply tent, preparing the evening's meals for their guests. They watch me stagger across the burn, wade through the last strip of tall yellow grass. I can barely talk, the words stick on my dry tongue.

The receptionist, a Tanzanian named Jonas, invites me to the dining tent for a bottle of water and a comfortable chair. Five hours have passed since I drove into the gully. It feels like ten minutes, it feels like three days.

Jonas waves goodbye as I climb up to the cab of the supply truck. Two Tanzanians will take me back to my Land Rover. We ride about eight feet off the ground—with a view like this, I would never have driven into the gully. But the suspension is so stiff that every bump bounces us off our seats. We move slowly. The drive seems infinitely longer than the walk.

After forty-five minutes, we spot the white Land Rover, tilting thirty degrees, now glowing pale orange as the sun sinks into the haze of grassfires from all around the periphery of the park.

We back the truck into position behind the car. They attach a thick steel towing cable. I hop in the Land Rover, throw it into reverse, and the truck yanks me out like a pull toy.

Simple.

The lesson for the day is perfectly clear: the next time I try to do something difficult and hopeless in a third world country, I need to find someone with a really big truck.

9 Uptow

GEORGETOWN, 16 SEPTEMBER 2013

"You asked me to fly all the way here to meet you," growled Luke Sidewalker. "You said you came to listen to me—but you haven't heard a word I've said!"

"If those pinheads at Fisheries and Wildlife ban lion hunting," Dave Barron shot back, "lions will be extinct in the next twenty years."

"Hunting has only helped a few small fenced populations—but nowhere else."

"That's not true."

"OK, so you tell me," Luke shouted, "if you know so much, tell me just one place. Go ahead, tell me: where?"

"Hunting helps lions in South Africa and Namibia."

"Almost all the lions that are being shot in South Africa come from lion farms—they aren't wild animals." Trying to calm down, Luke continued: "And Namibia received an enormous amount of aid money from USAID and the World Wildlife Fund to set up their conservancies—hunting can't take credit for that. Besides, Namibia's human population is tiny, and its government is the least corrupt in Africa."

Corruption. That's why Luke wanted to meet David Barron in the first place. Barron founded the International Conservation Caucus Foundation (ICCF), a bipartisan organization whose members include about thirty U.S. senators and a hundred twenty members of Congress, and Barron is hosting a dinner to honor President Kikwete the day after tomorrow. He lives on a large estate in South Carolina. He took a cab from the airport today with Vice President Biden.

Barron is used to impressing people: he was a surrogate speaker for President Reagan and hobnobs with movie stars. He graduated from Georgetown University and haunted a neighborhood restaurant, The Tombs, in his youth: he interrogates the student waiters, asking where they're from, which fraternity they've joined. When not being one of the boys at The Tombs, he has spent most of the past half hour showing off to his dinner companions: a senior member of his ICCF staff and to the U.S. government official who arranged the meeting with Luke Sidewalker.

Luke has had two other meetings today, neither of which led to anything concrete. He has dutifully acknowledged Barron's name-dropping, but now he wants to be heard.

Barron resumes the shouting match: "If you tell me that hunting is inherently harmful to lions, then we have nothing more to talk about, and this meeting has been a complete waste of my time. But if you tell me that hunting is harmful *as currently practiced*, then you need to tell me what can be done about it."

"It's the latter—current practices are driving everything down," Luke replies, half out of breath. Barron cuts him off. A fifth person is expected for dinner—a very wealthy man with stables along the Potomac—Barron and Mr. USG start talking about horseback riding. Luke sits on his hands, wishing he could rewind the evening and start over.

"If you'll excuse me," Barron says, as he stands and glares at Luke, "I am crossing the street now to attend mass. When I get back, we can have dinner, if you care to stick around."

Barron and Mr. USG depart. Luke sits alone with Barron's assistant.

"Well, that went rather poorly," Luke says, sighing.

The assistant is matter-of-fact: "You ended up agreeing with each other."

"We did?"

"You agreed that hunting should be reformed."

David Barron and Mr. USG arrive back at The Tombs at five past six P.M., and everyone orders dinner. A fifth person has joined the table, giving Barron the opportunity to swivel his head away from Luke for whole minutes at a time. The food is served, and Barron loudly says grace.

But as everyone's blood alcohol rises, Luke manages to ask Barron a few neutral questions about neutral topics, tensions start to fade, and they return to the matter at hand.

"The problem with hunting in Tanzania as it is currently practiced," Luke says, "is that only those people with direct access to the block allocation process are able to acquire land. And people like Pasanisi get all the land at cut-rate prices. This means that hunting can't possibly generate enough revenue to cover the costs of antipoaching or community conservation or anything else to protect the wildlife."

"I need you to sum this up for me in a sentence or two."

"Tanzania is the size of California, Nevada, and Oregon combined, and one-third of that land—three hundred thousand square kilometers—has been set aside for trophy hunting. All together hunting only earns fifteen million dollars a year."

Barron nearly chokes on his steak salad.

"*Fifteen* million dollars?"

"Yes, in a country of over forty-million people. So out of that paltry amount, local villagers only get about fifty cents apiece per year—that's the price of an egg."

"How much could be earned from all that land?"

"No one knows for sure, but we'd find out pretty quick if the Tanzanian government ever put the blocks up for public auction and let the market decide. There are plenty of billionaires out there who'd be happy to pay a fair price for a chunk of the Selous—and they could easily afford to hire an antipoaching staff."

Barron turns to Mr. USG and says, "We've got to get the Millennium Challenge Corporation to leverage the Tanzanians on this." The George W. Bush administration had established the Millennium Challenge Corporation as an alternative to USAID—a parallel development agency that forced recipient countries to reduce corruption. It was the threat of George W. Bush's visit to Tanzania in February 2008 that led Kikwete to sack Prime Minister Lowassa and Mama Meghji.

"We've got five hundred million dollars going over to Tanzania next year," Barron barks to Mr. USG. "We'll just make all that money conditional on the Tanzanians auctioning their hunting blocks."

Luke takes heart. If Pasanisi, Sheni, and the politicians were forced to bid against real money, they might have to find some other way to ruin the world.

But even if David Barron can convince his pals at the Millennium Challenge Corporation to coerce the Tanzanian government to earn an honest dollar and finally recognize the true value of these precious wildlife areas, the WD hasn't set the next round of block allocations until 2017—and the Tanzanians have resisted auctioning their hunting blocks ever since USAID first suggested the idea back in 1991.

MINNEAPOLIS, 7 OCTOBER 2013

"Ooh, those are *so* cool!" says Ali as she tears opens the large cardboard box to reveal a stack of pixilated posters of a Serengeti wildebeest. Look closely, and each tile of the photomosaic is a tiny camera-trap photo. The animal's dark flanks consist of night shots. The blue sky is made up of clear days on the open plains.

The wildebeest poster is a perk from our month-long Snapshot Serengeti fund-raiser. We sought to raise thirty-three thousand dollars and we timed the start of our online campaign to coincide with the official publication of the Serengeti lion story in the August 2013 issue of *National Geographic*. But our funding appeal generated little interest until my daughter Catherine started prodding her Facebook friends and Zooniverse notified its volunteers about "Save Snapshot Serengeti." Donations bumped up for a few days but faded again, and we had only raised a few thousand dollars by the end of the first week. How could we have managed to make the Serengeti so unappealing?

At the eleventh hour, Zooniverse warned its volunteers that Snapshot was in danger of shutting down forever, *National Geographic* posted a couple of links to the campaign on their webpage, and the donations started rolling in.

It was like standing in the surf and feeling the tide rush in.

By the end, over seven hundred people from all around the world had donated a total of $36,324 plus an additional seventeen thousand dollars from benefactors who preferred to give directly to the University of Minnesota.

Meanwhile, I had assembled a collection of greatest hits from the first few years of the camera-trap study: startlingly candid images of gazelle sex, porcupine sex, gazelle birth, and zebra death, and, since the cameras fired three shots in rapid succession, some of the sequences made hilarious animations: dancing warthogs, hartebeest ears flapping, backlit secretary birds preening like sun gods. The Expedition Council at *National Geographic* gave us another thirty thousand dollars to extend the project and collect enough material to produce publishable collections of intimate wildlife photography.

Today I'm helping Ali put together all the perks to send out to our hundreds of donors at the $55+ level, just like those volunteers at PBS or NPR, wrapping t-shirts and boxing coffee mugs. We have a few extra wildebeest posters, so we decide to pin one up on the Lion Lab bulletin board, making space by retiring Peyton and Karyl's old photos of Serengeti picnics, broken-down Land Rovers, lion dummies, and Grant Hopcraft looking like a high school student. We rearrange Ingela's spectacular sequence of photos of the Killers as they surrounded and severely injured a rival male called C-Boy.

It is a stark reminder of how the Killers earned their name: a collective willingness to destroy their enemies, a willingness that sparked a ferocity far greater than the sum of its parts.

Working together they were invulnerable—and *they knew it*.

And then I remembered something that David Barron had said to Luke Sidewalker back at The Tombs in Georgetown.

Before daring Luke to punch him in the chin, David wanted to know more about Luke's background, his motivations.

"All I know about you," he said, "is that you're some sort of Lone Ranger trying to do something about lions."

And, indeed, that was Luke's problem. He was ultimately alone. No one else saw any hope of moving that great mountain of corruption. No one else dared risk being deported from the paradise of wild Africa.

But no one else knew as much as he did about how the system worked.

And with great knowledge comes great responsibility.

He had to keep trying.

WASHINGTON, DC, 26 OCTOBER 2013

The African lion is currently under consideration for listing on the Endangered Species Act. If the lion is listed as a *T* (for "threatened"), greater care will be required to limit the impacts of sport hunting on lion populations, but most practical aspects of lion hunting will remain unaltered. Alternatively, if lions are listed as an *E* (for "endangered"), lion trophies can no longer be imported into the United States.

Luke Sidewalker originally investigated corruption in the Tanzanian Wildlife Division in hopes of encouraging sustainable harvest practices—he compiled his dossier on Severre on the assumption that Severre alone was responsible for overhunting. Severre kept raising the quotas to raise revenues: hunters were charged more for dead animals than for hunting opportunities, thus the WD ran an "industry soaked in blood."

Luke assumed that reforming the WD would allow honest hunters to hunt honestly.

But Severre was as much a symptom of the corruption as its cause. He was replaced by a well-meaning director, Erasmus Tarimo, who encouraged responsible harvest practices but was himself removed from office a month before he was due to retire—not only because he had put a damper on lion offtakes but because he had also tried to reform the block allocation system.

The WD had established a bulletproof consortium of ex-presidents, ex-prime ministers, and ex-ministers: all the best-connected and most revered families in the country received healthy incomes from subleasing "their" properties to expatriate hunting operators. Within this cozy mafia, a small number of extremely wealthy Asian business tycoons and a family of French colonialists gained the lion's share of hunting blocks and ran their businesses at a healthy profit—without worrying about the overheads from investing in antipoaching patrols or community conservation projects. They were strip miners, and like any other extractive industry, they only cared about short-term profits. Once all the animals were gone, they could happily move on and extract something else—like uranium, for example, or petroleum or precious gems.

The political elite prospered from the status quo. If some meddler like Tarimo tried to alter the system, he would just be brushed aside. Gerard Pasanisi had befriended all of Tanzania's presidents; the director of wildlife was a presidential appointment. Tarimo was eventually replaced by another one of Severre's trusted subordinates, Paul Serakike—the man who went to Parliament with suitcases full of cash to protect Severre's business empire.

Meanwhile, the hunting blocks had been so badly neglected that at least a dozen blocks were abandoned every few years because they no longer held any wildlife. But the well connected demanded more land, so the remaining blocks were subdivided—and each partial block retained the quota of the original—effectively doubling the quotas every few years.

Poaching is so badly out of control that most wildlife populations are in free fall; only a few hunting companies are willing to invest in protecting their blocks—but the rich wildlife populations in these well-protected areas only makes the land more attractive for the insiders. So the best blocks are systematically subdivided and re-assigned to the strip miners.

Every few years, some Arab country donates money to the ruling party in return for a hunting block. Half of a subdivided block was recently given to the Royal Family of Qatar. The original block was known as Saffron-North; the subdivisions became Saffron-North/North and Saffron-North/South. The Qataris were assigned North/North, but they wanted North/South instead. To keep the Arabs happy, the WD relabeled the blocks, announcing that North/North was actually *south* of North/South.

Luke asks the three women on the ESA committee in USFWS headquarters, "Sorry to go through all these details. Is corruption really relevant to you in considering whether lions should be an E or a T?"

"Yes," they confirm as they look up from their notes.

So he continues.

A core group of professional hunters had worried for years about the future of Tanzania's lions: they knew that lions were being overhunted—they could see for themselves how quotas were being filled with underaged "trophy" males—and they understood how infanticide could devastate whole populations. The wise men sought to reform lion hunting, but they were outnumbered and outmaneuvered by the people who really ran the show.

Even though the Tanzanian Parliament passed a law in 2010 to limit offtakes to six-year old males, underaged lions are still being shot in 2013. The law has never been implemented, and, in fact, the WD is preparing to overturn the law altogether—as soon as USFWS announces its decision on the ESA. Once the law is overturned, all the illegal lion trophies from the past three years can be safely exported.

"There has been no reform in Tanzania whatsoever," Luke paused. "The hunting industry regulates the government, not the other way around." Another pause. "Classifying the lion as a T on the ESA might seem appealing, because USFWS oversight might help modulate hunting practices."

He waved his hand, "But, no, the hunters will do whatever it takes to perpetuate the status quo. The Tanzanians already lied about their elephant population back in 2009; it will be even easier for them to lie about lion numbers. Give them an opportunity to keep hunting lions, and they'll just twist your rules or ignore them altogether."

"But," said the head of the ESA committee, "what would happen if Americans no longer came to hunt lions in Tanzania—wouldn't they just be replaced by wealthy hunters from somewhere else?"

"Look," Luke said, "this committee is being watched by the whole world. If you classify lions as Endangered on ESA, a lot of European governments will work with you to reclassify lions onto appendix 1 at CITES—and that would end lion imports everywhere. There might be a year or two when American hunting clients are replaced by other nationalities, but it would only be temporary."

Luke talked with the committee for over an hour. He had arrived feeling so brittle that he might shatter into a million pieces. But after exchanging banalities and commiserating over the recent shut down of the American government by the Tea Party Republicans, Luke felt unexpectedly calm and detached. He had gained unique access to the inner workings of the WD, he had conducted all these investigations into the machinery of corruption, and he hoped it might be valuable to the ESA committee deliberations. He just needed to unburden himself. He just wanted to be heard.

And, as far as he could tell, they did hear. They had only been slightly familiar with corruption in the WD, but now they were openly dismayed by how far Tanzania had fallen from the standards of well-regulated wildlife management.

Emboldened, he decided to stick his neck out all the way. "I presume you've all seen the op-ed in *New York Times* written by the new Tanzanian director of wildlife? It was called 'Saving Lions by Killing Them.'"

Alex Songorwa had expressed alarm that a listing on the Endangered Species Act would prohibit the importation of lion trophies into the United States. Over half of Tanzania's sport-hunting clients came from the United States; without lions on the menu, the hunting industry might collapse entirely.

"Songorwa is a good man, but he has to follow the party line." Luke paused, "And it is just possible that he's trying to send you a message."

The committee put down their pens.

"There is no way Songorwa can reform the system himself. If he is too open about it, he'll get sacked—or physically harmed.

"But he is telling you what you can do to shut the whole thing down: ban lion imports into the United States as soon as possible. Go ahead and let all those crooks in the hunting industry go broke. Once all the politicians and Asian businessmen decide to make money doing something else, a new block allocation system could be built up from scratch. You guys could even advise him how to set up an auction system and train his staff how to monitor their wildlife.

"If you can help him to start over, you can eventually revise the lion's ESA status, and then maybe trophy hunting really would be an effective conservation tool in Tanzania."

Luke Sidewalker had done all that he could. He had no idea how the committee might vote. There was more support for an ESA listing than he expected, but he knew that there was a long way to go before any decision would be final.

But *his* job was done. Finished.

CHICAGO, 2 NOVEMBER 2013

"Lions are in the balance." I pause, face the audience, and continue. "Lions are in the balance, and their future is far from certain."

I'm in front of about three hundred people, speaking at the Chicago Humanities Festival in an annex of the Fourth Presbyterian Church near Water Tower Place. Most of the festival speakers are playwrights, journalists, and gurus. I was asked to present a summary of my life's work, so I spent the past two weeks putting together a general presentation illustrated with dozens of Daniel's lion photos.

It has been decades since I was this nervous before a big talk. So many bad things have happened the past few months, and I don't want to come across as whiny or bitter. But I'm genuinely worn down, and I'm not sure if

I'll be able to summon the appropriate level of energy. I have confidence in the quality of my visuals, and the sound system here faithfully reproduces the sounds of a lion's roar. The key is to get off to a good start.

Inspiration strikes as I take the stage.

"Lions are in the balance, and their future is far from certain."

Then a few words about what to expect in my talk today, and we're off.

Briefly, I tell why lions live in groups, the fact that they can count when they hear each other's roars, why they have manes, the videos of Fabio and Julio, how we hope to measure their role in the ecosystem through Snapshot Serengeti.

I move to the reasons why lions are such a difficult species to conserve, their man-eating habits, the full moon, and the problem of getting people to fence their crops to keep out the bushpigs. Then the problems with shooting underaged males, the pictures of prepubescent "trophies," the unimplemented new law in Tanzania that is about to be overturned, and the loss of my research clearance for sabotaging the Tanzanian hunting industry.

The audience gasps.

"So much for Tanzania," I move on quickly, "what about the rest of Africa?"

Then the extraordinary costs of conserving unfenced lion populations and the value of fences and my appeal to the donor community to get serious about the real scale of the challenge. Then a quote from an anti-fence conservationist who once e-mailed to say that he would "rather see the Serengeti disappear completely than with a fence around it," and the statement by a major conservation agency that they will only promote the preservation of "virgin wilderness."

More gasps.

I had known in advance that the Chicago festival would want a more personalized account of my research than usual, and I knew that I was going to have to recite this dreadful list of developments, so I struggled for weeks trying to figure out how to engage an audience with a talk full of failure and despair.

My neighborhood back home in Minnesota was called Uptown. The most prominent local landmark was the marquee of the Uptown Cinema, which rose high above its surroundings. Whenever I walked to and from the local supermarket, I could see the green neon lettering of the Uptown marquee, but the bottom was often obscured by nearby buildings so that all you could read was:

U
P
T
O
W

So despite the constant wearying undertow of African corruption, global economics, and personal adversaries, I took heart from that sign every time I saw it.

It was up to each and every one of us to create our own up-tow.

And so today in front of three hundred people in a Presbyterian Church in downtown Chicago, I show a slide of the UPTOW sign and report the news from our Snapshot Serengeti crowd-sourced initiative. We sought thirty-three thousand dollars but raised fifty-three thousand dollars. Add the money from *National Geographic*, and we're good for another year.

The bad guys may control the Tanzanian hunting industry for now, but international regulations by ESA and CITES will have the final word on lion hunting all across Africa. If hunters can't reform themselves, they'll only have themselves to blame: populations of wild lions will keep falling and falling until lion hunting is eventually banned.

The western edge of the Serengeti will someday be fenced, whether the conservationists like it or not. When that day comes, someone should measure the effects. Establish a camera-trap grid to learn whether the fence was as good (or as bad) as people think.

I close by stressing the need for realpolitik. The solutions for lion conservation may not always be aesthetically pleasing, but if they work, they should be adopted.

Even more important, if anyone here today knows anyone who can get the World Bank and the other big donors to devote real resources to protecting the last few places where lions might conceivably thrive for the next century, please do what you can, because *lions are in the balance*.

MINNEAPOLIS, 15 NOVEMBER 2013

Back in the dark days of Costa Mlay when I was looking for a backup research site, I visited half a dozen wildlife reserves in South Africa. Some were small, some were as big as New Jersey, but they all had healthy lion populations. I had hoped to find a place where lions could be observed as easily as in the Serengeti, but they rarely came out into the open from the thick brush. I missed the short grass, stately acacia trees, and overall grandeur of East Africa. Starting over in scrubby South Africa would be a lot more trouble than it was worth, so it was a considerable relief to outlast Costa Mlay and be able to return "home" to Seronera.

But something stayed with me from those trips to South Africa.

Lions were being reintroduced to so many parts of the country—including places close to a lot of people. Lion habitat was being reclaimed from a century's use as cattle ranches; lions were moving to the edge of town—and no one seemed to mind.

Lions are dangerous. They can kill you. The reserve managers accepted full responsibility for keeping people safe and only brought back lions if neighboring communities were comfortable having them around.

But in East Africa, lions had been part of the woodwork continuously. Human fatalities and livestock losses were inconvenient for conservationists and government officials, but no one took responsibility for what happened to local communities. Of course, dangerous threats like small pox and rinderpest must be eradicated—even in the poorest countries on earth—but lions and elephants were a world heritage, and they brought revenues to a lot of powerful people. Somehow governments seemed to think that families would tolerate lions by paying a forty dollar consolation for the loss of a husband or a child. Somehow conservationists felt that modest incomes from wildlife would overcome the fact that *the villagers would never be safe*.

When asked, Maasai say they don't mind living with lions that don't eat their cattle. People around the Selous don't mind lions either, as long as they don't eat people.

But in the meantime, there is less and less for the lions to eat besides people and livestock. Kill this week's problem animals, and the good lions will turn bad soon enough.

Of course, South Africans haven't faced the challenges of protecting vast areas of land where wildlife graze among domestic stock during the annual ebb and flow of green grass. East African landscapes are dotted with semipermeable wildlife areas from which migratory herbivores emerge and mingle with livestock for weeks on end. When the grasslands dry up, the wildlife retreat back inside, and pastoralists try to graze livestock in the same dry-season refuges.

And where cattle live with wildlife, lions will always misbehave.

Maybe a Lion Guardian program can mediate the human-lion conflicts in Kenya and Tanzania for the next decade or so, and maybe Ingela will find some way to help the lions in the NCA. But how big of an area can these projects really cover? And what happens when the recovering lion populations eat more cows than before?

Lions rebounded in Namibia's Skeleton Coast after a decade of successful conservation, but then livestock losses became intolerable. Poison once more ruled the day.

It's hard to imagine these programs succeeding into the far future, especially when Africa's human population is expected to quadruple by the year 2100.

I drove to Lake Victoria from Seronera in 1978 and passed through the Grumeti Game Reserve, a hunting block that was intended to provide a buffer between the Serengeti National Park and the neighboring villages. I saw a fair number of animals on that first trip, but everything was suspiciously skittish, and when I went back to start our dog-vaccination program in 2003, the wildlife were all gone. The villages had grown, and more and more people lived along the park boundary.

Everyone knew that the Greek who had operated the hunting concession in Grumeti had made no effort to fight bushmeat poachers. Neither had the WD. But 2003 was also the year when Paul Tudor Jones obtained a thirty-year lease on the Grumeti and Ikorongo Game Reserves. Tudor Jones is a hedge fund manager, who makes about a hundred thousand dollars a minute. His first year in Tanzania he spent tens of millions of dollars on antipoaching and community conservation projects. He also aimed to fence the outer perimeter of the reserves.

But he was told "that's not the way we do conservation in Tanzania."

I visited Grumeti during my last days in the Serengeti in August of this year. The western boundary of the national park is shaped somewhat like the western side of Texas, and Grumeti and Ikorongo Game Reserves cover the equivalent of Carlsbad-Las Cruces and Hobbs-Clovis, New Mexico.

The drive from Seronera passes through Fort Ikoma, the site of the only prosecuted war crime following World War One: a German officer tricked a British officer with a white flag and murdered him. The fort was briefly converted to a tourist lodge in the 1960s, complete with swimming pool, but the bullet-scarred walls are crumbling and the pool has long been empty.

The road from Ikoma to Sasakwa winds through hills and mesas not unlike those of the American southwest, but thousands of topi and zebra once more grazed the bleached blond grass in August of this year.

I'm told that Sasakwa Lodge is spectacular; rooms cost two thousand dollars per person per night—Oprah Winfrey stayed earlier this year. Earnings from the lodge are meant to cover the costs of conserving the two game reserves, but the revenue shortfall amounts to millions of dollars each year, leaving Tudor Jones to make up the difference from personal funds.

I follow the path past the horse stables to the management offices, where I meet the reserve's chief ecologist, Peter Goodman. Despite the millions spent annually, the conservation pressures have only worsened: seventy-five thousand livestock enter the blocks illegally each year—as do poachers, who claim not to know the location of the reserve boundaries.

A physical barrier would reduce or even eliminate the pedestrian traffic from the surrounding communities. Goodman is eager to discuss the idea with several high-priority villages, get their agreement on materials and location and then slowly extend the barrier piecemeal until it spans the entire outer perimeter of the wildlife areas. There's nothing left outside the reserves: the wildebeest migration has nowhere left to go. Grumeti and Ikorongo would effectively become an annex of the national park.

What about setting out a second Snapshot Serengeti grid to document the before, during and after? A fence should reduce poaching and prevent illegal livestock grazing, but if wildlife suffers, tear it down. Fencing is contentious. The Tanzanian government will only agree if the conservationists don't make too much of a fuss.

In the papers this morning, I see a headline from the *Tanzanian Daily News*: "Serengeti to Erect Fence to Curb Rampaging Elephants." The paper reported on a recent meeting involving villages adjacent to Grumeti Reserve. Local people are up in arms about crop-raiding elephants that have been chasing people on their farms. The farmers have heard about wildlife fencing in South Africa, and they say it is time for Tanzania to do something similar.

I write to Pete Goodman, assuming that the WD had given Grumeti the green light, but he replies that they had merely scheduled a meeting so that the villagers could discuss the elephant problem. The villagers demanded that something be done, and Goodman mentioned that Grumeti was contemplating asking for permission to build a fence.

The villagers took the idea as their own.

MINNEAPOLIS, 18 NOVEMBER 2013

"Hello this is David Barron from the ICCF. I'm putting together a group of people to go to Africa in the next few months at the request of the Tanzanian Ministry of Natural Resources. They know they've got serious problems with their wildlife, and they're worried about the next set of elections. A lot of jobs out there depend on tourism, so they've come to think that if they do something meaningful about conserving their wildlife, they might improve their chances at winning that election—they want to do something soon, and they understand they'll have to do something big with those three hundred thousand square kilometers of hunting blocks.

"The committee I'm putting together will work with the minister of natural resources, and the president himself will be directly involved.

"I'm calling to ask if you'd join us. I like the way you think. I like your ideas. They're ambitious, but they're realistic."

Over the past decade, it has sometimes felt that entire years have passed when nothing has happened, but then sometimes days passed and whole years happened.

When I was in my early twenties, I went to see my doctor about some physical discomfort or other, and he couldn't find anything seriously wrong. At the end of the exam, he looked at me and said, "Your problem is that you thrive on uproar. You can't sit still—and the wear and tear of all that constant motion will eventually get you."

It's not the uproar that gets to me. It's those long gaps in between—and the fear that maybe the uproar is over.

But it's not over yet.

JULIUS NYERERE INTERNATIONAL AIRPORT, DAR ES SALAAM, 12 MAY 2014

"Excuse me, but I forgot to ask for your vaccination card," says the African woman from the ticket counter as she finds me at the Flamingo Cafeteria in the Departure Lounge.

"I'm afraid I don't have it—I haven't been asked for so long, I forgot to bring it . . ."

"Without evidence of yellow fever vaccination, we will have to remove you from the plane."

"But I've had all my shots—I must have left my card in Arusha."

"Without your yellow card, you will be unable to fly."

I don't want to stay in this country for another minute—and I'm not even sure if my yellow card is here in Africa.

She starts to make a call on her cell phone.

"But wait—"

"I'm calling my manager," she looks at me above her glasses. With her black jacket and white blouse, the red ribbon of her ID tag makes her look like royalty in the dingy café with its once-mirrored glass walls that are now scratched and foggy. Travelers at nearby tables edge away from us as if they, too, haven't carried their yellow cards in years.

She says a few words on the phone in Swahili, then says to me, "How long have you been in the country?"

"Only a few days—I'm on my way to Johannesburg . . ."

"Only three days?"

I nod as if confiding an important secret to a dear friend.

"And did you come here straight from America?"

Another nod.

"And you were only here for three days?"

Solemn nod, looking straight into her eyes.

More discussion on her cell phone in Swahili; the room has almost stopped reeling.

"Ok, you can go. But when you get to Johannesburg, tell them you came straight through from America."

I promise, but I'm not sure it would be such a good idea. After all, my passport shows I've been in Tanzania since the sixth of February, and just a few minutes earlier I had smiled sweetly while the immigration officer failed to notice that I had overstayed my ninety-day tourist permit by five days.

It wasn't the hassle of the vaccination card that put me on edge. It was the idea of being yanked out of the crowd that made me as anxious as Luke Sidewalker at the Econo Lodge when a fist banged on his door and death stood in the hallway.

I had always felt that if I was in a public space—a ritzy hotel or an airport—I would be safe. But any sense of refuge vanished on Saturday—two days ago—at the Kilimanjaro Hotel, my old office during the days of Severre and his tyranny.

I came back to Tanzania this February, hoping that the appeal for my research clearance had been approved, only to learn that the matter had been stalled at the desk of the permanent secretary, Mama Tarishi, for the past six months.

We had met before, and she remembered me well. When I walked into her office, she gave me the longest hug I've ever had in a government building. I said that I was so sorry to trouble her with my problems, and she assured me that there was no problem because we were friends, and Tanzania needed people like me to help conserve their wildlife. I thanked her and said I was so glad to see her, but I really did have a serious problem. She leaned back, holding me at arm's length, and asked what was my problem. I said that the appeal for my research clearance had been in her office for six months and could she please let me know her decision. She seemed surprised and called in her secretary, who looked around for five minutes and found my file. As I watched Mama Tarishi thumb through a two-inch thick dossier, the temperature dropped by about a hundred degrees.

"According to the file, you came to work in this country on a permit that allowed you to work as a researcher but instead you have become an activist."

I said, "An activist?"

"It says you are a member of Transparency International."

"Transparency *who*?"

"Transparency International."

And I said, "No, no, I'm not a member of anything—and I have never heard of Transparency International."

And to my dear friend Mama Tarishi, I was telling the absolute truth.

But she paid no attention as she turned the pages, and when she reached a certain document, she stared at me cold as ice and asked, "Why did you say that about *his excellency*?"

And now I'm thinking that I'm not here to ask for her approval—I'm here to get arrested.

I blurted out: "I didn't say that—I was just quoting an official at TAWIRI."

"Who?"

I told her.

"Why did you quote him? The director general of TAWIRI is the only person who represents the government in matters concerning wildlife research."

"But Dennis Ikanda is Tanzania's official representative on lions. He travels to Washington, DC, on behalf of the government. He goes to CITES meetings. He quoted the president to me in a way that sounded very official, like he was conveying a specific message."

But there was no use arguing. She said she must treat the matter with utmost seriousness, and she refused to look at me. As the meeting froze to a halt, and I staggered toward the door, I asked if she could please make her decision soon, as I had been waiting for a long, long time.

I went back to the Econo Lodge feeling that the world had come to an end.

The most familiar place—a room or a street—that has always been an anchor, a home away from home, can suddenly become a mere waypoint, an empty spot on the way to nowhere. Everything is torn away. No connection to anything. The exotically bad odor of a filthy African street that once felt like the road forward to some new hope now just stinks, and it all feels unpleasant, unnecessary, pointless. But where else is there to go? What else is there to do?

An all consuming anxiety had slowly built over those six months of waiting in Minnesota, and I had self-medicated myself to exhaustion. I changed my classroom teaching by "flipping" a small honors class so that the students watched lecture videos at home, and we spent class-time in discussion and group projects. I decided to flip all my teaching and as soon as the term ended, I went to a studio and presented all my lectures to a video camera day after day for nearly a month.

Profoundly tired, jet-lagged, and feeling old for the first time, I emerged from Tarishi's office feeling utterly defeated, and three days later, my appeal was officially rejected, on the grounds that I had made "unfounded allegations" about the integrity of the Tanzanian government.

But the shock just drove me back to work. First, I had to safeguard the lion studies. Daniel had better eyes for the lions; Ingela had a better voice for talking with the Maasai. I was best at writing, and I could probably be more productive if I didn't have to fiddle with any more high-lift jacks in the flat middle of nowhere.

In the beginning of the Serengeti lion study there was George Schaller, who begat a succession of scientists every four years until my turn came, and I clung to it for thirty-six years. Given a choice, my heart would have stayed in the Serengeti another dozen years, but this was no longer my decision to make; the projects had to pass to new hands.

Second, I had to focus on myself. I had gained dangerous knowledge, and I couldn't keep my big mouth shut. I was only guilty of telling the truth, but if I hadn't been so careless, I might have endured. Why had I been so obsessed with the evil hunters? Did I really think I could make a difference? Or was I just showing off?

The self-flagellation continued for days, but once back in Arusha, I worked out the succession for the lion project. By the middle of April, two new applications were sent to TAWIRI with new principal investigators. The Serengeti project would now be headed by Mike Anderson and based at Wake Forest University in North Carolina. The next-generation program would integrate the lions and the Snapshot camera-trap grid with new studies of grasses and soils.

The Ngorongoro project would now be based at the Swedish University of Agriculture in Umeå; the next-generation program would bring together Swedish reindeer herders with Maasai herdsmen to mitigate lion-pastoralist conflicts and restore genetic connectivity between the Crater and the Serengeti.

But what to do with myself? The director general of TAWIRI, Simon

Mduma, wanted me to appeal Mama Tarishi's decision, but he also warned that TAWIRI and the WD had both recommended I merely be given a warning. Mama Tarishi had instead sought advice from a higher level in the government than the Ministry of Natural Resources, and she had presumably overruled her technical staff under direct orders of his excellency, the president, who, as Ikanda had warned, would cancel my clearance on account of my meddling with lion hunting.

The Ministry of Natural Resources was in chaos. Minister Kagasheki had been sacked a few months earlier because of a botched antipoaching campaign, called Tokomeza (Swahili for "clean-up"). Tokomeza had been halted, it was said, because so many pastoralists had been beaten, raped and shot for illegally grazing their livestock in the protected areas. But, unofficially, everyone knew that the antipoaching drive had gotten too close to the big potatoes in the government who were directly involved in the ivory trade—including, it seemed, the president's son.

In January, Frankfurt Zoological Society announced the results of the first careful elephant census in many years, and the Selous—which once held over a hundred ten thousand elephants—now only had thirteen thousand. The numbers were announced while President Kikwete was attending a meeting in London on the ivory trade, and at one point he told CNN that he knew who ran the poaching trade in Tanzania. The ringleader was a rich businessman in Arusha. Kikwete knew the house where the man lived, but he was powerless to catch him.

Kagasheki's replacement as minister of natural resources, Lazaro Nyalandu, came into the job with fire and energy and sacked the director of wildlife, Alex Songorwa, a few days after I met with Tarishi in February. Songorwa's reaction when asked about losing office? "I feel enormously relieved."

There had also been repeated calls to fire Mama Tarishi, but she weathered the storm and became the only major figure left in the ministry of natural resources in the aftermath of Tokomeza, so I decided not to renew my appeal until David Barron brought out the troops in May.

But there is waiting and there's waiting. I could distract myself with teaching in the United States, and Arusha was another story. To be so near yet so far from the Serengeti, unable to stay in the lion house or drive my own car even if I entered the park as a tourist, was intolerable.

It was essential to stay out of trouble and be quiet, but on the seventh of April, Ingela rang to say that a collared Serengeti lion had been speared in the NCA. It was another ritual killing by a group of young Maasai warriors, and the lion had died near where Ingela had found MH106 and MH37 when they were speared in 2010.

Daniel heard the mortality signal of MH35 and found her carcass with spear wounds on the face and body. Parts of her skin had been taken as trophies. The transmitter had been cut off with a knife, and the rest of the collar was missing.

Ingela worried about what to do. She asked her Maasai informants, who told her about the morani, who they were and where they lived, but if she were to report everything to the authorities, she might lose the trust of the local communities. But she couldn't cover it up, or the morani would think they could get away with anything.

I encouraged her to act as if she had only heard about MH35 from Daniel and to urge the NCAA to make an official response. She started to conceive a plan, but a few days later her informants told her of something far worse.

National Geographic came to Tanzania in 2011–12 to produce a story on the Serengeti lions. The NG team was based along the rim of the Barafu gorge, located near the eastern boundary of the national park. They decided to focus the story on the Vumbi Pride, so-named because their dry-season range is truly barren and harsh (vumbi means dust). But in the wet season, the Vumbi lions have it easy, and they are a wildlife photographer's dream of yellow cats on a short green lawn with the occasional umbrella acacia framing the volcanic highlands of the NCA.

We gave NG a tracking box and told them the family histories of each lion in the pride. The Vumbi's had a large communal litter at the time, and the team got to know the five moms and eight surviving cubs, especially the radio-collared female, VU-M. VU-M was an avid hunter and a key figure in the eventual magazine spread, despite her rather unphotogenic necklace.

On the thirteenth of April this year, Ingela received more bad news from her Maasai informants: another lion had been speared inside the Serengeti Park, just a few miles from the eastern park boundary. But the lion had not been the only fatality.

Ingela called Daniel, and he drove out past the site of *National Geographic*'s campsite from two years earlier. Ingela's informant had been so specific that Daniel wound down between groves of spiky green Commiphora trees and clusters of red-flowering aloes to a particular spot on the bottom of Barafu gorge on one of those sun-drenched wet-season days of unnatural clarity. He bounced over deeply rutted game trails along the emerald grass of the valley floor until he reached a particular shallow pond, where he finally heard the fast-paced, high-pitched beeps of the mortality signal. He stepped out of his car, took off his shoes, and sloshed methodically through the muddy water until he felt a cylindrical metal object with his feet. The transmitter had been removed from the collar with a knife.

Following the odor of a dead animal a hundred meters away, he located VU-M's body underneath a low bush. She had been repeatedly speared, and several claws had been cut off from her hind paws.

Barely fifteen meters away lay the corpse of a dead Maasai, one arm under his chest, legs splayed and a spear wound through his skull.

I had managed to stay calm until now, but this was too much. The NCA had merely tolerated Ingela's conflict-mitigation project and had never made any meaningful contribution. She had managed to document the extent of the problem—the number of cows killed by lions each year, the number of goats killed by leopards—but it was time to start offering some practical strategies for improving livestock husbandry, for encouraging the Maasai to view lions as an asset rather than as a hazard or macho challenge, to start protecting their lions, and to stop killing each other with their errant spears.

If we didn't do anything now, nothing would ever happen—but I wasn't allowed inside the park, I wasn't allowed inside the NCA.

I couldn't stand it.

I called Simon Mduma, and said, "Please, Simon. A man has died. A famous lion—a *National Geographic* lion—has died. There is bound to be coverage by the Press. People will ask me what we are doing to help, and I'm going to have to say that I've been banned from conducting research. It's going to make Tanzania look bad if I'm not allowed to help. Can you please, please contact the permanent secretary and ask her to reconsider my appeal?"

I felt trapped in a cage.

When VU-M was first speared, she turned on one of her attackers and a spear aimed at the lion hit the young man in the head. His body had been abandoned because Maasai only bury their dead if the victim dies near home. A corpse in the bush is left for the vultures and hyenas.

Ingela drove out to the village nearest to Barafu and talked to the Maasai elders. They were unable to control their young men. Whole communities were punished by the NCAA whenever a lion was killed, but their youth had nothing better to do with themselves, and every Maasai morani knew where to find unwary lions from the Serengeti each rainy season.

Ingela explained about the Lion Guardians. The elders were supportive: such a program would require the participating warriors to control their peers. Ingela warned that the authorities would start flying over the NCA on a regular basis, ready to arrest any Maasai they spotted inside the national park, ready to detain any suspicious gatherings in the wildlife areas.

Even as she spoke, a Frankfurt Zoological Society plane zoomed overhead.

But there was a bigger issue to resolve.

In 2013, Ingela and I had had a brain wave. Ingela always lamented the lack of a performance payment scheme like she had seen in northern Sweden, where the Sami pastoralists were rewarded for the number of wolverines and lynx that co-existed with their reindeer herds. We had initially been attracted to the Lion Guardians model, but we had no way to pay for it or to assure its persistence beyond the duration of Ingela's PhD project.

I had always liked the idea of paying people for conservation—for the number of rare animals in their midst—rather than compensating them for their lost livestock. Compensation schemes have never worked for long anywhere in the world because of the lack of incentive to reduce livestock losses. Hey, if you had a sick cow, wouldn't you just leave it out for the wolves to kill so that someone would buy you a healthy new one?

Performance Payments might be appealing, but it's one thing to track wolverines and lynx in the snows of northern Sweden, how on earth would you know how many lions lived in the neighborhood of a Maasai boma?

Unless, of course, you had a camera-trap grid.

If grids were established around each village, the overall performance-payment fund could be based on how many cows are killed throughout the NCA, and funds could be distributed to each community based on the number of lion photos from their neighborhood camera traps. Maasai who live with lions would get more than those who don't; if lion numbers go up from one year to the next so would the payments.

Now we had a technique for establishing a Swedish-style system in the NCA.

But there was just one problem.

Who would put up the cash?

We met with the top management of the NCA in July 2013. They liked the fact that they wouldn't have to verify the cause of death of every last cow and goat in an eight thousand square kilometer area. But they lacked an official conservator at the moment, and no one wanted to stick their neck out for a new idea—could we try to get approval from higher up in the ministry?

In early August 2013, I met with Alex Songorwa in Dar es Salaam. Alex loved the idea until we got to the payments.

"The NCA gets thirty-million dollars a year in gate receipts," I suggested. "The fund would only cost about two hundred fifty thousand."

"But the NCAA has committed all those funds to other activities." The director of wildlife paused. "Can't you get *National Geographic* to pay for it?"

I gripped my chair as tightly as possible. "An outside donor might help cover the costs for a pilot project—but they'll look at your revenues and ask why you can't pay for it yourself."

In mid-August 2013, I was with Ingela in the NCA. We heard that the permanent secretary was on an official visit and would be dining at one of the lodges on the Crater rim. We waited for our moment, introduced ourselves, and outlined the performance payment idea. Mama Tarishi grasped the advantages immediately. We didn't get as far as to how it would be financed, but when I saw her again this February, I raised the idea as a good reason why I hoped the government would restore my research clearance. The performance payments would help the NCAA by reducing administrative responsibilities, encouraging the Maasai to become active conservationists, helping to restore the lion population . . .

Meanwhile, the NCAA had done nothing for Ingela or the Maasai; the few remaining lions in the pastoralist areas were still killing livestock and still being killed in retaliation.

And now the young warriors were invading the Serengeti National Park.

Daniel mentioned the latest massacre in Barafu in an e-mail to one of the *National Geographic* photographic team, who then called David Quammen, the author of the *National Geographic* article. David was eager to write a story, but I wanted to hold off until we had tried some friendly persuasion.

A new man had recently been appointed conservator of the NCA. Dr. Freddy Manongi had taught at Mweka, the College of African Wildlife Management, for years, and he was known for his outspoken integrity. He would be in town on the twenty-sixth of April—the golden jubilee of the union between Tanganyika and Zanzibar—and he agreed to meet us in Arusha.

Freddy is a bit overweight, with a distinctive goatee and dark-rimmed glasses—he looks a lot like jazzman Charlie Mingus circa 1957. We sat together in the comfortable leather chairs of the lobby of the Arusha Hotel, and Ingela and I took turns reviewing the need to restore the ailing Ngorongoro Crater lion population to good genetic health, showing Freddy the extent of livestock losses in the NCA and discussing the broad concept of performance payments.

Like Mama Tarishi, Freddy immediately saw the advantages—but he also raised a concern: "If the rewards from good conservation are given to the community rather than to individuals, people may not make the connection with conservation; they may think it is just one more form of community development that has nothing to do with lions."

Yes, we said, Ingela planned to work closely with Swedish conservationists who have grappled with exactly this challenge in the Sami communities—we had applied for funding from the Swedish government to bring some Sami reindeer herders to the NCA to discuss performance payments with the Maasai and to take some Maasai up to the ice and snow.

The idea, he said, was worth considering—at least for a pilot project—and Ingela asked if the NCAA could help fund the pilot by covering the costs of eight Maasai scouts who would eventually become Lion Guardians and maintain the camera traps.

"I can easily find that amount of money in my budget," said Freddy, and I felt such intense relief that I almost burst into tears.

We continued our discussion in broad terms until the time was ripe. I apologized to Freddy in advance for what I was about to show him, then presented Daniel's photos of MH35, VU-M and the dead morani. "*National Geographic* wants to write a follow-up story on the Vumbi pride, but I asked them to wait until we had talked with you," I said. Then the pitch: "Rather than have Geographic focus on the bad news, we were hoping that you might talk to them directly, and maybe even announce your support for the performance payments and Lion Guardians—and mention that the NCAA will pay the salaries for some of the Maasai on the project."

Freddy readily agreed, and David Quammen titled his piece, "Can Good Come from Maasai Lion Killings in the Serengeti?

It is pretty to think so.

The pastoralist problem is much bigger than just the NCA. Between the boundary of the Conservation Area and the Kenyan border is the Loliondo Game Controlled Area—the hunting concession controlled by the Arabs. But the neighboring Maasai have decided to overrule all conservation boundaries, and the hunting block is now overrun with pastoralists: the Arabs didn't even try to hunt in Loliondo in 2013. Over three hundred Maasai families have recently built bomas in a five hundred square kilometer section of the Serengeti National Park—just north of our long-term study area—and eliminated the local lions, thereby drawing our study prides northwards to fill the void and reducing the number of lions in our woodland prides by half.

We identify every nomadic lion that passes through our study area;

taking mug shots, drawing ear notches and whisker spots. Because loose lions follow the migration all around the ecosystem, the ratio of nomads to residents provides a measure of how the lion population might be faring in the rest of the Serengeti, but nomads have become increasingly scarce over the past fifty years.

Between the Maasai to the east and the poaching to the west, the balance is tipping and lions are at risk in one of the best-protected, best-loved parks in Africa.

Meanwhile the hunters have been on the offensive. Earlier this year, the Tanzanian Tourist Operators had issued a public appeal for a permanent ban on all trophy hunting in the country. Tanzania's wildlife populations were declining. Overhunting was contributing to the problem, and most tourists didn't approve of hunting. Hunting, in short, gave the country a bad name.

The Tanzanian Hunting Operators Association shot back that photo tourists should be forced to pay a surcharge to cover the costs of protecting the hunting blocks, since hunting didn't generate enough revenue. . . .

But weren't hunting fees supposed to raise so much revenue that they single-handedly protected all the wildlife in their blocks?

The recent elephant survey was a pretty good clue that hunters had failed to protect the Selous in any meaningful way. But these were the guys with the guns, and they only relied on logic when it suited them.

In mid-March, a few days before Bernard Kissui was scheduled to give a lion talk in Arusha, he received an anonymous call from the WD, warning him not to talk about trophy hunting. Benny's talk had been organized by the Tanzanian Tourist Operators, which opposed trophy hunting. Lion hunting was critically important to the hunting industry.

Benny assured the voice over the phone that he only planned to discuss human-lion conflicts around Tarangire and to show how he had been reinforcing Maasai cattle bomas with chain-link fencing. But the next night Benny received another anonymous call and yet more warnings not to talk about hunting.

About two hundred people attended his talk, mostly tour guides and drivers with a fair number of expatriates, a handful of hunting operators, and two people from the WD, who, presumably, were there to make sure that he didn't say anything negative about lion hunting.

Meanwhile, the International Union for Conservation of Nature had asked me to help update the Red List status of the African lion. Lions had been classified as vulnerable for the past few cycles, but with the potential ESA listing and recurrent considerations of the lion at CITES, we needed to be as comprehensive and up to date as possible. New surveys in West Africa found only 404 lions in all of Benin, Burkina Faso, Cameroon, Côte d'Ivoire, Ghana, Liberia, Niger, Nigeria, and Senegal. We updated the dollars-and-fence analysis and added new data from Botswana, Zambia, and Zimbabwe.

Once all the numbers were in, we divided Africa's last lions into four categories.

First, the fenced populations were fine, with about three thousand lions safely protected in Kenya, Namibia, South Africa, and Zimbabwe.

Second, lions in Mozambique's unfenced Niassa Reserve held a growing population of about a thousand lions, but ivory poachers had overrun Niassa, and thousands of dead elephants meant abundant meat for the local lions despite an accelerating loss of smaller prey to bushmeat poachers—not a situation that could persist for very long.

Third, unfenced populations in the empty desert areas of southern Africa—Botswana, Namibia, and Zimbabwe—were holding steady at about two thousand lions.

But fourth were the unfenced reserves everywhere else. These showed a 62 percent decline over the past twenty-one years. All the West African countries plus Kenya, Rwanda, Tanzania, Uganda and Zambia had lost nearly two-thirds their lions in three lion generations. This was enough to qualify lions as endangered on the Red List in every country outside of southern Africa, and we reckoned that there were now only about twenty thousand lions left across the whole continent.

Although our final category was based on two dozen populations in a dozen countries, we had no idea what was happening in large parts of the Selous and western Tanzania. The Tanzanian government had forbidden Henry and Kirsten from continuing their long-term lion study in the photo-tourism area of Selous and blocked anyone else from conducting surveys in any of the hunting blocks. So in the absence of real numbers we were forced to exclude them.

Maybe lions in the Selous hunting blocks had thrived on poached elephant as happily as in Niassa. Maybe lions elsewhere in Tanzania were safe from poaching, poisoning and pastoralists. But the government had committed itself to a massive cover-up, so now it was stuck. If Tanzania's lions didn't truly qualify as "Endangered," someone would have to prove it—and no one would believe the hunters if they tried to do it themselves.

*

On the morning of the ninth of May, I walked up to Mama Tarishi in the middle of the coffee shop at the Kilimanjaro Hotel in Dar es Salaam.

"Good to see you again," I said, gripping her hand.

"Good to see you, too," she said without showing any obvious displeasure, but she was a bit too quick to ask, "Are you attending the meeting?"

"Yes, I've been invited by ICCF. Will you be here the whole time?"

"Yes, yes," she said, looking off to the side.

"Would there be a chance for us to revisit my research clearance? I would be very grateful for some time to talk with you again."

"Yes, I'll be here," she said, annoyed. "We can talk. Just look around, and you'll find me."

I walked through the back patio, crossed the footbridge above a fake stream lined with shade trees and potted plants, and found my purple-ribboned nametag at the welcome desk outside the Kilimanjaro Hotel marquee, a vast air-conditioned tent with glass walls on two sides. Inside, tables had been arranged in a large empty square that would allow film and TV crews to take close-ups of honorable guests as they spoke on matters of great importance.

The circus had come to town. Under the big tent, we had boy scouts and girl scouts, journalists, photographers, religious leaders, ambassadors, high commissioners and the ministers of natural resources, livestock and local government, Tanzanian soldiers, police officers and game rangers all in their best uniforms. The International Conservation Caucus Foundation had brought an ex-senator from Wisconsin. Scattered around the marquee were dozens of people from FZS, TANAPA, USAID, USFWS and the World Bank.

On one side, John Jackson, Sheni Lalji and Eric Pasanisi.

On the other side, Markus Borner, Benson Kibonde, Freddy Manongi and Simon Mduma.

On yet another side, we have the WD: Paul Serakike, Julius Kibebe and, the man himself, Emmanuel Severre—his dull teeth browner than ever, the rims of his glasses still gleaming golden, and his hair still dyed black, but his face showing the signs of advanced age and prolonged illness.

The only notables missing are Dennis Ikanda and Alex Songorwa. I have no idea about Ikanda's whereabouts, but Songorwa's absence is the scandal du jour. The new minister of natural resources, Lazaro Nyalandu, sacked Songorwa at the beginning of March and appointed Severre's old henchman, Paul Serakike, as a temporary replacement. But even though Songorwa initially professed relief at being removed, he changed his mind and appealed to Mama Tarishi, who then announced that since only the president could sack the director of wildlife, Lazaro's action had been illegal. So history repeats itself: first Severre clung to power when Diallo tried to oust him, and now Songorwa refuses to budge.

Minister Lazaro and Mama Tarishi are at war. Lazaro insists that Serakike is the one true director, and only Serakike has been allowed inside the big top.

The ICCF meeting had originally been intended to involve only about forty people so the Americans could quietly negotiate behind the scenes, but the Tanzanians have turned it into an extravaganza with a cast of thousands.

Minister Lazaro Nyalandu is a handsome man with a strong jaw and prominent cheekbones. He *looks* the role of the heroic statesman, though he is still only in his early forties. He addresses the honorable guests by assuring everyone that his government is committed to fighting poaching and finding new strategies to sustain its wildlife.

He introduces the Tanzanian vice president, who expresses profound dismay that the Selous now has only 13,084 elephants. The government has stepped up antipoaching, the vice president says, but it needs more money, as the protected areas are so large, and the problem is so severe.

Then Lazaro retakes the floor and reminds us that "the slave trade was in Tanzania because of poaching. Slaves were porters for ivory. This problem runs deep through our nation's history. We have a stockpile of ivory that is over 104 tons. We do not want to sell it until the poaching crisis has been resolved.

"We have made progress: poaching has been cut in half the past year. I can also announce that we have finalized the steps for launching TAWA—the Tanzanian Wildlife Agency. TAWA will look after the hunting blocks as TANAPA looks after our national parks. We have already hired 430 new rangers. We will hire five hundred more by the end of the year. We will soon implement a new disciplinary code for our rangers to avoid the [human rights] problems of Tokomeza."

He pauses briefly, avoids looking at anyone in particular, and says, "And there will be zero tolerance for corruption at all levels of the government."

A few more dignitaries give a few more remarks before Benson Kibonde takes the floor, saying, "We have doubled our boots on the ground and poaching has more than halved. By doubling the boots, we have seen good results so far. We have faced this crisis before, and we succeeded. But now we are back at the same place. We can't subject ourselves in this sinusoidal way. We must stop it. Let me say that I'm sometimes disappointed coming to these meetings and these trainings and seeing all the money that is spent. The poachers haven't been to conferences or taken the training.

They haven't been to Mweka or the University of Dar es Salaam, yet they are winning. Why? I think it is because they are committed, they are hard working, and they know what they are doing. We must show the same commitment."

Benson said a lot more, but the most interesting part was when he talked about prioritization. He discussed the need to prioritize law enforcement efforts, and he almost certainly started to say that efforts should be focused on certain *people* who were most powerful in the ivory trade, but he caught himself in time. Instead, he mentioned that resources should be focused on the *places* that still had the most wildlife, since 20 percent of the areas had 70 percent of the wildlife.

Prioritization on places remained a popular topic for the rest of the morning, but most of the afternoon ended up as a feel-good session about how all the remaining wildlife areas outside the parks and reserves and WMAs should also be protected, and how everyone should be trained to manage the massive influx of money that would come with wildlife conservation and that the local law courts and schools and village infrastructures would all be transformed by community conservation.

And yet very few parks and WMAs make any money.

I gave up hope for any serious discussion about reform or restructuring. The best use of my time would be to try to catch Lazaro for a moment and point out how I had generally been a decent fellow, but I needed help. Simon Mduma had suggested that the minister might be able to help overturn Mama Tarishi's decision, but I should first talk to him about safer topics.

So I found a moment during tea break to introduce myself and asked if we could meet sometime because it might be useful to think about adding wildlife fencing to Tanzania's conservation toolkit. The minister was intrigued but busy with newspaper reporters and TV crews, and he suggested we meet for a few minutes the next day.

I thanked him, walked outside and decided, what the hell, Sheni is here—let's talk about fencing. Sheni says fencing would be difficult unless the government supports the idea. I remind him that, with his political connections, he is well placed to help shift government policy, and he soon starts coming around to the idea—presumably because he has inspired so much ill will in the communities around his own blocks. It would be rather amusing if Sheni were to become known as the king of fences, but as long as he helps keep Tanzania's rural communities safe from lions, elephants, and people like himself, I'd be the first to congratulate him.

As it happens, Sheni speaks to the gathered assembly later that afternoon, complaining about the lack of management skills in the WMAs and the fees he is forced to pay—nearly a million dollars a year, he claims, to a single WMA where he does everything for them. He makes it sound like he personally delivers each of their kids to school, darns their socks and mows their lawns. He is the greatest benefactor in the Eastern Hemisphere, a true humanitarian.

Then Eric Pasanisi takes the floor. Pasanisi, the younger, has spent the past year trying to gain a toehold in the Serengeti through an illegal contract with a neighboring WMA. But by forcing out a well-financed American company, he overplayed his hand, and an impromptu coalition of hunting operators threatened to leave TAHOA if he didn't drop his claim.

Eric Pasanisi may be under a cloud, but he still has plenty of pawns on his payroll in the Tanzanian government, and he is the current president of TAHOA. In his heavy French accent he reads a written statement that recounts the virtues of the hunting industry, which has generously purchased three Land Cruisers for the WD this year. He blasts the American government for banning the import of elephant trophies into the United States from Tanzania. Hunting brings value to wildlife. Without hunting, the situation would be even worse. Very few elephants are shot by trophy hunters each year. Hunting is part of the solution, not the problem. Every third sentence mentions Valery Giscard d'Estaing's patronage of TAHOA. No one can understand a word he says, but he looks the part of the French aristocrat with his wavy hair, silver at the temples, his navy blue suit, and a face that belongs in a James Bond movie.

On Saturday morning, the religious leaders discuss ways to persuade their flocks to be better stewards of the land. Muslims and Christians stand together, extolling the glories of Creation and love for all living things. During tea break, I meet with Lazaro again, and we are just getting down to business when I see Mama Tarishi scowling at the minister and me apparently conspiring together, and then someone from ICCF "poaches" Lazaro away from me so the minister can meet with a Tanzanian billionaire who is about to leave the meeting.

I return to my seat and am only there for about three minutes, when a lackey from the WD comes up and says, "I'm sorry to disturb, but I have been told that you came to this meeting without an invitation."

"No, no, I was invited by ICCF."

"I have been told that you were not invited, so please do not make a scene. You must leave immediately."

"That's not true; look at my nametag with the purple ribbon. Why would I have been given an official nametag if I hadn't been invited?"

"Please, I have orders to ask you to leave."

"Orders from whom?"

"Very high," he says, raising his hand above his forehead.

"Wait, ask anyone from ICCF; they'll tell you that I was invited."

He leaves for about five minutes before coming back to say, "I have again been informed that you must leave immediately."

The next session has begun. I stand up straight and sidle out between two rows of seats with the Frankfurt contingent looking at me quizzically. "I've been kicked out," I say half-dazed.

I am in the same funk as when I left Mama Tarishi's office. Except this time I am frightened. I'm only a few hundred yards from the site of Luke Sidewalker's attack. I'm on the ninety-third day of my ninety-day tourist pass; Mama Tarishi could have me arrested. The Tanzanian Wildlife Research Institute isn't scheduled to approve the next-generation lion projects until next week. Everything is in jeopardy.

At first I just want to leave—to get as far away as possible from this nest of vermin, but then I enter the coffee shop in the Kilimanjaro, my "office" from the old days with Severre.

I start bursting with defiance.

If this were the end, the WD would just have to arrest me in full view of USFWS and USAID and half the Tanzanian Cabinet. The minister of livestock and the speaker of the Parliament both worked with me in the Serengeti. Just about everyone here knows me.

A few hours later, I am at lunch with an official from USAID. Like most other officials, she had only been in Tanzania for two or three years and would soon cycle to another posting in Asia or Africa. Susan and I had met her a few days earlier, discussing how Savannas Forever might help USAID move forward with some of their Feed-the-Future projects. We didn't discuss our rejected proposal, but word was that the Beltway bandits' efforts at monitoring and evaluation had been less than stellar. There was no hope that we could get the contract now, but our survey experiences in northern Tanzania could help guide USAID with some of their smaller efforts.

The past two years, Savannas Forever had conducted a series of disconnected surveys for various aid agencies interested in famine relief, cotton production, soybeans, child labor and family violence. Aid funding had dropped in the aftermath of the 2008 financial crisis, agencies were stingier and stingier with their M&E budgets, and we had lost hope of getting a large "anchor grant" that would pay our rent in Arusha and the salaries for Susan's staff.

Though we no longer collected the holistic data that had inspired the concept of the Whole Village Project, we had learned enough to discern a recurrent flaw in foreign aid.

Someone might have the best idea in the world—some new way to improve subsistence agriculture or to reduce infant mortality—and it might work beautifully on a small scale. When piloting a project with a dozen or so beneficiaries, the practitioners are always on hand to provide in-depth training and to follow up with intensive supervision.

But in scaling up from ten people to ten thousand, the original beneficiaries are meant to train yet more trainers. Imagine being a subsistence farmer who has just learned some new skill but then has to leave the family farm to go teach his neighbors who are scattered over a wide area with no roads or transport—except for the bicycle provided by the aid agency.

Development projects promise fast results, but rapidly moving from small scale to large scale always seems to fail. Even if the Whole Village Project had received all the money needed during our la-la land days, we might only have shown the consequences of this one basic flaw over and over and over again.

Aid can certainly help by building highways, dams, and power plants. Construct the infrastructure and commerce will flourish in even the most miserable setting, and the Millennium Challenge Corporation makes funding such extensive construction projects contingent on recipient countries reducing their corruption.

This is the half billion dollars that Tanzania may risk losing if their happy talk about waging war on elephant poaching today has just been a smoke screen.

After lunch, David Barron comes up and asks, "Have you sorted yourself out with the Tanzanians?"

"Yes, I refused to leave, and they've stopped chasing me for the moment."

He drains his coffee and looks off in the middle distance. Clenching his jaw, he says, "I have to tell you that I'm worried about the future of big game hunting. I don't think you can be a proper man unless you've hunted. This younger generation is missing out on an essential part of what it means to be a man. Without that, there will be a general deterioration in the human race—a genetic legacy that will have been lost. These boys will all grow up to become metrosexuals with no idea how the world works or how to behave."

Night has fallen in Dar es Salaam. The tables and chairs have been removed from the marquee, and we have gathered for the closing ceremony. Ambassadors have brought their spouses. Women are dressed in long gowns, men in suits. A Tanzanian military brass band is playing polkas from the days of the kaiser when the mainland was still Tanganyika. Waiters deliver canapés, wine and beer to the assembled guests as we await the arrival of Prime Minister Pinda. I'm in my usual polo shirt and khaki pants, hoping I can linger long enough to say hello. I haven't wanted to bother him with my problems. He has taken a lot of flak from the press the past few years about far more urgent matters of state, but I would like to see him one last time.

Time is crawling, but then the crowd stirs, and the prime minister makes his entrance, running the gauntlet of dignitaries and dutifully shaking hands with each one in turn.

He finishes with the ambassadors and spots me, standing by myself across the room. He leaves his entourage to greet me.

"I wasn't expecting to see you here." We hold hands in front of the entire crowd and ask after each other's families and health. He is soon called back to duty.

Pinda is not a tall man, but he has a fierce dignity that makes him seem as big as anyone in the room. He climbs up to the dais, positions himself behind the podium and delivers a rather stilted address about the importance of wildlife and the government's commitment to winning the war on poaching.

But then he puts down his script, looks out at the room, and his voice suddenly comes alive. "I have to tell you that I wasn't looking forward to coming to this meeting tonight. Parliament is in session, and we have many important matters to resolve, and I attend so many of these conferences just to shake hands and say a few words.

"But when I came in here tonight, I was so happy to see my friend, Craig Packer. It has been too long since we last saw each other. And I want you all to know that he is the greatest conservator in this country; he has done so much for us."

And he went on in that vein for I don't know how long. My face was burning; the world had started spinning in such an unlikely direction.

I can't remember if the crowd applauded or if he just returned to his written speech without any break in the action.

It didn't matter. Traditional African culture is utterly hierarchical, and one of the most important men in the country had called me out in the most remarkable way imaginable.

I mostly remember this amazing feeling of warmth and gratitude as I thanked Pinda from across the room, bowing my head and pressing my hands together as if in prayer. Mama Tarishi stared at me from her spot in the crowd, her face like a popped balloon.

10 Object Permanence

DINOKENG, 21 MAY 2014

"OK, let them out."

Kevin opens the door and two female lions walk through the grass, heading toward the heavy black metal box firmly staked to the ground. Inside the box are two pieces of meat about the size of large pork chops. Each lion learned how to open the box when it was equipped with a single rope, but now there are two. Livy and Ginny know there is food inside, and they quickly find the ropes. They pull at the same time but not hard enough, and nothing happens.

Livy then pulls hard on her rope—but by herself—and nothing happens. Ginny chews on the other rope, and finally pulls it, but Livy has given up on the rope. Ginny gets discouraged, too, and tries to find a way into the box from the top. Livy starts digging, trying to get to the meat from below. Both lions get so frustrated that they walk off a few meters and lie down, as if to collect their thoughts.

After a minute or so, Livy comes back to the box, tugs at a rope, digs underneath for a moment then goes to the second rope. Ginny returns to work, too, and, by chance, they both happen to tug the two ropes at the same time.

The door opens.

"How long did that take?" asks Kevin.

"Thirteen and a half minutes," says Natalia.

"Not very impressive."

"They seemed to get discouraged when it didn't open the first time they pulled together."

"Let's do it again."

Livy and Ginny come back to the re-baited box. They seem more focused this time. They sniff around, play with the ropes and finally tug together. The door opens within six minutes.

Reload the box, re-latch and repeat.

But by the third trial Ginny has had enough.

"Come on, girl," Kevin encourages, but Ginny has lost the thread. Livy continues pulling, digging and climbing for awhile before giving up and joining Ginny in the shade.

Next come Amy and Megan, and their initial tug is successful, but their performance deteriorates in successive trials: on the first occasion they pull the ropes out of synch, they lose the connection between work and reward. By the third test, Amy has developed an obsession with the rope, and she prefers to eat the sisal rather than to pull on it.

Third, we have Bobcat and Gabbie. Bobcat is a splendid black-maned male, and Gabbie is his only wife. Once again, the pair has initial success, but these two seem to get the point of working together and rapidly open the box on the second and third trials. But then the fourth trial is like a replay of the two sets of females: the couple arrive slightly out of synch, Bobcat tugs the rope too hard and lifts the box a few inches off the ground. The resultant crash startles Gabbie, and she loses her confidence around the whole contraption. They finally open the door after about ten minutes, but the sequence of trials doesn't indicate any increasing degree of coordination—of waiting until the other lion has also started to pull.

Natalia Barrego studies animal intelligence—specifically social intelligence. Lions hunt and defend territories together but can they solve a novel problem together?

Natalia came to test the Serengeti lions in 2012. But during her five weeks at the lion house, she couldn't find a reliable source of meat inside the park, and the Serengeti lions wouldn't eat beef or goat. Even when she did have the right bait, she couldn't control how many animals might approach the box during each test. Sometimes it was only one, sometimes a half dozen. Without any way to control the tests, the results were meaningless.

Natalia moved her project to South Africa and started working with Kevin Richardson, who keeps his lions in a series of one-hectare enclosures in the Dinokeng Reserve. Each lion group spends about a week per month in a thirty-hectare enclosure, and Kevin occasionally takes them for walks in the 180-square-kilometer reserve where they sometimes catch wild prey. His lions already received a fair amount of stimulation, but he welcomed Natalia's experiments as a form of behavioral enrichment.

Virginia Quinn is filming Natalia as part of a documentary comparing the intelligence of lions versus spotted hyenas. Captive hyenas passed the double-rope test in an earlier study in Berkeley, but Natalia is the first person to test lions. Even though none of the lion pairs improved with practice, the new box presented a fair number of technical difficulties (e.g., the tactile pleasure of chewing sisal). And whereas the Berkeley study allowed the hyenas to pull the ropes in any direction, this black metal box only opens when the ropes are pulled at a precise angle. So the lions have to master two problems at once: pull and direction.

Virginia is on a tight schedule, so after the first three pairs have cycled through the double-rope test, it's time to film the next experiment. Five metal buckets have been welded together in a steel frame; each bucket has a hinged metal lid. Two racks of five are positioned in parallel, so the lion must learn to find food in one of ten covered buckets. Kevin will train Ginny to find a piece of meat by opening the lid of a particular bucket. After training, she will come back with Livy, who has never seen the setup before.

In some species, the trained animal will misinform its companion by going to the wrong bucket. So instead of opening the bucket of food, Ginny might try to lead Livy to one of the nine empty buckets.

Kevin trains Ginny to find the food in the middle bucket at the front. After two or three trials, Ginny immediately returns to the same bucket each time.

When Kevin releases Livy, Ginny goes straight to the middle bucket at the front rather than mislead Livy by going to a different bucket. Ginny is a lion, where possession is nine points of the law. By arriving first, she guarantees her access to the food. Animal species that engage in deliberate deception are hierarchical: the devious subordinate would otherwise lose the meat to its dominant companion. But female lions are egalitarian: a latecomer cannot supplant the first.

Basic lion biology may undermine the lions' performance while manipulating ropes and buckets, but I came to Dinokeng to see if the captive lions had the same instincts as Serengeti lions. I first wanted to measure their responses to strangers, so Kevin went out with Bobcat for a walk in the park one morning, and I broadcast the roar of another male from Kevin's captive colony. These lions roar all night long, and Bobcat should have found it normal to hear a near neighbor while out for a stroll. Sure enough, he barely looked up at Rafiki's roar.

However, a Serengeti male should evoke a more serious response, since a complete stranger would represent a challenge to the status quo, and the roar barely finished before Bobcat advanced straight toward the speaker like a torpedo, ready for trouble.

Next, three Dinokeng lions, again familiar, again no big deal, and Bobcat barely noticed.

But three Serengeti males, in contrast, presented a significant threat to Bobcat's world. One against three would have meant certain death. The roars immediately captured his attention, and he reacted just like a typical Serengeti lion when confronted with an existential threat. He clearly wanted to investigate the challenge, but caution won out, and he shot off in a wide arc around the speaker.

Bobcat is a big black-maned cantankerous captive, and he can count as well as any wild lion, and when we play the roar of three Serengeti males to a trio of Kevin's males, they head determinedly toward the speaker, convinced that the strangers are somewhere off in the distance. Three amigos against three enemies just like in the Serengeti.

Peyton always tried to restrict her dummy tests to single-sexed groups of males or females. Her goal was to find out the kind of mane that females preferred and males avoided. And in most of the single-sexed tests, the lions went up to the dummies, sniffed a few times and lost interest once they smelled the fabric.

But on a few occasions, the intended subject was with an unseen member of the opposite sex. In most of the mixed-sex experiments, the male felt compelled to prove himself to his wife by rushing up to one of the dummies and tearing out its stuffing.

So today we are working with Mr. Yak, a taxidermically mounted male with fake teeth and plastic lips set in a permanent snarl. In real life, Mr. Yak was a subadult with a weedy blond mane, so Virginia commissioned the manufacture of a yak-hair wig, which provides a reasonable facsimile of a darkish mane, but nowhere near as impressive as Bobcat's.

Our first couple today is Ginny and her husband, Vayetse (Vie-YET-see). Vayetse is the sort of male your mother wished you would marry. He has good manners and an even temperament, he's respectful, and he has such a nice long blond mane.

Kevin takes Ginny and Vayetse for a walk. When they are about a hundred meters away, I play the roar of a lone Serengeti male. Vayetse looks startled, but Ginny is intrigued. She makes her way toward Mr. Yak, and Vayetse looks like he's thinking, "Maybe we should just stay here." Ginny keeps coming forward, and Vayetse is saying, "I don't think this is such a good idea. I really don't think we should go over there."

But she keeps coming, and ("Dammit, Janet!") Vayetse has to keep up with her. She arrives at the dummy and starts checking it out. Vayetse finally arrives—and timidly sniffs Mr. Yak as Ginny watches (judges?) her husband. The lions sniff Mr. Yak together, accidentally tipping him over, and Vayetse jumps backward as his wife reassesses her marital status.

Kevin tries to get the lions to resume their walk, but they keep sniffing

Vayetse and Livy meet Mr. Yak

Bobcat and Gabbie dispatch Mr. Yak

around the dummy. Finally, one of his assistants drives up with a pickup truck, Kevin throws Mr. Yak in the back, and the assistant starts to carry it off.

Vayetse is relieved to see the stranger heading away, but Ginny runs after the truck as if to say, "Come back, come back—don't leave me here!"

Now we try Bobcat and Gabbie. Again with the walk in the park with Kevin, again with the recorded roar, but Bobcat heads straight up to Mr. Yak, reaches around its back and bites it on the face. Gabbie follows her husband up to the dummy, and as soon as Bobcat throws it over, she bites down on its throat. Bobcat's full black mane is ten times as impressive as Mr. Yak's, and Gabbie's devotion to her husband is steadfast. Over the next few minutes, they trade places with Bobcat now throttling Mr. Yak's throat as Gabbie disembowels the Styrofoam stranger. The couple cannot be separated from their quarry until Bobcat rips off its head and struts around, carrying it like a trophy in the Roman Coliseum.

I have to leave in a few days to join Ali Swanson for a weeklong leopard survey, but there's one more test I want to try with the buckets. Ginny is still trained on the middle bucket at the front, but the next time she comes to the rack, Kevin will remove the food and place it in the rightmost bucket at the back, showing Ginny where it went.

Ginny watches Kevin move the food but then goes straight back to the middle bucket at the front. She saw the food go to a different bucket, but her training tells her that the food is in the middle bucket at the front.

After finding that the usual spot is empty, she seems confused. Kevin reopens the right-hand bucket at the back and shows her the food before closing it again.

Now she goes to the right-hand bucket at the back and opens the lid.

Kevin switches buckets again, putting the food in the second bucket from the left in the front, again making sure she sees it before closing the lid.

Ginny goes to the right-hand bucket at the back.

This phenomenon is known as object permanence, where experience trumps the evidence of our eyes. This turns out to be a surprisingly difficult test for animals to master. A recent study compared a wide variety of species: pigeons, dogs, monkeys, and apes. And who did worst on object permanence? The Asian elephant, which is nobody's idea of a dummy.

And maybe we humans don't always fare so well ourselves. Sometimes we spend so much time in the place we love most and worry so much about how we can stay on forever to enjoy the same familiar sights and feel the same exalted emotions, and maybe sometimes we have to be shaken out of our dreams and face up to the fact that it's time to move on.

SABI SANDS, 3 JUNE 2014

Word comes through over the two-way radio that a pack of wild dogs has just caught an impala on the east bank of the Sand River. Ali has never seen a wild dog before, so our driver tells us to hang on, and we race along, accelerating across ridgetops, haring around tight curves of the two-track road until we reach the west bank in time to see four yellow-splotched black dogs running straight across the shallow water with their feathery white-tipped tails held high.

Vultures retreat from the impala kill, and we pass reeds and sand banks, traversing the water atop a bed of stones. We park next to another open-backed Land Rover with tiered seats and awestruck tourists. The dogs tear into the carcass as the sun sinks below the horizon. The vultures remain at a respectful distance, and the evening light grows dim. The dogs see a spotted hyena skulking down the opposite bank—they cross to the east again, squeaking and chattering, splashing through the water. At least three hyenas have appeared. The dogs run up on shore, the hyenas turn squealing, and the dogs give chase somewhere out of view.

We sit in silence in the near darkness. Birdcalls and stream babblings are the only sounds on earth. About a hundred yards across the river, beyond a stand of trees and scrub, an unseen hyena gives out the most unearthly sound—a low loud angry *hoooooo*, like the cry of a banshee, a defiant sound that could shake the gates of hell.

And as the eerie call echoes between the opposing riverbanks, I am transported to another world and look out at the landscape of timeless Africa where there is far more mystery than we can possibly know and where so much remains to be discovered.